Basic Pathology

Fifth edition

Basic Pathology

An introduction to the mechanisms of disease

Sunil R Lakhani BSc MBBS MD FRCPath FRCPA
Professor and Head, Molecular and Cellular Pathology, School of Medicine,
University of Queensland and State Director, Anatomical Pathology,
Pathology Queensland, Brisbane, Australia

Susan A Dilly BSc MBBS FRCPath
Emeritus Professor of Pathological Sciences, Barts and The London Medical
School, Queen Mary University of London, UK

Caroline J Finlayson MBBS FRCPath
Formerly Honorary Senior Lecturer and Consultant in Histopathology,
St George's Hospital Medical School,
London, UK

Mitesh Gandhi MBBS MRCP FRCR FRANZCR
Princess Alexandra Hospital and Queensland X-ray, Woolloongabba,
Brisbane, Australia

CRC Press
Taylor & Francis Group

CRC Press
Taylor & Francis Group
6000 Broken Sound Parkway NW, Suite 300
Boca Raton, FL 33487-2742

© 2016 by Sunil R. Lakhani, Caroline Finlayson, Susan A. Dilly, and Mitesh Gandhi
CRC Press is an imprint of Taylor & Francis Group, an Informa business

No claim to original U.S. Government works

Printed on acid-free paper by Ashford Colour Press Ltd.
Version Date: 20160201

International Standard Book Number-13: 978-1-4822-6419-7 (Pack - Book and Ebook)

Visit the Taylor & Francis Web site at
http://www.taylorandfrancis.com

and the CRC Press Web site at
http://www.crcpress.com

CONTENTS

PREFACE

'What is the use of a book', thought Alice, 'without pictures or conversations.'

Lewis Carroll

Any artist will tell you that in drawing objects, you cannot ignore the spaces in between: the picture ceases to exist when only one aspect is viewed in isolation. Musical pieces composed entirely of notes and without pauses would be nothing more than a noise and an irritation. Yet when it comes to teaching, we may ignore this fact and fail to put our own specialty into the context of the whole curriculum.

Over the last decade, there has been a trend towards a more integrated approach to medical education. We are delighted that such an approach, which we have always used in our teaching, is now widely adopted throughout the world.

Our aim in this book has been to create a tutorial on the mechanisms of disease over a background of history, science and clinical relevance. The goal is to give the student a sense of belonging to a movement, the movement from past to present and from cell to patient.

This book has been written in the hope that the student will read the text fully and at leisure. This not only contains detail about the disease processes but also historical anecdotes and clinical scenarios. The cartoons are intended to amuse as well as illustrate the importance of certain topics, and we sincerely hope that students reading the book will be able to shed the dull, dreary image of pathology that they all seem to be born with. Pathology is one of the most fascinating and fun subjects students are likely to encounter during their undergraduate training. If you understand the basic principles of disease, then the interpretation of clinical symptoms and signs, the rationale behind investigation and treatment, and the unravelling of complex cases

become more logical. Time spent building a framework of mechanisms will assist your clinical practice.

This fifth edition of *Basic Pathology* can be used in conjunction with a companion volume called *Pathology in Clinical Practice: 50 Case Studies*, which is designed to help you use your pathology knowledge in clinical settings. Some extracts from *Pathology in Clinical Practice* are included in this book as well as links to the full cases highlighted with a 'link' symbol.

 Read more about bacterial infection in Pathology in Clinical Practice Case 7

When you access the e-book of *Basic Pathology*, either via the unique code printed in your hard copy or if you have purchased the e-book only, you will have access to the full text of both titles, and hyperlinks from *Basic Pathology* will take you directly to relevant cases in *Pathology in Clinical Practice*. We hope that you will find these stimulating, fun and complementary to this book because they contain more advanced, clinically relevant information.

This edition introduces radiological images in place of some of the traditional autopsy-based photographs of diseased organs. We recognise how significantly technology has changed everyday clinical practice and students are expected to be familiar with the increasing repertoire of scanned images. Over the next decades, the same may become true for genomic information, and we have extensively revised and expanded these sections.

This book is primarily intended for medical students, but should also be useful to students of dentistry, human biology and other health professions. Postgraduate students studying for pathology or surgical exams may also wish to consult these books.

ACKNOWLEDGEMENTS

When we set out more than 20 years ago to write a textbook in which we would take the students on a journey, from past to present and from patient to the cell and back again, the phrase 'integrated curriculum' was not in common usage. Today, most medical schools have switched to the new-style curriculum in which pathology is taught as an integrated subject with other disciplines. With each version of this book, there has been a move towards greater integration, especially with our colleagues in radiology who share our key role in diagnosing and monitoring disease in patients. We have also been influenced considerably by the many clinicians who attend joint clinico–pathological meetings to discuss individual patients and the best approach for their clinical management. All these should be acknowledged as informing and shaping this book.

In addition, our endeavours to produce this fifth edition have been considerably aided by help, comments and constructive criticism from a number of people, and we would like to acknowledge their time and effort. In particular, thanks to Peter Riley, Alison Milne and Alex Roche for their useful comments. This book has evolved through each edition and all contributors to earlier editions have influenced this latest version. They include Professor Philip Butcher, Professor Mike Davies and Dr Grant Robinson at St George's Hospital; Dr Ahmet Dogan, one of our co-authors from the third edition, and Dr Barry Newell, the lead author on the companion volume of *Pathology in Clinical Practice*. Dr Peter Simpson and Dr Amy McCart Reed from Brisbane made a useful contribution to the molecular sections of the current edition.

We would like to thank the staff of Queensland X-ray and the radiology department of the Princess Alexandra Hospital, Brisbane for their help in providing the radiology images which have been used.

Finally, and most importantly, we would like to thank our families for their encouragement and support. But for their understanding, it would be difficult to get away with the disproportionate amount of time that such projects consume.

SRL, SAD, CJF, MG
2016

PART 1

DISEASE, HEALTH AND MEDICINE

INTRODUCTION

DISEASE, HEALTH AND MEDICINE

Medicine, to produce health, has to examine disease.

Plutarch (*c.* 46–120)
Greek biographer and essayist

Medicine is the science and practice of diagnosing, treating and preventing diseases and it depends on understanding the mechanisms of disease – the topic of this book.

Diseases have causes (aetiology) and mechanisms (pathogenesis). They may produce symptoms (experienced by the patient) and signs (elicited by the physician). There may be structural changes that are visible to the naked eye (gross or macroscopical appearances), or only detectable down a microscope (microscopical appearances). Functional changes may be apparent to the patient or only detected by clinical or laboratory tests. All of this is pathology, and pathology is the study (logos) of suffering (pathos).

Our approach to investigating mechanisms has, however, changed significantly over the last 25 years from a simple linear cause-and-effect approach (reductionism) to an appreciation of the significance of complex interactions (holistic approach). Huge advances in computer technology have allowed modelling of complex systems that would not previously have been possible and this is opening up whole fields of new research (the 'omics'). The problem for a doctor or student is that this is mind-boggling and, in day-to-day life, we rely on simpler concepts that have practical application to the patient in front of us, so we acknowledge the complexity but adopt the simplistic.

Generally in this book we use a mechanistic model that regards the body as a machine with repairable or replaceable parts. It places a high emphasis on the scientific evidence base for untangling cause and effect in both the disease and its treatment, because this is important for patient care and prognosis. We also describe 'associations' that have been identified between various factors and diseases but where no clear mechanism has been discovered to explain the link. These are areas of intense research and worth reading about in research journals to learn the latest ideas.

So how would you define 'disease'? In fact, this is fraught with problems because disease definitions change with time and across cultures. Do you, in fact, need to suffer from a disease as part of the definition? Does a person have the disease before the symptoms appear? This is particularly important now that there is screening for many diseases and people may have abnormal 'disease markers' without any symptoms, or the 'disease marker' occurs in people who will never have the disease. This can lead to over-treating people who have lesions considered as premalignant but who would never develop invasive cancer, or people who have raised blood pressure but might not have a stroke or heart attack.

With those caveats, we offer a biomedical definition of disease – 'Disease occurs when homeostasis fails'. Homeostasis is the concept of a *regulated* system that maintains equilibrium (a balanced state) within the body, despite changes in the internal or external environment. If you accept this definition then understanding the mechanisms of disease will involve understanding the processes for maintaining homeostasis, identifying the agents and events that disrupt homeostasis, trying to determine why homeostatic mechanisms fail, and testing treatments that can alter the sequence of events and restore health. A disease should have the potential to produce some impairment of function, but may be detected while asymptomatic. It may also be treated or heal through the body's normal processes so that no permanent damage is produced.

Let us take the example of lobar pneumonia (see Chapter 4, fig 4.25) and the key concept of homeostasis and introduce a diagram that helps to illustrate the various components of the disease process (Fig. 1).

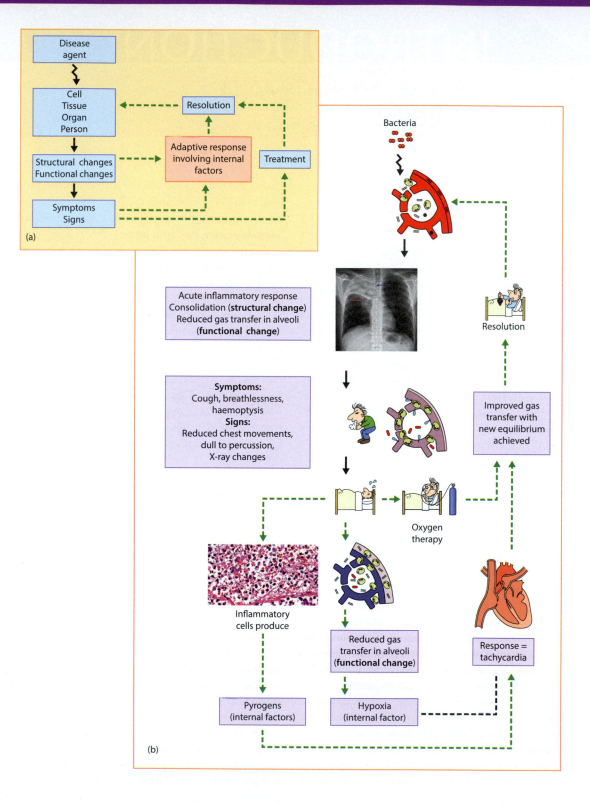

Figure 1 Disease occurs when homeostasis fails: (a) homeostatic mechanisms restore equilibrium; (b) some homeostatic mechanisms in lobar pneumonia (green dotted arrow = homeostatic mechanism).

Figure 2 Don't be confused by terminology!

In its simplest form, an intrinsic or external factor (the cause) acts on a cell, tissue, organ or whole person to produce structural or functional changes and a response. If an adaptive response is 100% successful, then homeostasis is maintained and no symptoms or signs result. If unsuccessful, then the disease manifests and the structural and functional changes may have an impact on another cell, tissue or organ to produce another set of reactions. Thus, in lobar pneumonia, the *extrinsic* cause is the bacterium, *Streptococcus pneumoniae*, which affects the lung. This stimulates an acute inflammatory response that can produce the structural changes of consolidation and functional changes of reduced gas transfer in alveoli.

The patient has *symptoms* of a cough, breathlessness and, even, haemoptysis (coughing up blood).

The physician may detect the *signs* of reduced chest movements, an area that is dull to percussion and radiological opacity reflecting lung solidification. However, this is not the end of the story.

We can produce another 'disease sequence' where the reduced gas transfer results in hypoxia (reduced oxygen saturation in the blood), which acts as an *internal* factor that leads to the heart responding with an increased heart rate (tachycardia). More blood is pumped through the part of the lung, which is not consolidated, so increasing gas transfer, and this may successfully compensate so that the hypoxia is corrected, i.e. a new equilibrium is achieved. But before you relax, look at page 139 where we explain about pyrogens that may accompany the inflammatory response to bacteria and the effect that they may have on the heart. Yes, that is another route for producing a tachycardia and so our diagram could become almost infinitely complex as we add the different pathways.

By now, you will have realised why we started by comparing reductionist and holistic approaches for understanding mechanisms. In the simplest situation, we just identify the primary mechanism, treat the patient with antibiotics against *S. pneumoniae*, remove the primary cause and return the patient to health. If the patient is very unwell and the secondary mechanisms are operating, we may need to provide oxygen by facemask to reduce the hypoxia, i.e. treat the secondary factor. We could consider using drugs such as paracetamol that have an antipyretic action and will help counteract the effect of the pyrogens; and so it goes on! Where should we stop? Are we able to analyse intelligently all the processes that may be operating?

The answer is no but we are making huge leaps forward. We mentioned the 'omics', which is the term used for studying the totality of activities. Genomics is about analysing the functions and structure of the genome (i.e. the interactions of all the genes). Proteomics studies the interactions of all the proteins in a cell, lipidomics all the lipid-based actions.

If you start off healthy and successfully maintain homeostasis, surely you should remain healthy. But do you start off healthy? That leads us to the concept of *predisposition*. We each have around 25 000 genes present in every cell but expressed differently, depending on the cell type and its response to its environment. The likelihood of our suffering from a disease depends on interactions between our genes and the environment, so called G×E interactions. The environment, however, starts working from the moment of conception so even genetically identical twins may have different predispositions for diseases by the time they are born. This continues on through life, with our personal 'exposome' being the life-long impact of the various environmental exposures on our genome and our health. However, things are rarely inevitable and so you may be able to identify your higher predisposition for disease, understand the risk factors that will make matters worse, avoid those risk factors and so not get the disease! Preventive medicine in the future may be much more tailored to individual circumstances and susceptibility than was previously possible. Drug treatments may be used only where they will be truly beneficial. This could be the beginning of an era of personalised medicine based on individual genome analysis so understanding the mechanisms of disease is fundamental – read on!

CHAPTER 1
WHAT CAUSES DISEASE?

Those who are enamoured of practice without science are like a pilot who goes into a ship without rudder or compass and never has any certainty where he is going. Practice should always be based upon a sound knowledge of theory.

Leonardo da Vinci (1452–1519)
Italian artist, sculptor, architect and engineer

Every day we eat food, breathe air, walk, talk and perform the tasks of daily life before going to bed unscathed. Some days, however, we may encounter and be harmed by viruses, bacteria, unusual antigens, extremes of temperature, chemical pollutants or fast-moving vehicles. We are harmed by environmental (or extrinsic) factors. The cause is usually obvious if it happens immediately but may be hard to discern when the effect takes days or years to manifest, e.g. tobacco smoking causing lung cancer. In some situations, the environment plays no part and the cause is entirely genetic (intrinsic), e.g. the anaemias related to abnormal haemoglobin (see page 279).

There is no particular way of classifying the causes of diseases but it is useful to have a checklist that you find helpful when talking with patients and thinking about their 'differential diagnosis', i.e. the possible conditions and the causes that would explain their current symptoms. Table 1.1 shows such a list of causes, which are presented in an order that moves from the relatively simple to the more complex, so that you could

Table 1.1 Categorising the causes of disease

Category	Example of causes and diseases
Physical	Heat, cold, radiation, electrical, mechanical
Chemical	Tobacco smoke, alcohol, drugs, poisons, air pollution
Structural	Congenital neural tube defects, vascular occlusion, bowel perforation
Infectious	Bacteria, viruses, fungi, protozoa, prions
Inflammatory	Peanut anaphylaxis, dust-mite asthma, autoimmune conditions
Nutritional	Obesity, malnutrition, diabetes
Degenerative	Osteoarthritis, dementia
Genetic	

identify and treat the straightforward cases quickly. Depending on your location and specialty, you could use this approach to construct a flowchart of the most likely causes for your patients.

We have put genetic causes at the bottom of the list because the genes interact with most of the other causes in complex ways. It is possible to produce a GxE (genes x environment) card to illustrate this, as shown on page 215. Mostly these causes are based on statistically derived associations rather than precise knowledge of the mechanisms but we attempt to highlight the main mechanisms through which genes affect health.

PHYSICAL CAUSES

Let us start with a **physical cause** of disease and look at extremes of temperature.

CLINICAL SCENARIO: FROSTBITE

A 29-year-old skier lost his way after drinking a few beers in a mountain restaurant. He was discovered, unconscious, without a hat, 20 hours later in an exposed location on the mountain face. He had obvious frostbite to his nose and fingertips, manifest as hard white areas. A weak, thready, but regular pulse was present and he was breathing shallowly. He was wrapped in a space blanket and rushed by helicopter to hospital. His rectal (core) temperature was found to be 30°C. He was warmed with blankets and hot water bottles. He suffered a cardiac arrest but was immediately resuscitated. His mild metabolic acidosis was corrected. As he regained consciousness he began to shiver violently and his limbs became reddened and swollen. He tried to raise himself up but promptly fainted. Over the next couple of days his fingertips and nose turned black and the skin began to slough off. At one point it appeared that he would lose the tips of most of his fingers, but after the skin had fallen off only two fingertips were lost and his nose was saved (Fig 1.1).

What has happened here?

Our patient was hypothermic, i.e. his core temperature was <35°C (normal temperature 37°C). He contributed to his condition by drinking alcohol (see below). He was not wearing a hat, and 50% of the body's heat loss may be through the head. There is no mention of a facemask, which might have served to warm the freezing air before he inhaled it, losing core heat when the lungs warmed the air before exhalation.

The onset of hypothermia is dependent on many factors, including body build, ambient temperature, whether the individual is dry or wet, the presence of

Figure 1.1 Frostbite is caused by the freezing of tissues. Sludging of cells in capillaries causes ischaemic damage. Endothelial cells are vulnerable to both cold and hypoxia; on thawing of the tissues, there is leakage of blood through the damaged vessels, and the tissue appears blackened over a larger area than is actually necrotic, so delaying amputation is advised.

protective clothing and whether or not alcohol has been ingested (even small amounts can greatly increase the risk of hypothermia).

Alcohol plus exercise is probably the worst combination, because not only is his blood glucose depleted by exercise but also the alcohol interferes with the generation of new glucose from body stores by reducing the amount of pyruvate available.

Once he became hypothermic he would have been less able to make rational decisions or take sensible measures such as seeking help, finding shelter or covering exposed areas of his body.

How does the body respond to cold?

Initially the heat gain centre in the hypothalamus directs the body to shiver violently and shut down peripheral blood vessels by vasoconstriction, but if this fails to restore core temperature the cooling of the heart interferes with cardiac output. Below 26°C the cardiac output is too low to sustain life (Fig. 1.2).

Oxygen combines more strongly with haemoglobin at low temperatures, further depleting tissues of oxygen. Anoxic effects on the heart include arrhythmias. Patients whose body temperature is below 30°C are at risk of cardiac arrest and should be monitored using ECG. Respiration is diminished, usually in proportion with tissue requirements, but slight CO_2 retention may

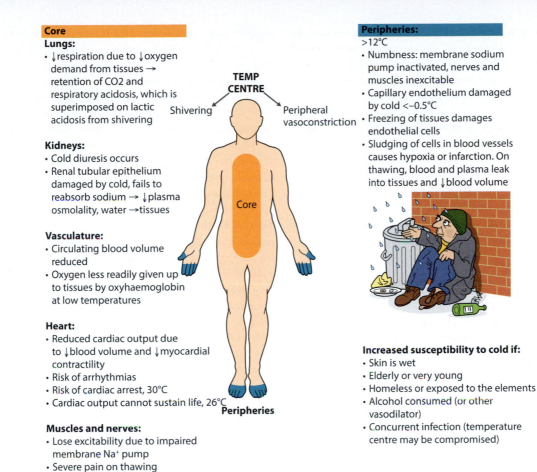

Core

Lungs:
- ↓respiration due to ↓oxygen demand from tissues → retention of CO2 and respiratory acidosis, which is superimposed on lactic acidosis from shivering

Kidneys:
- Cold diuresis occurs
- Renal tubular epithelium damaged by cold, fails to reabsorb sodium → ↓plasma osmolality, water →tissues

Vasculature:
- Circulating blood volume reduced
- Oxygen less readily given up to tissues by oxyhaemoglobin at low temperatures

Heart:
- Reduced cardiac output due to ↓blood volume and ↓myocardial contractility
- Risk of arrhythmias
- Risk of cardiac arrest, 30°C
- Cardiac output cannot sustain life, 26°C

Muscles and nerves:
- Lose excitability due to impaired membrane Na+ pump
- Severe pain on thawing

TEMP CENTRE

Shivering

Peripheral vasoconstriction

Core

Peripheries

Peripheries:
>12°C
- Numbness: membrane sodium pump inactivated, nerves and muscles inexcitable
- Capillary endothelium damaged by cold <−0.5°C
- Freezing of tissues damages endothelial cells
- Sludging of cells in blood vessels causes hypoxia or infarction. On thawing, blood and plasma leak into tissues and ↓blood volume

Increased susceptibility to cold if:
- Skin is wet
- Elderly or very young
- Homeless or exposed to the elements
- Alcohol consumed (or other vasodilator)
- Concurrent infection (temperature centre may be compromised)

Figure 1.2 Changes associated with cold and reperfusion.

cause a respiratory acidosis, compounded by the metabolic acidosis that occurs when lactic acid accumulates as a result of shivering.

Why did the patient faint as he began to recover?

Blood volume falls in the hypothermic patient due to a combination of a 'cold diuresis' which occurs in response to a drop in core temperature, and damage to the renal tubular epithelium as a result of the cold, preventing sodium reabsorption; this causes a drop in plasma osmolality and water moves out of the vascular compartment into the tissues to balance this. When the patient is warmed and vasodilatation occurs the blood volume is insufficient for demand and there is hypotension.

What is frostbite?

Frostbite is the result of the freezing of tissues, which occurs at temperatures below -0.54°C. Before freezing, cooling to <12°C causes paralysis of muscles and nerves by interference with the membrane sodium pump. Lack of sodium ions renders nerves and muscles inexcitable. Damage is reversible after a few hours, but not if left longer. If the tissue freezes, the tissue proteins become denatured and the cell dies. Vascular endothelial cells are particularly susceptible. When they thaw, plasma leaks out of the small vessels through the damaged endothelium and the retained red blood cell sludge, obstructing the lumen and causing local infarction of tissue.

Although it is dangerous to warm the body of a hypothermic person too quickly, if a patient is suffering only from frostbite it is best to warm the affected area as

Chapter 1: What causes disease?

quickly as possible, so that sludging and infarction are reduced to a minimum due to quick restoration of blood flow. It is harmful to thaw an area of frostbite and then allow it to refreeze, as may happen on a mountainside when colleagues treat the area by warming and then the patient with frostbite is carried back through the same harsh environmental conditions which caused the problem.

Although damage may seem extensive, the actual area of necrosis may be less than it first appears and it may be worth waiting for a few days before amputating an affected area. The process of thawing out is very painful and requires analgesic support, plus elevation of the affected area to reduce tissue swelling.

CHEMICAL CAUSES

The most common chemical causes of diseases are in the air that we breathe, whether we are indoors or outside. The contaminants in the air vary considerably

depending on the season and the location, but included among them are substances that cause life-threatening lung inflammation and cancer. Indoors, the major hazards are tobacco smoke and wood smoke. Outside they can be a mixture of pollutants, most of which come from industrial production, home heating and car exhaust fumes (Fig. 1.3). In cities, ozone is a particular problem because it is produced by the action of sunlight on nitrogen oxides from car exhausts, and so can be present in large amounts. Ozone generates free radicals (see section on cell damage) which damage the lining cells of the lung, so decreasing lung function and increasing airway reactivity and inflammation. Particles are also damaging, especially if they are carried down to the alveoli rather than being trapped and cleared by the mucus. In the alveoli, they can stimulate macrophages and provoke chronic inflammation. These effects can limit the activities of healthy people and cause hospitalisation for people with asthma or chronic lung conditions (Fig. 1.4). It is increasingly recognised that air pollution also causes lung cancer, yet environmental controls fail to

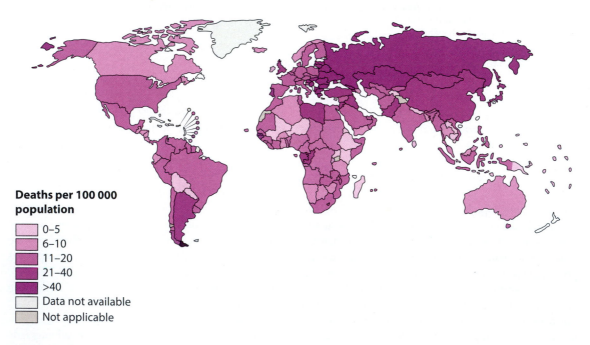

Deaths attributable to outdoor air pollution, 2008

Deaths per 100 000 population

- 0–5
- 6–10
- 11–20
- 21–40
- >40
- Data not available
- Not applicable

Figure 1.3 Deaths attributable to outdoor air pollution. (From WHO global observatory map. © WHO 2011. All rights reserved.)

adequately reduce air pollution. In Europe, the limit is set at 25 mg/m^3 but this is higher than the level causing damage and studies on 360 000 city residents show a 7% increase in mortality rate for each 5 mg/m^3 increase in fine particulates.

Indoors, wood smoke is an irritant that contains carcinogenic polycyclic hydrocarbons, which cause recurrent lung infections and bronchial tumours. The most significant, readily avoidable, chemical air pollutant is tobacco smoke. Tobacco smoke is a complex mix of substances, some of which cause immediate damage by irritating airways and stimulating inflammation and mucus production, resulting in cough and reduced exercise tolerance. In the medium to long term, the recurrent inflammation causes chronic obstructive pulmonary disease (COPD) and emphysema, which are irreversible. The risk of lung cancer depends on the number of cigarettes smoked in a roughly linear relationship and the overall survival is reduced by around 7 years for smokers compared with those who have never smoked. Most of the harmful effects are on the lungs and cardiovascular system; however, there are also organ-specific carcinogens in tobacco smoke causing tumours in the oesophagus (N'-nitrosononicotine or NNN), pancreas [4-(methylnitrosamino)-1-(3-pyridyl) -1-butanone or NNK], bladder (4-aminobiphenyl and 2-naphthylamine) (Fig. 1.5) and oral cavity (polycyclic aromatic hydrocarbons, NNK and NNN). Combining tobacco smoke with alcohol significantly increases the risk of laryngeal, oral and oesophageal cancer.

Some of the components of tobacco smoke and their effects are listed in Table 1.2.

The list in Table 1.2 does not explain the major effects of tobacco on the cardiovascular system, which is believed to be the cause of a third of heart attacks. This is also discussed in the section X CVS atheroma (page 239) but, briefly, it is likely that the combination of damage to vascular endothelial cells, hypoxia and increased platelet aggregation

Outdoor air pollutants

Ozone   Acid aerosols
Nitrogen dioxide   Particulates
Sulphur dioxide

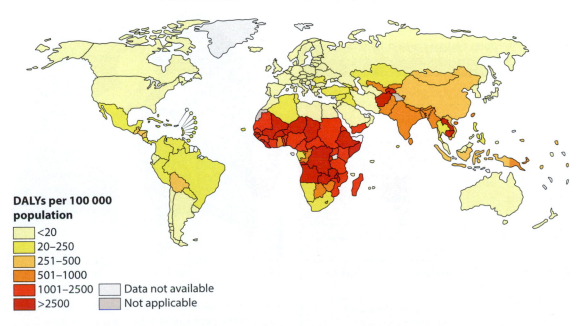

DALYs attributable to household air pollution, 2004

DALYs per 100 000 population
- <20
- 20–250
- 251–500
- 501–1000
- 1001–2500
- >2500
- Data not available
- Not applicable

Figure 1.4 Disability-adjusted life-years (DALYs) attributable to household air pollution. (From WHO global observatory map. © WHO 2011. All rights reserved.)

Chapter 1: What causes disease?

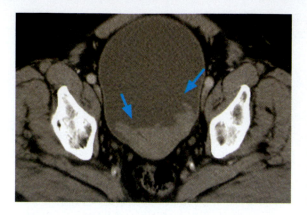

Figure 1.5 An axial CT scan showing a soft-tissue mass arising from the floor of the bladder (blue arrows). Tobacco smoke is one of the main culprits in the development of bladder cancers. Some occupations, such as bus driving and working with rubber, are also associated with exposure to toxins, such as benzidine and 2-naphthylamine, implicated in the development of bladder cancer.

contribute to the production and progression of atherosclerotic plaques. The estimated number of smoking-related deaths worldwide is more than 4 million and, sadly, despite educational and public health measures, the number of people smoking continues to increase (Fig 1.7).

ASBESTOS-RELATED LUNG DISEASE

Asbestos is a crystalline silicate mineral that has been mined for centuries and used in building insulation because of its advantageous physical properties. Exposure and inhalation of asbestos fibres lead to a variety of disease processes in the lung, resulting in death or considerable disability (Fig. 1.8). Asbestos is now banned in many countries.

We have included drugs in the category of chemical causes. This is clearly true when thinking about drugs of abuse (e.g. amphetamines) or deliberate overdoses of therapeutic drugs. It is rather more complex when considering the 'side effects' of medications. These are very important because almost all treatments have some unwanted effects and patients are commonly on several different types of medication. However, the detail is beyond the scope of this book and we just highlight a few exemplars in the relevant chapters (Fig. 1.9).

Table 1.2 Effects of some tobacco smoke elements	
Component	Action
Polycyclic aromatic hydrocarbons	Carcinogenic
Benzopyrene	Carcinogenic
Nitrosamine	Carcinogenic
Nitrogen oxides	Damage cilia and irritate mucosa
Formaldehyde	Damage cilia and irritate mucosa
Carbon monoxide	Reduce oxygen delivery to tissues
Nicotine	Stimulate ganglia and tumour promotion

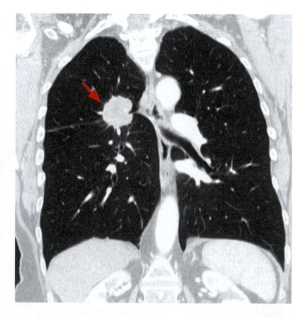

Figure 1.6 Coronal CT scan showing a speculated mass in the right upper lobe – a primary lung cancer.

Case study: coughing up blood

A 64-year-old woman presents to her GP with a 3-month history of breathlessness and occasional episodes of haemoptysis. She is known to have chronic obstructive pulmonary disease but thinks that her shortness of breath is worse than she normally expects from her COPD. She is also very worried by the haemoptysis.

Her previous medical history includes COPD and hypertension. Her hypertension is controlled with an angiotensin-converting enzyme (ACE) inhibitor. She smoked 15 cigarettes a day from age 20 to age 60 (when her COPD was diagnosed).

Examination of the cardiovascular and abdominal systems is unremarkable. There is a mild reduction in expansion in the right lower zone; the percussion note is a little dull and the breath sounds are quiet relative to elsewhere.

Question: What diagnosis must be considered?

Answer: In a patient of this age, who is a smoker or ex-smoker, and presents with haemoptysis, bronchial carcinoma is a definite possibility.

The GP arranges an urgent referral to the local chest clinic but receives a request for a home visit from the patient 4 days later. Her dyspnoea has worsened and she now has a cough that is productive of green sputum. She has pain on inspiration on the right side of her chest. Examination reveals decreased expansion on the right side of the chest which is of a greater degree than 4 days previously. The percussion note is dull over the right lower zone, again to a greater degree than when the patient was first examined. The breath sounds over the right lower zone are quiet and have the quality of bronchial breathing. The breath sounds elsewhere in both lungs are normal.

Question: What has happened and how do all the patient's symptoms at this stage relate to her suspected diagnosis?

Answer: The majority of bronchial carcinomas develop as masses in the proximal bronchi, near to the hila of the lungs. As with other tubular structures/organs in the body, if a tumour develops, from either the mucosal lining of the tube or the deeper tissues of the wall, the tube may become partially or completely obstructed by the neoplasm. Narrowing of a bronchus impedes airflow in and out of the part of the lung distal to that airway. If the volume of lung affected is sufficiently great, this can manifest as dyspnoea, especially in a patient who has COPD and therefore has decreased lung function anyway. In extreme cases, the obstruction is complete and causes collapse of the affected part of the lung. The findings in this patient on presentation suggested an element of collapse.

If a region of the lung is obstructed, the drainage of mucus and other airway secretions from that lung is impaired. Stagnant fluid is an advantageous environment for bacteria and they can colonise the fluid. In the lung, this leads to pneumonia in the occluded region. The immune system reacts with an acute inflammatory response, leading to the generation of pus which is coughed up as purulent sputum.

Extension of the pneumonic process to the pleura induces inflammation there. Movement of an inflamed parietal pleura over the visceral pleura, as occurs naturally during respiration, produces pain.

Haemoptysis may result from two mechanisms: in addition to its ability to obstruct the lung, a bronchial carcinoma can ulcerate the mucosa and will invade deeper structures. This damage may be accompanied by small quantities of blood loss. The other process that generates haemoptysis is cavitation. Malignant tumours are frequently rapidly growing and in some instances grow faster than their ability to induce adequate angiogenesis. When this happens, part of the tumour becomes necrotic and may undergo cavitation. This destructive collapse can be accompanied by haemorrhage.

(Continued)

Chapter 1: What causes disease?

    This is part of Case 6 taken from the book *Pathology in Clinical Practice: 50 Case Studies*, which is complementary to this book as it contains more advanced clinically relevant information. Have a look at that so that you can decide if it is useful for your programme.

🔗 Read more in Pathology in Clinical Practice Case 6

Percentage of tobacco use among adults, 2005

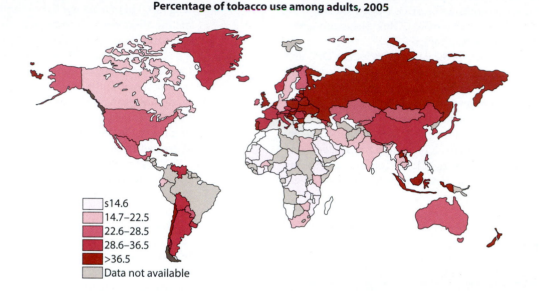

	≤14.6
	14.7–22.5
	22.6–28.5
	28.6–36.5
	>36.5
	Data not available

Figure 1.7 Percentage of tobacco use among adults. (From WHO global observatory map. © WHO 2008. All rights reserved.)

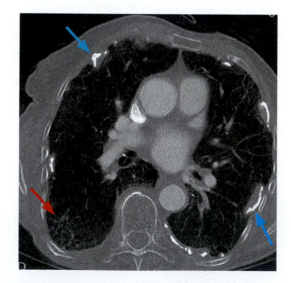

Figure 1.8 Axial CT scan showing calcified pleural plaques (blue arrows) and a fibrotic process with increased interstitial thickening at the bases (red arrow). The latter is also known as asbestosis and is associated with progressive respiratory dysfunction.

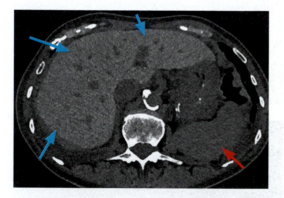

Figure 1.9 Amiodarone and the liver: an iatrogenic chemical cause of liver disease. Amiodarone is a drug used in the treatment of cardiac arrhythmias. It can potentially have side effects on the liver, ranging from abnormal liver function tests to, rarely, cirrhosis. Over the long term amiodarone is deposited in the liver, resulting in a change in the density of the liver on a CT scan. The normal liver should be the same density as the spleen (red arrow). However, this patient's liver has a higher density (blue arrows) in keeping with amiodarone accumulation.

STRUCTURAL CAUSES

Structural causes include conditions that are congenital and produce obvious physical disability (Fig. 1.10), such as neural tube defect, or are acquired, such as acute obstruction or rupture of blood vessels or parts of the bowel. The cardiovascular structural conditions are covered in Chapters 7–8 which describe thrombosis, embolism and atherosclerosis.

The congenital conditions are often of unknown cause but some have clear genetic or infective causes.

Genes are extremely important during normal development and mistakes that occur during the production of gametes, fertilisation of the ovum and formation of the embryo can result in malformation or death. A key concept is 'body patterning', which provides a basic anatomical organisation through creating gradients of gene products. For example, the four *HOX* gene clusters create a gradient along the anteroposterior axis of the embryo. Not all of our organs are paired and symmetrical; as an example we only have one heart and this should be situated on the left. To achieve this requires a right–left differentiation which is dependent on functioning cilia creating an asymmetrical flow of fluid (Fig. 1.11).

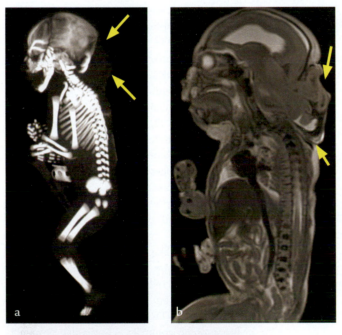

Figure 1.10 (a) Postmortem CT and (b) MRI on a fetus born with a large occipital encephalocele – a defect in the posterior skull bones with herniation of the brain and meninges through the defect (yellow arrows). This is one of the types of neural tube defect. Others include spina bifida and anencephaly.

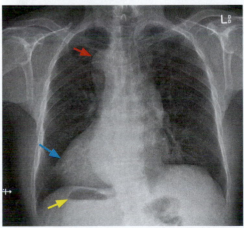

Figure 1.11 Kartagener's syndrome is due to abnormal cilia that affect embryonic development – chest radiograph showing the situs inversus or 'mirror imaging' component of the syndrome, with the apex of the heart pointing to the right (blue arrow), a right-sided aortic arch (red arrow) and a gastric air bubble (yellow arrow) on the right side.

Chapter 1: What causes disease?

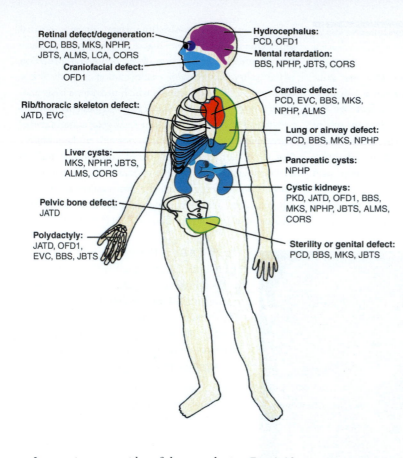

Retinal defect/degeneration:
PCD, BBS, MKS, NPHP,
JBTS, ALMS, LCA, CORS

Craniofacial defect:
OFD1

Rib/thoracic skeleton defect:
JATD, EVC

Liver cysts:
MKS, NPHP, JBTS,
ALMS, CORS

Pelvic bone defect:
JATD

Polydactyly:
JATD, OFD1,
EVC, BBS, JBTS

Hydrocephalus:
PCD, OFD1

Mental retardation:
BBS, NPHP, JBTS, CORS

Cardiac defect:
PCD, EVC, BBS, MKS,
NPHP, ALMS

Lung or airway defect:
PCD, BBS, MKS, NPHP

Pancreatic cysts:
NPHP

Cystic kidneys:
PKD, JATD, OFD1, BBS,
MKS, NPHP, JBTS, ALMS,
CORS

Sterility or genital defect:
PCD, BBS, MKS, JBTS

Figure 1.12 Almost every organ in the body shows vulnerability in the ciliopathies. Most ciliopathies have overlapping clinical features in multiple organs. Cystic kidney and retinal defects are frequently observed. Skeletal dysplasia is predominantly seen in JATD, OFD1 and EVC. ALMS, Alström's syndrome; BBS, Bardet–Biedl syndrome; CORS, cerebro-oculo-renal syndrome; EVC, Ellis–van Creveld syndrome; JATD, Jeune asphyxiating thoracic dystrophy; JBTS, Joubert's syndrome; LCA, Leber's congenital amaurosis; MKS, Meckel's syndrome; NPHP, nephronophthisis; OFD1, oral–facial–digital syndrome type 1; PCD, primary ciliary dyskinesia; PKD, polycystic kidney disease. (Reproduced from Lee and Gleeson. *Genomic Medicine* 2011;**3**:59.)

Just to give you an idea of the complexity, Fig. 1.12 is a picture of some congenital abnormalities associated with ciliopathies and the genes involved. Please don't attempt to memorise these; just be grateful that you can look them up on genetic databases when required.

WHICH ARE THE MOST IMPORTANT DISEASES TODAY?

The World Health Organization (WHO) 2012 mortality statistics show that, of the 56 million people who died in 2012, 'non-communicable diseases' caused 68% of all deaths: cardiovascular diseases (CVDs – mainly ischaemic heart disease and strokes), cancers, diabetes and chronic lung diseases were the most important of these. 'Communicable', i.e. 'infectious', maternal, neonatal and nutritional conditions together caused 23% of global deaths, and injuries caused 9%. The use of tobacco is thought to be at the root of 10% of all deaths, largely from respiratory tract (and other) cancers and cardiovascular disease.

What makes troubling reading is the fact that, in low- and middle-wealth countries, the average age at which deaths occurred is considerably lower than in the wealthy countries. Of deaths in affluent countries 87% are from CVDs and cancers, compared with 37% in low-income countries. Deaths due to infective causes and poor hygiene are far more common in low- and mid-income countries, of which tuberculosis, diarrhoeal diseases, malaria and HIV/AIDS are major contributors. Tuberculosis (TB) is second only to HIV/AIDS as the greatest killer worldwide due to a single infectious agent and TB causes a quarter of all deaths in HIV/AIDS patients.

BIOLOGICAL AGENTS AS CAUSES OF DISEASES

Infectious causes are especially important because of the number of people affected worldwide and the fact that many are treatable with antimicrobials.

Read more about bacterial infection in Pathology in Clinical Practice Case 7

changes in the classification of some medically important microbes.

Bacteria are generally classified according to their shape and whether they stain blue–purple with the Gram stain (Fig. 1.13 and Table 1.3). Endotoxins and exotoxins are important mechanisms by which they cause damage and these are described in detail in Chapter 2 page 72.

Advances in molecular biological techniques have led to some significant changes in medical microbiology. Most importantly there have been

TAXONOMY

Microbes were traditionally classified on the basis of their microscopical morphology, culture requirements, biochemical and serological properties – in other words their phenotypic characteristics. Nucleic acid amplification and genome sequencing mean that organisms can now be speciated by their genotypic properties. In bacteriology this has resulted in the reclassification of many

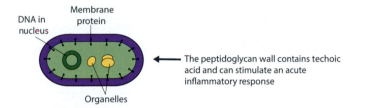

Gram-positive bacteria stain blue with crystal violet, which lodges in their thick outer wall. Inside they have a double-layered phospholipid membrane, with membrane proteins, surrounding cytoplasm with organelles and a circular nucleus composed of double-stranded DNA

Gram-negative bacteria have an extra, double-layered coat that traps crystal violet before it can reach the peptidoglycan coat (which is thinner than in G+ bacteria). The violet crystals are washed out by alcohol as part of the Gram staining pocess and the bacterium is visualized using a red counter-stain

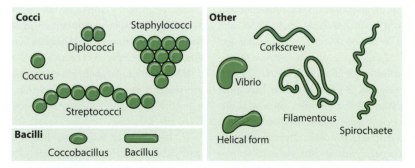

Bacterial shapes – bacteria are classified by their shape and groupings as well as their Gram staining. Bacteria can be circular, rod shaped, spiral, curved and/or have flagellae. They can occur singly, in pairs (diplo), clusters (staphlo) and strings (strepto); eg streptococci.

Figure 1.13 Gram-negative and Gram-positive bacteria.

Chapter 1: What causes disease?

Table 1.3 Simple summary of the ways to classify bacteria

	Gram-positive	Gram-negative	Acid fast
Examples of aerobic and anaerobic bacteria			
Obligate aerobes	*Bacillus cereus*	*Neisseria* spp.	Mycobacteria
Facultative anaerobes	*Bacillus anthracis*	*Escherichia coli*	
	Staphylococci	Salmonellae	
Microaerophilic bacteria	Streptococci	Spirochaetes	
Obligate anaerobes	Clostridia	*Bacteroides* spp.	
Examples of morphologically distinct Gram-positive and Gram-negative bacteria			
Cocci (spherical)	Streptococci (in chains).	*Neisseria* spp.	
	Staphylococci (in clusters)		
Bacilli (rod shaped)	Corynebacteria	*Haemophilus* spp.	
	Clostridia	*Bordetella* spp.	
	Bacilli	*Klebsiella, Proteus* and *Shigella* spp., *E. coli*	
	Listeria spp.	*Vibrio, Salmonella* spp.	
Spiral		*Treponema, Borrelia, Leptospira* (spirochaetes)spp.	
		Helicobacter, Campylobacter spp.	
Pleomorphic		*Rickettsia* spp. (intracellular obligates)	
Branching	*Actinomyces, Nocardia* spp.		

medically important organisms and, in some cases, this confusingly means that the name of the organism has changed, often with the generation of a new genus to comply with taxonomic rules. A good example is in the genus *Pseudomonas* where the organism *Pseudomonas maltophilia* (a cause of bacteraemia in immunocompromised patients) became *Xanthomas maltophilia*, and is currently known as *Stenotrophomonas maltophilia* and where the organism *Pseudomonas cepacia* (a cause of respiratory tract infection in patients with cystic fibrosis) is now *Burkholderia cepacia*. Some organisms have now been found to be made up of a complex of several different species. The genus *Streptococcus* is a good example of this: *Streptococcus bovis* has now been divided into two species – *Streptococcus gallolyticus* and *Streptococcus pasteurianus*. Is this important? In some cases it has little practical relevance, but in the case of the former *Streptococcus bovis* it does. It was known in the past that there was an association with colonic carcinoma in patients who had a bacteraemia with this organism. It has now been shown that this association is really with *Streptococcus gallolyticus* not with *Streptococcus pasteurianus*. More profoundly, some microbes have been found to have been completely misclassified. *Pneumocystis carinii*, which can cause respiratory infection in immunocompromised hosts, is no longer classified as a protozoan and is now known to be a fungus and has been renamed *Pneumocystis jirovecii*. Molecular techniques such as 16-S ribosomal DNA polymerase chain reaction (PCR) have allowed previously unculturable organisms to be identified, e.g. the bacterium *Tropheryma whipplei*, the cause of Whipple's disease.

Although most infectious disease is due to bacteria and viruses, there are other categories that are just as important, and that have evolved their own mechanisms for bypassing host defences and causing disease. Briefly, these are the fungi, protozoa, parasites, helminths and prion proteins.

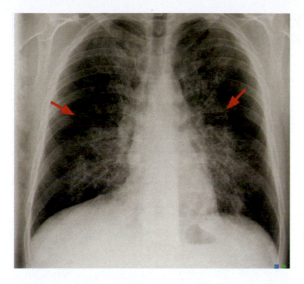

Figure 1.14 A radiograph in a young patient of 32 with AIDS showing bilateral perihilar infiltrates due to a pneumocystis infection (pneumocystis pneumonia). This fungus causes infection in immunosuppressed patients.

Table 1.4 Outline classification of viruses (see Figs 2.30 and 2.31)

Nucleic acid	Symmetry	Envelope	Strand	Family	Example
RNA	Icosahedral	No	SS1	Picorna	Polio, Coxsackie
			DS	Reo	Rotavirus
		Yes	SS	Toga	Rubella (rubivirus)
				Flavi	Dengue fever
	Helical	Yes	SS1	Corona	Colds
			SS2	Orthomyxo	Influenza A, B and C
				Paramyxo	Mumps, measles
				Rhabdo	Rabies
	Complex	Complex	SS1	Retro	HIV, HTLV
DNA	Icosahedral	No	SS linear	Parvo	Red cell destruction
			DS circular	Papova	Papillomavirus
			DS linear	Adeno	Colds
		Yes	DS linear	Herpes	Herpes simplex
					Varicella-zoster
					Cytomegalovirus
					Epstein–Barr virus
			DS circular	Hepadna	Hepatitis B
	Complex	Complex	DS linear	Pox	Smallpox

SS, single strand; DS, double strand.

Chapter 1: What causes disease?

PROTOZOA

Protozoa are free-living, single-celled eukaryotes with nuclei, endoplasmic reticulum, mitochondria and organelles. They ingest nutrients through a cytosome and can reproduce sexually and asexually. Most are able to form cysts when in hostile environments (Table 1.5).

HELMINTHS

Helminths or worms can usually be seen by the naked eye. They can enter from contaminated food or water, penetrate the skin or be transmitted by insects (Table 1.6).

FUNGI

Fungi are eukaryotic cells requiring an aerobic environment. They are more likely to cause significant diseases if a patient is immunocompromised (so-called opportunistic infection) or the normal flora of the mouth, gut or vagina are altered by antibiotics. They are generally classified as superficial, cutaneous, subcutaneous and systemic (Table 1.7).

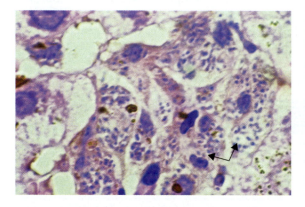

Figure 1.15 *Leishmania* sp., a parasitic flagellate protozoan (arrows).

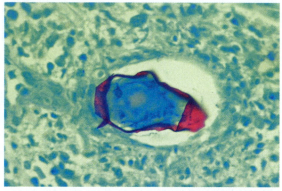

Figure 1.16 Ovum from a schistosome, one of the trematodes.

Table 1.5 Examples of protozoa that cause human disease	
Organism	Disease
Toxoplasma gondii[a,b]	Cerebral, ocular and lymphoid damage
Plasmodium (*falciparum, vivax, ovale, malariae, knowlesi*)	Malaria
Leishmania spp. (various)	Cutaneous and visceral leishmaniasis (Fig. 1.15)
Trypanosoma spp. (various)	Sleeping sickness and Chagas' disease
Entamoeba histolytica	Diarrhoea
Giardia lamblia	Diarrhoea
Cryptosporidia[a]	Diarrhoea
Isospora spp.[a]	Diarrhoea
Acanthamoeba spp.[a]	Keratitis
Naegleri fowleri	
Trichomonas spp.	Vaginal discharge
[a]People with defective immune systems are more liable to have significant problems with these organisms.	
[b]Problems often relate to reactivation because the immune defences are reduced rather than there being a primary infection.	

Table 1.6 Examples of helminths that cause human disease

Helminth	Disease
Trematodes (flukes)	
Schistosoma spp. (various) (skin penetration)	Schistosomiasis (liver, lung, gut and bladder damage) (Fig. 1.16)
Cestodes (tapeworms)	
Echinococcus	Hydatid disease
Taenia solium, Taenia saginata and *Diphyllobothrium latum*	Tapeworm infestation from pig, cow and fish
Nematodes (roundworms)	
Necator, Trichuris spp.	Hookworm, whipworm (gut infestation)
Wuchereria, Onchocerca spp. (insect bites)	Elephantiasis, river blindness, filariasis

Table 1.7 Examples of fungi that cause human disease

Fungus	Disease
Superficial	
Malassezia globosa	Tinea versicolor
Cutaneous	
Microsporum, Trichophyton spp. (dermatophytes)	Ringworm, athlete's foot, etc.
Systemic	
Histoplasma capsulatum	Pneumonia or disseminated disease
Aspergillus spp. (various)	Pneumonia or disseminated disease (Fig. 1.17)
Candida, Cryptococcus spp.	Disseminated disease in immunosuppressed individuals
Pneumocystis jirovecii	Interstitial pneumonia

PRIONS AND TRANSMISSIBLE SPONGIFORM ENCEPHALOPATHIES

This is a short section but we should flag up our ignorance concerning certain diseases and some 'infectious' particles. There is great interest in transmissible spongiform encephalopathies which can produce progressive and fatal brain damage in humans (kuru), sheep (scrapie) and cows (bovine spongiform encephalopathy [BSE] or 'mad cow' disease). These diseases are experimentally and naturally transmissible with no viruses or bacteria detectable. Prion proteins have been proposed as the cause. They are naturally occurring proteins in most mammals. Their role is not entirely clear, but it seems that they are important in the differentiation of neurons. Prion proteins may be induced to change shape, either spontaneously (as in sporadic Creutzfeldt–Jakob disease [CJD]) or if a mutant or foreign prion protein enters the cell (e.g. variant CJD, thought to be the human equivalent of BSE, caught from infected cattle) (Fig. 1.18). The particle is protein without any evidence of nucleic acid. Its neurotoxic effects are thought to be due to abnormal folding, brought about by a mutation of just a few amino acids, which inactivates the normal relationship between the cytoskeletal components and the proteasome. The misfolding seems to be catching and, once it has begun, the normal protein also becomes misfolded. The effects are to cause degeneration of the brain, which develops a sponge-like texture ('spongiform encephalopathy'). The disease is characterized by problems with coordination, rapidly progressive dementia and death.

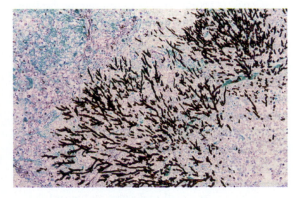

Figure 1.17 *Aspergillus* spp.

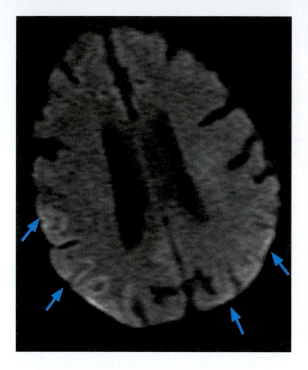

Table 1.8 Examples of inflammatory conditions

Disease	Target antigen	Hypersensitivity type
Asthma	Pollen, dust mite	I
Myasthenia gravis	Acetylcholine receptor	II
Pernicious anaemia	Intrinsic factor of gastric parietal cells	II
Systemic lupus erythematosus	Nuclear antigens	III
Reactive arthritis	Bacterial antigens, e.g. *Yersinia* spp.	III
Type 1 diabetes mellitus	Pancreatic islet β cells	IV
Rheumatoid arthritis	? collagen	IV

Figure 1.18 Diffusion-weighted MRI of the brain showing 'bright' areas representing cytotoxic oedema in the cortex of a 60-year-old patient with rapidly progressive dementia. This is a variant of prion disease called Creutzfeldt–Jakob disease.

INFLAMMATORY CONDITIONS

These cover a broad range of causes and mechanisms. The mechanisms are based on the different types of hypersensitivity reactions (see pages 80–86). The causative allergens can be environmental, genetic or a mixture (Table 1.8).

COELIAC DISEASE: GENETIC PREDISPOSITION AND HYPERSENSITIVITY

Coeliac disease is an example of a genetic *predisposition* to disease. In such patients, inherited characteristics that are dictated by the major histocompatibility complex (MHC) class I or II molecules on the surfaces of all nucleated cells (type I) or antigen-presenting cells (APCs, type II) mean that a person is at increased risk of developing a particular disease. Coeliac patients almost always have either MHC molecules of HLA-DQ2 or -DQ8 type displayed by their APCs.

If gluten molecules enter the lamina propria of the small intestine in an undigested state they may trigger an immune response. Why this should happen has not yet been fully elucidated, but gliadin, the antigenic component of gluten, is said to be particularly resistant to digestive enzymes.

The key to the genetic element is that people with MHC-DQ2 or DQ8 MHC class II molecules on the surfaces of their APCs can easily bind and display deamidated gliadin peptides, altered by the tissue enzyme tissue transglutaminase (tTG). The immune response causes enterocyte destruction, local inflammation in the lamina propria and formation of specific coeliac antibodies.

Affected people develop coeliac disease when sufficient enterocytes disappear. This causes microscopically visible flattening of the intestinal villi and reduces the available absorptive surface. Patients with coeliac disease have malabsorption of all types of food and often complain of diarrhoea, extreme fatigue or abdominal pain. However, presentation is very variable, and often it is the incidental finding of anaemia or investigation for skin rashes or irritable bowel syndrome (IBS)-like symptoms that leads to diagnosis. Osteoporosis may develop, and occasionally osteomalacia, because vitamin D is poorly absorbed. A particular kind of blistering, itchy skin rash, dermatitis herpetiformis, is virtually always associated with coeliac disease.

Treatment is by withdrawal of the stimulus, gluten, and this is usually sufficient to reverse the process.

Patients with coeliac disease often have other 'auto-immune' diseases, such as diabetes mellitus (type 1) and thyroid diseases (Fig. 1.19, and see the clinical scenario below).

Normal duodenum

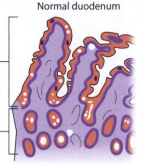

Normal slender villi with few intra-epithelial lymphocytes

Crypt to villus ratio 1:3

People with type II MHC antigens DQ2 or DQ8 are at risk of developing coeliac disease if wheat protein enters the lamina propria, possibly following an infection.

Some enteroviruses are thought to mimic wheat antigens.

An anti-inflammatory enzyme, tissue transglutaminase (tTG) deamidates gliadin, changing its three dimensional shape. The altered wheat protein 'fits' a groove in the DQ2 or DQ8 MHC antigens.

Intestinal epithelial cells present the wheat antigen associated with MHC II antigens to T cells. The T cells delete the epithelial cells bearing the foreign antigen, causing villous atrophy.

Removal of wheat from the diet restores normality in 3-6 months. Relapse occurs within days if wheat is eaten again.

Gluten-induced enteropathy (coeliac disease)

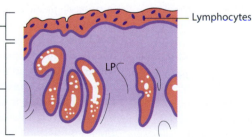

Near total villous atrophy — Lymphocytes

Crypt elongation, inflammation in lamina propria (LP) and increased lymphocytes in the surface epithelium

LP

Figure 1.19 Coeliac disease occurs in genetically predisposed individuals who are exposed to gluten molecules that enter the lamina propria in an undigested state. A reaction to gluten in these patients causes enterocyte destruction and local inflammation in the lamina propria. Removal of gluten from the diet is curative and prevents long-term complications of the disease. APC, antigen-presenting cell; MMP, matrix metalloproteinases; tTG, tissue transglutaminase enzyme.

Chapter 1: What causes disease?

Case study: coeliac disease

Clinical

A slightly built 25-year-old woman consulted her family doctor. She complained of fatigue and breathlessness on exertion, worse over the last 6 months. On examination, she was of medium height (1.65 m), low weight (50 kg) (see BMI chart in Fig. 1.20) and had pale mucous membranes. Glancing through her medical notes, the GP observed that she had been diagnosed with IBS a few months ago, having complained of abdominal pain and bloating. He also noticed that she had some bruising.

Investigations were as follows:

- Hb: 9.6 (normal 11–13.5) g/dL
- MCV: 86 (normal 78–95) fL
- WCC: 3.8×10^9/L (normal $3–5.5 \times 10^9$/L)
- Platelets: 330×10^{12}/L (normal 150–400) $\times 10^{12}$/L
- Red cell folate: 152 (normal 150–750) μg/L
- Vitamin B_{12}: 68 (normal 150–1000) ng/L
- Iron: 13 (normal 14–30) μmol/L
- International normalised ratio (INR): 2.5 (normal 0.8–1.1)

Provisional clinical diagnosis: malabsorption, probably due to coeliac disease (gluten-induced enteropathy)

Management and progress

She was referred for investigation at the gastroenterology clinic in her local hospital. The results of the investigations were as follows:

- Upper gastrointestinal (GI) endoscopy and duodenal biopsy: macroscopically normal, but biopsy showed subtotal villous atrophy and increased intraepithelial lymphocytes.
- Serology for anti-tTG IgA antibodies was positive.
- Bone density scan showed early osteoporosis.

Pathology

Pallor and symptoms suggest anaemia. Iron deficiency due to menorrhagia is the most common cause of anaemia in a young woman, but bruising would be unusual. Of patients with coeliac disease, 40% have IBS. This is a common diagnosis; however, 35–40% of patients with IBS will have coeliac disease if investigated.

Microcytic anaemia is the hallmark of iron deficiency and macrocytic anaemia typically indicates folic acid and/or vitamin B_{12} deficiency. This patient has a low serum iron and low normal red cell folate, and her anaemia is normocytic, which means that the average red cell size is normal.

Prolonged prothrombin time (measured as the INR) is likely to be due to decreased vitamin K causing a deficiency of clotting factors (see Table 1.9).

The incidence of coeliac disease in the UK is approximately 1:100 people. Many patients have subclinical disease.

Subtotal villous atrophy with crypt hyperplasia and increased intraepithelial lymphocytes is the typical microscopical appearance in coeliac disease, due to increased destruction of enterocytes by a T-cell-mediated reaction.

IgA anti-tTG will diagnose 95% of patients with untreated coeliac disease. IgA anti-endomysial antibodies are even more specific (99%) but slightly less sensitive and the test is more expensive. (Of patients with coeliac disease 2–4% are IgA deficient, compared with 0.2–0.5% of the general population.) IgG antibodies to tTG or endomysial antibodies are slightly less sensitive but are used if the patient is IgA deficient. Testing for deamidated gliadin is fractionally less sensitive.

Her diagnostic endoscopic biopsy in coeliac disease showed subtotal villous atrophy, increased lymphocytes among the surface enterocytes, lamina propria inflammation and crypt hyperplasia.

The mild osteoporosis noted on the bone density scan was due to vitamin D malabsorption, with resultant poor calcium absorption.

The suspected diagnosis of coeliac disease (gluten-induced enteropathy) was confirmed serologically and by endoscopic duodenal biopsy, and she was advised to exclude wheat, barley and rye from her diet (a gluten-free diet). Oats contain avenin, to which some people with coeliac disease are sensitive. She felt symptomatically improved after 1 month.

After 6 months she returns to the clinic. She is bored with the gluten-free diet. She feels well and is putting on weight. She requests a return to a normal diet.

Repeat duodenal biopsy after 6 months showed greatly improved microscopical appearances, which were almost normal.

She is warned to remain on a strict gluten-free diet, to prevent long-term sequelae.

Other members of her family are screened for occult coeliac disease.

Should the general population be screened for coeliac disease?

A gluten-free diet is difficult to sustain in western countries: wheat is present in bread, cakes, biscuits and sauces. Any trace is sufficient to spark a recrudescence of disease. Social occasions are fraught with difficulty.

Short-term problems due to coeliac disease include vitamin deficiencies, growth retardation (in children), steatorrhoea, malnutrition and anaemia, and osteoporosis. These are relatively quickly improved on a gluten-free diet.

She must not give up the gluten-free diet, however, or the problem will recur.

The duodenal biopsy may take up to 1 year to return to normal (but just days to show subtotal villous atrophy again after a gluten challenge!).

Long-term problems include an increased incidence of malignant gastrointestinal tract tumours, e.g. patients with coeliac disease are at 50–100 times the normal risk of developing malignant lymphoma and there is a moderately increased risk of small and large bowel adenocarcinoma, and squamous carcinoma of the oesophagus. It appears that the increased risk can be averted by adherence to a gluten-free diet.

The prevalence of coeliac disease in relatives is as follows:

- First-degree relatives: 10%
- HLA-identical siblings: 30%
- Dizygotic twins: 25%
- Monozygotic twins: 70–100%.

There are strong associations between HLA-DQ2 (95% of patients) and HLA-DQ8 and coeliac disease. All patients with coeliac disease inherit a particular HLA type, which renders them more likely to develop the disease – if negative it is not coeliac disease. However, HLA-DQ2 is common, present in 40% of the world's population, of whom only 3–5% will have coeliac disease, so serology for coeliac antibodies and endoscopic biopsy are recommended for diagnosis.

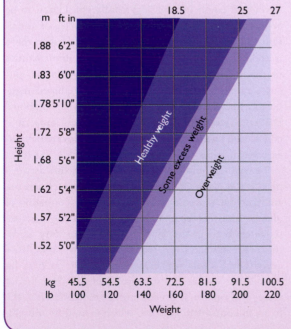

Figure 1.20 Body mass index (BMI) is calculated as follows: weight (kg)/height (m)^2. Values from 18.5 to 25 are healthy. Our coeliac patient's BMI is 18.3, so she is fractionally underweight.

Chapter 1: What causes disease?

Table 1.9 Clinical consequences of malnutrition

Nutrient deficiency	Clinical effect
Calories	Fat loss
	Muscle wasting
	Organ atrophy
	Growth failure in children
Protein (kwashiorkor)	As above but without fat loss and with oedema and fatty liver
Fat-soluble vitamins	
Vitamin A	Epithelial changes affecting eyes, skin and viscera
Vitamin D	Rickets and osteomalacia
Vitamin E	Neuromuscular degeneration
Vitamin K	Haemorrhagic disease of the newborn
	Effect on anticoagulants
Water-soluble vitamins	
Vitamin C	Scurvy (bleeding, poor wound healing, bone lesions)
Vitamin B_1 (thiamine)	Beri-beri (neural, cardiac and cerebral problems)
Vitamin B_{12} (cyanocobalamin)	Pernicious anaemia
Niacin	Pellagra (dermatitis, dementia, diarrhoea)
Folate	Megaloblastic anaemia
Vitamins B_1 and B_2 (riboflavin)	Ocular lesions, glossitis, stomatitis
Vitamin B_6 (pyridoxine)	Infant convulsions, anaemia, dermatitis, glossitis
Minerals	
Iron	Microcytic, hypochromic anaemia
Copper	Nerve and muscle dysfunction
Iodine	Goitre
Zinc	Growth retardation and infertility
Selenium	Myopathy and cardiomyopathy

NUTRITIONAL DISEASES

For centuries, the major problems in nutrition were centred around deficiencies. Too little protein resulted in kwashiorkor, whereas too few calories in the first year of life produced marasmus. Other diets lacked one or more vitamins or minerals because the variety of foodstuffs was not available. These problems, sadly, still exist in some developing countries but, in the developed world, malnutrition is more likely to be a consequence of bowel problems, other illnesses or psychosocial issues, although *nutritional imbalance* has become a major public health problem for otherwise healthy people. The key point is that nutritional imbalance for normal adults is a choice, and does not have a biological cause.

The most common imbalance is ingesting too much. The consequence of excess calorie intake is obesity, with its long-term effects on cardiovascular disease, diabetes, osteoarthritis and some cancers. The other most commonly over-ingested substance is alcohol. Chronic alcoholism is now known to produce most of its damage through direct toxic effects rather than secondary to nutritional deficiencies and results in central nervous system (CNS) atrophy, cardiomyopathy, peptic ulcers, pancreatitis, liver damage, varices, testicular atrophy and upper GI tumours. We are now becoming much more aware of the effects of other imbalances in our diet. Too much salt worsens hypertension, too much fat increases the amount of atheroma and the risk of heart attack and strokes, and certain foods (e.g. smoked foods) may increase the occurrence of GI tumours.

Metabolic syndrome is a common condition in developed countries and is generally considered to be at least partly due to a mismatch between our nutrition and our biochemistry. It is diagnosed when there is abdominal (central) obesity and two abnormalities of raised blood pressure, raised fasting plasma glucose, raised serum triacylglycerols or reduced high-density lipoprotein (HDL) levels (criteria International Diabetic Federation 2006 http://www.idf.org). It is considered further in the section on cardiovascular disease because it is a major risk factor.

The pathophysiology of metabolic syndrome is complex, but it can be induced by eating too much sucrose and fructose (sucrose is the disaccharide consisting of glucose and fructose). It is thought that the fructose is metabolised in the liver to short-chain fatty acids; this overloads the liver, leading to elevated blood triacylglycerol levels which induces ectopic fat (i.e. fat around the viscera and in organs not designed for fat storage). Fructose is predominantly metabolised in the liver, in contrast to glucose, which is used by muscle.

Overprovision of dietary calories compared with energy used in physical activity could also conceivably create an excess of mitochondrial oxidation products and result in insulin resistance, which is a key feature of metabolic syndrome. Other risk factors for metabolic syndrome are increasing age, lack of physical activity and stress.

The amount and type of food that we eat is influenced by social and psychological factors beyond the scope of this book. There are also genetic factors, including leptin coded for by the *LEP* gene. This is secreted by adipocytes and acts by reducing food intake and increasing energy expenditure. It is part of a complex system including ghrelin, peptide YY and insulin which acts on the brain to influence appetite. Through feedback loops, this aims to balance intake with output and is an example of homeostasis.

The food that we eat and the level of exercise are lifestyle choices for most people but some are unfortunate and have specific problems with ingesting, absorbing, metabolising or controlling excretion of essential nutrients (Table 1.10).

These conditions mostly affect general nutrition and this is what you will most commonly see, especially in those with chronic illness. Specific deficiencies are less common, with iron, folate, vitamin D and vitamin B_{12} probably being most important. If you are working in areas where nutrition is poor then make sure that you are aware of the clinical effects of all of them (Table 1.9).

GENETIC CAUSES OF DISEASE

Genes are so fundamental to the functioning of every cell that it would be hard to think of a disease in which alterations in gene activity did not occur. However, that would not mean that the disease was caused by an abnormal gene; so in this section we are interested in diseases that result from specific gene abnormalities and what might cause those abnormalities.

First let us look at some common ones (Table 1.11).

INCIDENCE

Table 1.11 lists the incidence per 1000 live births of the most common genetic disorders. It is helpful to subdivide them into abnormalities of chromosomal structure or number. The single-gene disorders are due to abnormalities of structure, which will be inherited in a mendelian fashion. Abnormalities of chromosomal number are not normally inherited.

There are several points to highlight. The first is that the incidence relates to live births, which means that genetic abnormalities causing intrauterine death will be under-reported. This principally influences the figures for chromosomal abnormalities, because their incidence in spontaneous abortions and stillbirths is 50% whereas the incidence in live births is 6.5/1000. In spontaneous abortions with chromosomal abnormalities, around 50% will have a trisomy, 18% will be Turner's syndrome (XO) and 17% will be triploid.

The most common condition is X-linked red–green colour blindness, which, fortunately, is only a very minor handicap (and is not an excuse for avoiding histology sessions!). Klinefelter's syndrome is due to an extra X chromosome in males (47,XXY). Affected individuals are generally of normal intelligence and are tall, with hypogonadism and infertility. XYY syndrome also produces tall males. They may have behavioural problems, especially impulsive behaviour.

Table 1.10 Conditions affecting general nutrition	
Mechanism	**Examples**
Reduced ingestion	Psychiatric illness
	Anorexia, e.g. linked to malignancy or chronic illness
	Food allergy
	GI disorders
Reduced absorption	Gut hypermotility
	Inflammatory bowel damage
	Pancreatic or biliary disease
	Achlorhydria
Abnormal metabolism	Malignancy
	Hypothyroidism
	Liver disease
Increased excretion	Diarrhoea
Increased demand	Fever
	Pregnancy and lactation
	Hyperthyroidism

Chapter 1: What causes disease?

Table 1.11 Common genetic disorders

Condition	Estimated frequency/1000 live births	Abnormality
Red–green colour blindness	80[a]	X
Total autosomal dominant disease	10	AD
Dominant otosclerosis	3	AD
Klinefelter's syndrome (XXY)	2[a]	N
Familial hypercholesterolaemia	2	AD
Sickle cell disease	2	AR
Total autosomal recessive disease	2	AR
Trisomy 21 (Down's syndrome)	1.5	N
XYY	1.5[a]	N
Adult polycystic kidney disease	1	AD
Triple X syndrome	0.6[b]	N
Cystic fibrosis	0.5	AR
Fragile X-linked learning disability	0.5[a]	X
Non-specific X-linked learning disability	0.5[a]	X
Recessive learning disability	0.5	AR
Neurofibromatosis	0.4	AR
Turner's syndrome (XO)	0.4[b]	N
Duchenne muscular dystrophy	0.3[a]	X
Haemophilia A	0.2[a]	X
Trisomy 18 (Edwards' syndrome)	0.12	N
Polyposis coli	0.1	AD
Trisomy 13 (Patau's syndrome)	0.07	N

AD, autosomal dominant; AR, autosomal recessive; N, disorder of chromosomal number; X, sex-linked disorder.
[a]Number/1000 male births.
[b]Number/1000 female births.

In familial hypercholesterolaemia, patients have increased plasma low-density lipoprotein (LDL) levels and a predisposition to developing atheroma at an early age, which gives them an eightfold increased risk of ischaemic heart disease. The primary defect is a deficiency of cellular LDL receptors, so that the liver uptake is reduced and plasma levels are two to three times normal. Around 30 different mutations of the LDL-receptor gene have been identified. About 1 in 500 people are affected and they are heterozygotes who have half the normal number of LDL receptors. One in a million people is a homozygote and he or she usually dies from cardiovascular disease in childhood.

The incidence of *sickle cell disease* varies significantly in different populations and is greatest in groups that have originated in malaria-infested areas. In US African–Americans, the incidence of disease is 1 in 500 and the carrier heterozygous state is 1 in 12. For details on sickle cell disease see page 279.

Adult polycystic kidney disease is due to a defect on the short arm of chromosome 16 which is inherited in an autosomal dominant fashion. Both kidneys are enlarged, with numerous fluid-filled cysts and may weigh a kilogram or more (normal is 150 g). The patients develop symptoms of renal damage and hypertension in their third or fourth decade (Fig. 1.21).

Figure 1.21 Adult autosomal dominant polycystic kidney disease. Gradually cystic dilatations of renal tubules and collecting ducts accumulate and distend the kidney, causing pressure atrophy of the renal tissue.

Triple X syndrome produces tall girls who may have below average intelligence; in addition, although gonadal function is usually normal, there may be premature ovarian failure. Fragile X syndrome was first described in 1969 and is now recognised as the second most common cause of severe learning disability after Down's syndrome. Affected males have a reduced IQ, macro-orchidism, and a prominent forehead and jaw. Heterozygote females can show mild learning disability but counselling is difficult because not all female carriers show the chromosomal abnormality on testing. Turner's syndrome (monosomy X, i.e. 45,X) is a common cause of fetal hydrops and spontaneous abortion. About 95% of affected pregnancies will abort. Those surviving to delivery will be less severely affected and generally show short stature, webbing of the neck, normal intelligence, infertility, aortic coarctation and altered carrying angle of the arm (cubitus valgus).

These disorders all relate to genes in the germ cell lines that give rise to all the cells in the body. Genes can also mutate during life and this is most obvious in various cancers. The mechanisms operating at the gene and cell level in cancers are covered on page 324 of Chapter 13.

Although we have listed the common genetic disorders and their inheritance pattern, this does not answer the question: 'How are they caused?' This is really two questions:

1 How does the genetic abnormality produce disease?
2 How does the genetic abnormality arise?

Basic biology

Chromosomes, DNA and genes

Just to remind ourselves of the basic biology: there are 23 pairs of chromosomes with one of these pairs being the sex-determining chromosomes, X and Y. Each chromosome is formed of two chromatids, joined by a centromere and with telomeres at each end. The appearance of these chromosomes, stained with Giemsa during mitosis, is the basis of classic cytogenetics and karyotyping, and the simplest way to detect changes in chromosome number or structure (see below).

Each chromatid is cleverly packaged DNA. If the human genome were constructed as a single straight strand, it would be 6 feet long, yet it fits into a tiny nucleus! DNA is composed of purine (adenine and guanine) and pyrimidine (cytosine and thymine) bases, arranged as pairs in a helical ladder. This DNA helix forms double loops around *nucleosomes* (globular aggregates of histone proteins) to produce a beaded string structure. This beaded string is coiled into a *solenoid*, composed of five or six nucleosomes per turn. This is chromatin in its relaxed state. During mitosis and meiosis, there is further condensation involving non-histone proteins. For those who like numbers, the DNA in each human cell contains around 7000 megabase-pairs (Mbp) (1 Mbp = 1 million base-pairs). Only 3% of the genome codes for genes (the *exons*). There are around 25 000 nuclear genes and a small but important set of 37 genes in the mitochondria that are inherited exclusively from the mother. There is much interest in the 97% non-coding DNA and its functions, some of which are related to gene regulation. You will appreciate that breaks in chromosomes, deletion of segments and point mutations in DNA will have different impacts depending on whether they affect exons or non-coding sections.

(Continued)

Chapter 1: What causes disease?

(*Continued*)

Haploid cells have only one set of chromosomes, i.e. '*n*' as occurs in gametes.

Diploid cells have a normal number of chromosomes, i.e. 46= '2*n*' as in all somatic cells.

Polyploid cells have extra sets of chromosomes, i.e. '*xn*'.

Aneuploid cells have an abnormal number of chromosomes but not an exact multiple of the haploid '*n*' state.

Mosaicism is the presence of two or more cell lines that are both karyotypically and genotypically distinct but are derived from the same zygote. These generally arise from post-zygotic, mitotic non-dysjunction. Sex chromosome mosaicism is more common than autosomal mosaicism.

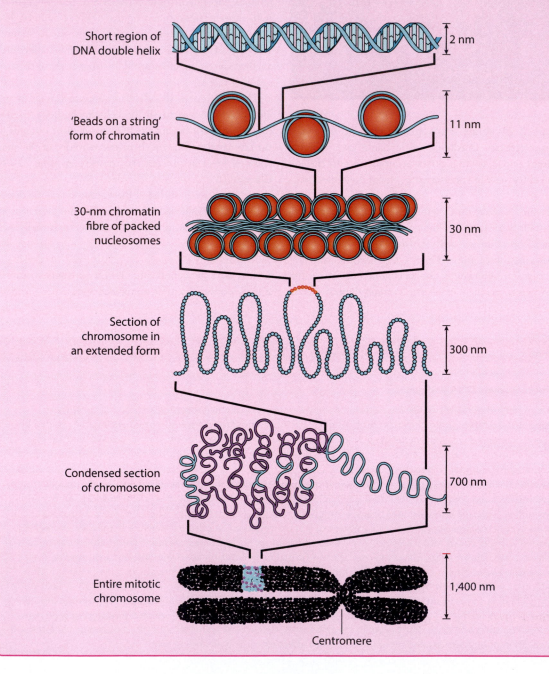

Short region of DNA double helix — 2 nm

'Beads on a string' form of chromatin — 11 nm

30-nm chromatin fibre of packed nucleosomes — 30 nm

Section of chromosome in an extended form — 300 nm

Condensed section of chromosome — 700 nm

Entire mitotic chromosome — 1,400 nm

Centromere

HOW DOES THE GENETIC ABNORMALITY PRODUCE DISEASE?

Genes code for proteins and so the disease patho-physiology will depend on what the normal protein does and how the changed protein functions. Going back to our list of common genetic disorders, we can now list the abnormal proteins and their function (Table 1.12).

Sometimes molecular biology can seem daunting with a vast amount of detail that can be difficult to relate to clinical practice. Often a way to simplify this is to decide into which category the abnormality falls and then to understand the typical mechanisms operating for that category. The majority of genes code for proteins and proteins can be:

- structural
- receptors or ion channels
- enzymes
- growth regulators.

In Chapter 2, we look at the mechanisms for how genes operate in abnormalities affecting the embryo, our biochemistry and the immune system. We choose examples in the embryo that affect cell differentiation which is *structural*; in our biochemistry the examples are linked to *enzyme* abnormalities and in the immune system the example is the generation of diverse surface *receptor* molecules to detect antigens. The topic of abnormalities in genes affecting *growth regulation* is covered in Chapter 11.

HOW DOES THE GENETIC ABNORMALITY ARISE?

We need to consider abnormalities of chromosome number separately from abnormalities in chromosome structure or genes, because different mechanisms operate.

Abnormal chromosome number

This most commonly occurs because of problems at the anaphase stage of either meiosis 1 or meiosis 2 during production of the gametes. This leads to unequal sharing of the chromosome material so that, after fertilisation, one daughter cell will have an extra chromosome (trisomy) whereas the other is missing a chromosome (monosomy) (Fig. 1.22). This is called *non-dysjunction* and can affect a pair of homologous chromosomes in meiosis 1, or sister chromatids in meiosis 2. It can also occur through delayed movement (anaphase lag) of chromosomes, so that one is left on the wrong side of the dividing wall. The cause is unknown but the incidence increases with maternal age, as discussed when considering Down's syndrome (page 38 or case study). If non-dysjunction occurs in mitosis of normal tissue, it can result in mosaicism. Polyploidy means that the cell contains at least one complete extra set of chromosomes. Most commonly, this is one extra set, i.e. 69 chromosomes, or triploidy. Affected fetuses usually die *in utero* or abort in early pregnancy. It can result from fertilisation by two sperm (dispermy) or from fertilisation in which either the sperm or the ovum is diploid because of an abnormality in their maturation divisions.

Table 1.12 Genetic abnormalities and their dysfunctional proteins

Disease	Abnormal protein	Function
Familial hypercholesterolaemia	Low-density lipoprotein receptor	Receptor transport
Neurofibromatosis type 1	Neurofibromin 1	Growth regulation
Adult polycystic kidney disease	Polycystin 1	Cell–cell and cell–matrix interactions
Cystic fibrosis	CF transmembrane regulator	Ion channel
Sickle cell disease and thalassaemia	Haemoglobin	Oxygen transport
Haemophilia A	Factor VIII	Coagulation
Muscular dystrophy	Dystrophin	Structural support: cell membrane
Fragile X syndrome	FMRP	RNA translation

Chapter 1: What causes disease?

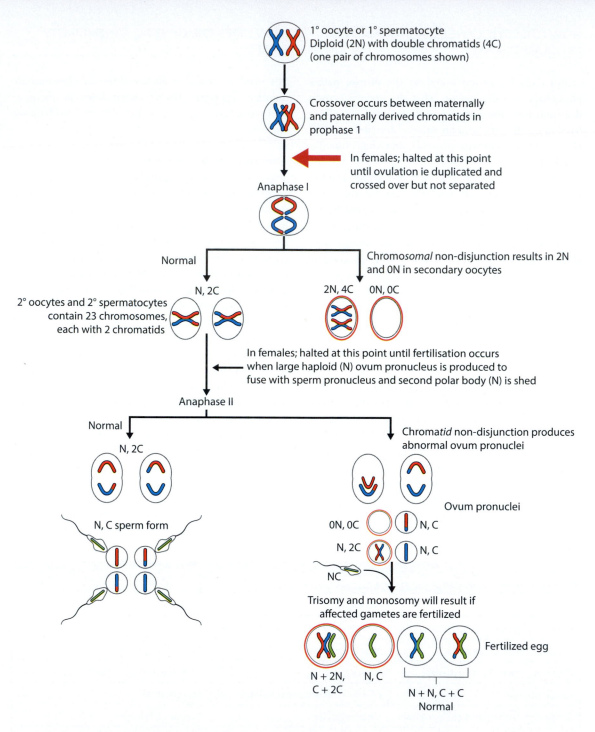

Figure 1.22 Abnormal chromosome numbers may arise through problems in Anaphase I and Anaphase 2 of meiosis. Resting cells (and fertilised eggs) have 23 pairs (2N) of single stranded (2C) chromosomes. Dividing cells (and 1° oocytes/ spermatocytes) have double strands (4C) known as sister chromatids. Highlighted in red are the abnormal secondary oocytes produced by chromosomal non-disjunction in Anaphase 1, abnormal ovum pronuclei produced by chromatid non-disjunction at Anaphase 2 and abnormal fertilized eggs. The left hand side shows normal meiosis for comparison.

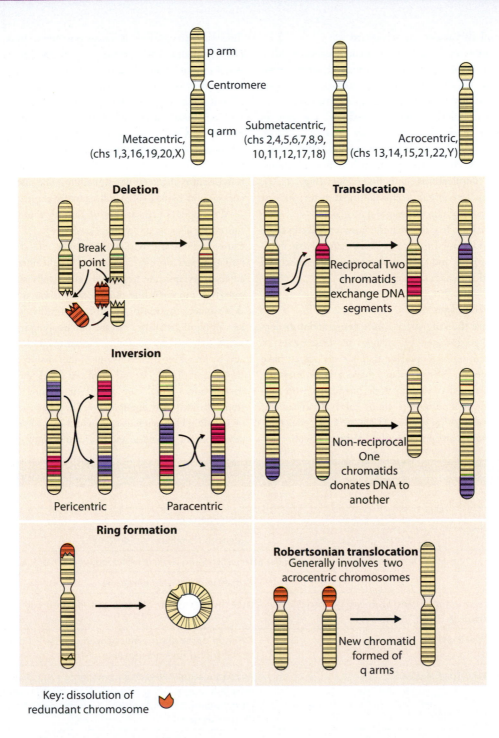

Figure 1.23 Structural chromosomal abnormalities.

Abnormal chromosome structure

Abnormalities in chromosome structure occur when chromosomes are inaccurately repaired after breaks have occurred. Chromosomal breakage can happen randomly at any gene locus, but there are some areas that are particularly liable to breakage. The rate of breakage is markedly increased by ionising radiation and certain chemicals, and some rare inherited conditions. Structural abnormalities, such as translocations, deletions, duplications and inversions (Fig. 1.23), occur when two break points allow transfer, loss or rearrangement of chromosomal material.

Abnormalities in chromosome number and structure have been identified for many years using classic cytogenetic methods and various karyotypes are associated with specific diseases (Table 1.13).

Gene level changes

Once we are thinking about single genes and disorders that may be produced by changes in those genes, it is worth reminding ourselves that we do not all have the same genes. At any locus, there may be multiple possible alleles that vary in their frequency with ethnic origin, depending on their survival advantage. This is normal *polymorphism*. Over 30% of genes coding for proteins are polymorphic. When patients react differently to diseases, it is now possible to look at their DNA sequence and compare it with the known genes listed in worldwide databases. If there are differences, then it may be relevant because of the known function of that gene, e.g. resistance to HIV occurs in people with a naturally occurring different gene for the T-cell receptor CCR5, due to a 32-bp deletion (Δ32) causing a frameshift mutation. Homozygotes can become HIV positive but do not develop AIDS because the new protein disrupts intracellular signalling. The allele is most common in Finns (16%) and virtually absent in Africa.

Many agents are capable of causing mutations, and mutations also occur spontaneously (Fig. 1.24). *Physical* causes include ultraviolet (UV) light, electromagnetic radiation and atomic radiation. UV light normally links adjacent pyrimidine bases and causes only somatic cell mutation not germline mutation. It is a major cause of skin cancer, especially in red-haired individuals whose lack of pigment allows release of damaging free radicals when exposed to sunlight. Electromagnetic radiation acts by knocking electrons out of their orbits and causes single- and double-strand breaks in DNA and base-pair destruction. This can result in major deletions, translocations and aneuploidy. They have most effect on dividing cells such as sperm and in early embryonic development. *Chemical* and *viral mutagens* often result in

Table 1.13 Examples of karyotypes

46XY	Normal male
46XX	Normal female
47XXY	Male with Klinefelter's syndrome
If there is a change in chromosomal number, then the affected chromosome is indicated with a + or -, e.g.	
47XX+21	Female with Down's syndrome
If there is a structural rearrangement, the karyotype indicates the precise site affected and the nature of the abnormality, e.g.	
46XXdel7(p13-ptr)	Deletion of the short arm of chromosome 7 at band 13 to the end of the chromosome
46XYt(11;14)(p15.4;q22.3)	A translocation between chromosome 11 and 14 with the break points being band 15.4 on the short arm of chromosome 11 and band 22.3 on the long arm of chromosome 14
Mosaicism indicates that two different cell lines have derived from one fertilised egg and the karyotype specifies both cell lines	
46XX/47XX+ 21	Down's syndrome mosaic
46XX/45X	Turner's syndrome mosaic

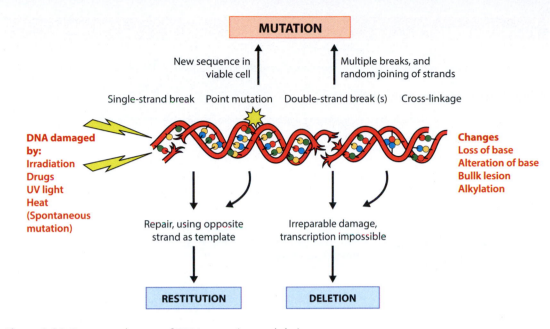

Figure 1.24 Causes and types of DNA mutation and their consequences.

cancer and the molecular biology of their mechanisms can be found on pages 323 and 335.

Mutations at the gene level can either involve millions of bases or be a point mutation. The factors that affect chromosome structure can also occur at a single gene level, i.e. segment deletions, duplications and translocations. Point mutations are usually spontaneous and of unknown cause, but are probably mostly the result of copying errors. Bases may be substituted, inserted, deleted and replicated (amplification).

Substitution of one base within a codon may lead to a different amino acid being inserted into the protein and major pathological effects, e.g. sickle cell disease. However, this is not inevitable because there are only 20 amino acids but 64 possible codons (4 × 4 × 4), which is the basis of the degeneracy of the genetic code, e.g. an mRNA sequence of GAA or GAG will code for alanine, so some point mutations can alter the codon but have no effect on the amino acid sequence. These are called *silent mutations* and approximately 25% of point mutations have no effect.

As well as coding for amino acids, codons also act as start and stop instructions. UAA, UAG or UGA is used as a stop codon by mRNA. If a point mutation produces a stop codon, the amino acid chain will terminate too early and is the effect of about 5% of point mutations.

If one or two (but not three) bases are inserted or deleted, this will produce a *frameshift mutation* with much greater effect, because the code is based on triplets of bases (codons) and so changes dramatically if the groups of three are shifted along to produce a nonsense message.

Amplification of sequences of three bases (triplet repeats) is called 'dynamic mutation' and is recognised as the cause in about 20 diseases, 3 examples of which are Huntington's disease, fragile X disease and myotonic dystrophy type 1.

In each of these diseases, there is 'anticipation' which means that the disease often becomes more severe with each generation or occurs earlier. If the genes are investigated, it is apparent that the number of triplet repeats correlates with the severity. The number needed for severe disease is different in the different conditions, with Huntington's disease requiring more than 40 CAG repeats, whereas fragile X disease requires several hundred repeats and is more severe in males. The mechanism is also different with the abnormal gene for Huntington's disease coding for a protein toxic to neurons, whereas the fragile X mechanism blocks transcription by methylating the promoter.

Chapter 1: What causes disease?

Basic biology

Gene disorders are most commonly produced by mutations in the DNA sequence. These can occur in the germ-line or the somatic cells.

Mutations (Fig. 1.25) take various forms:

- Substitution of single nucleotide bases, producing *missense mutations*, e.g. sickle cell anaemia
- Insertion or deletion of fewer than three bases, causing a *frameshift mutation*
- Creation of a *stop code* by a point mutation
- Amplification of a sequence of three bases
- Copy number variations (CNVs).

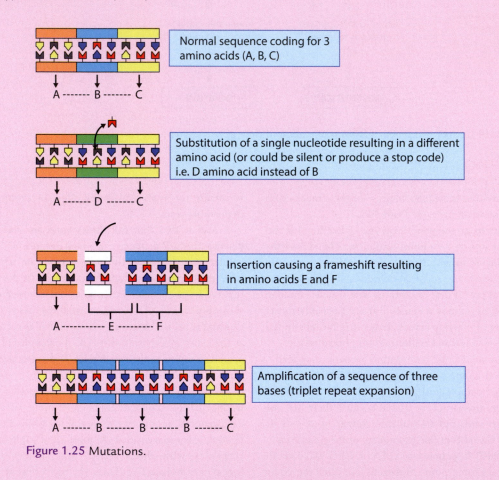

Normal sequence coding for 3 amino acids (A, B, C)

Substitution of a single nucleotide resulting in a different amino acid (or could be silent or produce a stop code) i.e. D amino acid instead of B

Insertion causing a frameshift resulting in amino acids E and F

Amplification of a sequence of three bases (triplet repeat expansion)

Figure 1.25 Mutations.

Epigenetic change: this is modulation of genes without alteration of the DNA sequence.

Imprinting: this is an epigenetic process during gametogenesis that switches off an allele so that there is monoallelic gene expression.

Single nucleotide polymorphism (SNP) is a variation in single nucleotides anywhere in the DNA. They can be useful for DNA fingerprinting and other investigations.

Copy number variations

CNVs are probably responsible for most of the diversity in normal humans and at least half are involved in gene-coding sequences. As they are so common, it is thought that some may convey an evolutionary advantage. They have also been identified in a variety of disorders, such as epidermal growth factor receptor (EGFR) copy number in non-small cell cancer and high CCL3L1 copy number in people with lower susceptibility to HIV infection.

Single nucleotide polymorphisms

Over 6 million SNPs have been identified in humans but only 1% are in coding regions and so they may have less biological effect, although they have been useful in studying inheritance if located near (and inherited with) a known disease-associated gene. This method is known as *linkage disequilibrium*.

Epigenetic change

Modulation of genes without alteration of the DNA sequence (*epigenetic change*) is important for homeostasis and development, and allows regulation of gene and protein expression through alteration of gene promoters and so on, e.g. by methylation of cytosine residues. It is important in cancer progression and treatment. If it occurs as a part of normal development, it is known as *imprinting* (see later).

Imprinting

Imprinting silences genes and occurs to some genes during gametogenesis, so that either only the father's or only the mother's allele will be expressed. The mechanisms involved in imprinting differ for different genes, but they are being intensely investigated because imprinting disorders occur in the assisted conception techniques of *in vitro* fertilisation and intracytoplasmic sperm injection. The most common disorders of imprinting are Prader–Willi, Angelman's and Beckwith–Wiedemann syndromes.

Alterations in non-coding RNAs

These are exciting areas of new research because non-coding RNAs are involved in regulation of genes and, hence, cell pathways. There are estimated to be around 1000 genes in humans for micro-RNAs (miRNAs) which are crucial for post-transcriptional silencing of genes. There are even more long non-coding RNAs (lncRNAs) that may be important in diseases such as atherosclerosis and cancer.

MULTIFACTORIAL DISORDERS

Every new patient is asked about his or her 'family history', the idea being that if the parents and siblings have a particular disease then the patient is at increased risk. Unfortunately, for most diseases it is not known how great that increased risk may be because the inheritance does not follow simple mendelian principles but is rather multifactorial. It is likely that there will be a variety of genes involved that interact with a number of environmental factors. Research into multifactorial disorders adopts a similar approach to single-gene problems. First, it is necessary to identify the diseases with a significant genetic component by comparing the incidence in family groups with that in the general population. This genetic contribution is termed heritability (Table 1.14) and some examples are listed below.

The next step is to look for genetic, biochemical and immunological features that affected individuals have in common. It has been well established that certain HLA types are associated with particular diseases and this may be helpful in counselling affected families, e.g. in a family with ankylosing spondylitis, a first-degree relative has a 9% risk of developing the disease if HLA-B27 positive but less than a 1% risk if HLA-B27 negative.

Table 1.14 Heritability or the genetic contribution to the aetiology of the disorder

Disease	Estimate of heritability (%)
Schizophrenia	85
Asthma	80
Cleft lip and palate	76
Coronary artery disease	65
Hypertension	62
Neural tube defect	60
Peptic ulcer	35

Chapter 1: What causes disease?

The ultimate goal is to identify the gene or genes and the environmental factor(s), so that those at particularly high genetic risk could attempt to avoid the relevant environmental hazard. At a simple level, this would mean giving vitamin supplements to pregnant women at risk of producing babies with neural tube defects, or advising individuals with potential 'arteriopathy' to modify their diet and not smoke.

Type 1 diabetes is a disease with a wide geographical variation, in which a genetic predisposition and environmental factors seem to interact to produce the clinical manifestations of pancreatic islet cell damage. It is associated with HLA and two haplotypes have a particular association with the disease – DR4-DQ8 and DR3-DQ2 – which are present in 90% of children with type 1 diabetes. It is clear, however, that the environment plays a significant role, with viruses being the major culprit. Enteroviruses, rotaviruses and, in particular, rubella appear to be the principal candidates.

Case study: an abnormal baby

A 41-year-old mother gives birth to a live male infant. She has had two previous normal pregnancies with the same father. Shortly after delivery, the baby seems to be relatively floppy, although he is moving all four limbs and has been crying and feeding successfully. Examination reveals that the baby's head is particularly round, the palpebral fissures are upward sloping and the medial epicanthic folds are prominent. The tongue is enlarged and heavily fissured. The baby's hands are broad and possess only a single palmar crease, whereas the little fingers are short and curved inwards. These features persist over the next few months and there is a delay in the child reaching the early motor milestones. Later, the child does not learn to walk until the age of 3 and his speech is delayed until 54 months, although by this time the diagnosis has already been made.

Question 1: What is the diagnosis?
Answer: Down's syndrome – the physical features are characteristic.
Question 2: What is Down's syndrome?
Answer: Down's syndrome is a chromosomal disorder in which the affected individual has trisomy 21. It should be noted that not all of chromosome 21 needs to be present in excess to produce Down's syndrome and that trisomy of a certain portion of the long arm alone will cause the disease.
Question 3: Other than confirming the diagnosis in the child, is there any other role for postnatal genetic investigation in this family?

Answer: Yes. Occasionally (around 3% of cases), Down's syndrome results from a balanced (robertsonian) translocation in one parent. In this situation, the parent has an abnormal karyotype in which the crucial part of one of the chromosome 21s is translocated on to another chromosome. The parent is phenotypically normal because, although the translocated portion of the chromosome is not in the right place, it is still present and can function normally. However, this means that the germ cells also carry the balance translocation and, when they undergo meiosis, a gamete results in which a normal chromosome 21 is accompanied by the augmented other chromosome, so that gamete has a double dose of the long arm of chromosome 21. (Chromosome 14 seems to be susceptible to acquiring the extra part of chromosome 21.) If a parent has a balanced translocation, they are at particular risk of having another child with Down's syndrome because the underlying cause persists. This is in contrast with the majority of cases in which the mutation is sporadic.

Read more in Pathology in Clinical Practice Case 26

Radiology

The role of imaging in identifying disease in the unborn baby

Imaging, in particular ultrasound, is extensively used in developed countries to identify prenatal disorders.

The nuchal translucency scan (NTS) (Fig. 1.26) uses ultrasound to non-invasively measure the amount of fluid collecting subcutaneously in the nape of the fetal neck at 11–13 weeks' gestation. The measurement is combined with blood tests to give a risk estimation for the presence of aneuploidy (an abnormal number of chromosomes). If the pregnancy is high risk, the mother can elect to have a more invasive test such as chorionic villous sampling or amniocentesis to detect a chromosomal disorder in the fetus. One of the most common disorders that NTS may help to identify is Down's syndrome (trisomy 21), which may otherwise have a normal morphological fetal ultrasound.

At 18–22 weeks the morphology scan is performed. By this gestational age the fetus is virtually fully developed and large enough for a detailed ultrasound examination of the brain, facial bones, spine, heart, abdomen and limbs to be performed. An abnormal morphology ultrasound can trigger further invasive investigations such as amniocentesis, usually at a tertiary referral centre.

Some disorders may be surgically treatable at birth and prenatal identification of these disorders, such as sacrococcygeal teratoma (Fig. 1.27), allows the obstetrician and surgeon to closely monitor the pregnancy at a specialist centre, and plan an elective delivery and subsequent surgery.

There are other disorders that may be identified which are untreatable genetic disorders. This allows parents to make informed choices whether to continue or terminate the pregnancy.

Some centres will perform fetal MRI to help identify disorders. Imaging in the form of MRI or CT is also used in fetal or neonatal demise (see Fig. 1.10) to identify genetic skeletal and visceral anomalies that may explain the death of the child. This also has implications for genetic counselling of the involved families, particularly with regard to future pregnancies.

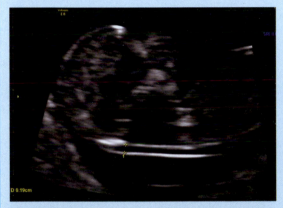

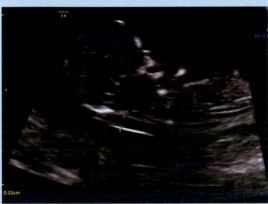

Figure 1.26 (a) A normal nuchal translucency measurement is seen; (b) an abnormally thick nuchal translucency.

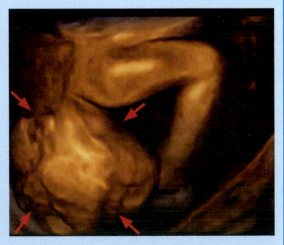

Figure 1.27 Three-dimensional fetal ultrasound showing a mass arising from the lower back and buttocks of an 18-week-old fetus. This is a sacrococcygeal teratoma, which can be benign or malignant, and is treated with surgical excision at birth.

Chapter 1: What causes disease?

CHAPTER 2

WHAT ARE THE COMMON MECHANISMS OF DISEASE?

In considering the Origin of Species, it is quite conceivable that a naturalist, reflecting on the mutual affinities of organic beings, on their embryological relations, their geographical distribution, geological succession, and other such facts, might come to the conclusion that each species had not been independently created, but had descended, like varieties from other species.

Charles Darwin (1809–1882)

INTRODUCTION

This is the moment to move from the whole patient and delve down into the tissues, cells and molecules. We have to assume that you are familiar with the basics of genetics, molecular biology and cell biology, and so we won't cover these in detail but will remind you of some key concepts, terminology or interesting examples in 'basic biology boxes'. To begin gently, we will consider Mendel and his pea experiments.

Mendel was born in 1822 and, at the age of 21 years, joined the Augustinian order. He studied at the University of Vienna for 10 years and then entered a monastery at Brunn. He started experimenting with peas in 1856, looking at the probability of inheriting certain characteristics. He published his results in 1865

but they didn't come to prominence until 1900 when other botanists were doing similar work.

Mendel looked at seven distinct characteristics and only bred plants that differed in one characteristic. For the sake of discussion, let us consider violet and white flowers. He crossed plants with violet flowers with those bearing white flowers to produce the next generation, called the F1 generation. He found that the F1 generation plants all had the same colour flowers – let us say that they were all violet. The F1 plants were then self-pollinated (inbred) to produce the next generation, called F2. Interestingly, there were three plants with violet flowers for every one plant with white flowers. He took this one step further and self-pollinated the white plants, which gave rise to an F3 generation of plants that all had white flowers. Self-pollination of the violet plants produced an intriguing result: some plants produced only violet plants whereas others produced a mixture of white and violet plants in the ratio 1:3 (Fig. 2.1).

As Mendel correctly deduced, although the violet plants in the F2 generation all looked the same, they had different inheritance factors. He postulated that each plant must possess two factors that determine a given characteristic, such as colour of the flower. If two plants are crossed, each will contribute one factor to the next generation and it is purely random as to which

Part 1: Disease, health and medicine

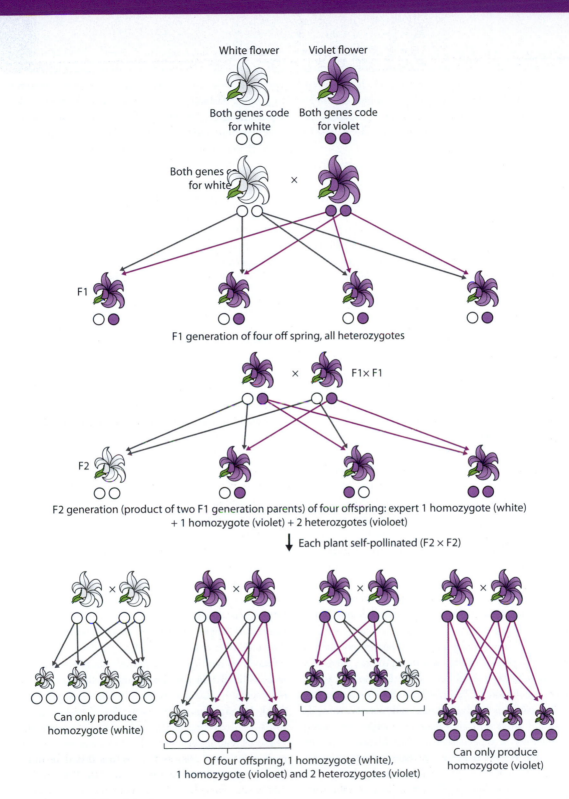

Figure 2.1 Mendelian inheritance.

factor is passed on. This is the law of segregation, also known as Mendel's first law. We now know that these 'factors' are genes on chromosomes, which are paired, and the two genes on the two chromosomes are alleles of each other. In Mendel's experiment, violet is the dominant allele and white is the recessive allele. The F1 generation has one white plant that is homozygous for the white allele, one violet plant that is homozygous for the violet allele and two violet plants that are heterozygous, i.e. they have one white and one violet allele but the violet one dominates. In simple examples such as these, one allele completely dominates, i.e. if the plant has at least one violet allele then all the flowers will be violet. Sometimes the situation is more complicated because there will be variable penetrance, i.e. the 'dominant' allele dominates only in a percentage of cases.

The physical basis for Mendel's 'factors' started to be discovered in the 1880s when Walter Fleming, working at the University of Kiel, stained nuclei and observed the changes in chromatin and chromosomes during cell division, which he called *mitosis*. He realised that cell nuclei came from the division of precursor nuclei with distribution of their chromosomes, but knew nothing of Mendel's work and did not make the link to inheritance.

Friedrich Weismann was another German biologist and he recognised the significance of *meiosis* for reproduction and inheritance in 1890. He also led on the idea that the characteristics bred true and were not altered during life, in contrast to Lamarck who believed that inherited characteristics could be acquired. To 'prove' the point, Weismann removed the tales of 68 mice over five generations producing 901 offspring who all had normal tails! Weismann was also unaware of Mendel's work but was strongly influenced by Darwin.

There was a huge amount of interest in inheritance in the early part of the twentieth century as Mendel's work was rediscovered and several people reached similar conclusions. Chromosomes were recognised as the physical basis for inheritance and these were studied as the science of *cytogenetics*. As more data were accumulated, it was recognised that Mendel's law of segregation wasn't true for all characteristics – many

characteristics were inherited together. What we now call genes were referred to as 'units' and it was postulated that these units were positioned on chromosomes in a linear fashion, were always in the same position (locus), and the likelihood of inheriting characteristics together depended on first whether they were on the same chromosome and second how far apart they were on a chromosome. The American Alfred Sturtevant worked on drosophila flies, between 1915 and 1928, to divine the principles of genetic mapping by realising the significance of unequal crossing-over and genes that formed 'linkage groups' and were inherited together. This was the phase of *classic genetics* based on cytogenetic experiments.

Avery, MacLeod and McCarthy, in 1944, demonstrated that DNA was the biochemical basis of genetics; this marked the beginning of the *molecular genetics* era (however, their ideas were not well received at the time and had relatively little impact and no Nobel Prize). Watson and Crick, in 1953, used crystallographic images of DNA provided by Wilkins and Franklin to discern the structure of DNA, and most crucially its helical nature. In 2007, James Watson became the second person to publish his fully sequenced genome online.

I am putting my genome sequence online to encourage an era of personalised medicine in which information contained in our genome can be used to identify and prevent disease and to create individualised medical therapies.

James Watson (2007)

Between the discovery of DNA and the widespread availability of whole genome sequencing, there are two crucial technological advances. The first came from Fred Sanger's laboratory in 1977 when they used the 'dideoxy' chain termination method (the Sanger method) to rapidly and accurately sequence long sections of DNA (in contrast to the earlier use of DNA polymerase). This was ultimately used to sequence the whole human genome in 2003. The second is ongoing and is collectively called 'next-generation sequencing'. These techniques are dependent on sophisticated automation and huge data processing capability as well as

History box

Frederick Sanger (1918–2013) (Fig. 2.2)

Sanger was a British biochemist who won two Nobel Prizes. His father was a general practitioner and a Quaker. Sanger was a pacifist and conscientious objector during the Second World War and started his PhD in 1940, investigating whether an edible protein could be obtained from grass. His first Nobel Prize, in 1958, was for determining the complete amino acid sequence of the two chains of bovine insulin, at a time when protein structure was not understood and thought to be amorphous. He correctly concluded that the sequences were precise and different, and that every protein might have a unique amino acid sequence. This was crucial to Crick's later ideas on how DNA codes for proteins. The second Nobel Prize was in 1984 for the Sanger method for sequencing DNA molecules.

Figure 2.2 Frederick Sanger.

the fundamental biochemical methods. At present, 'massive parallel signature sequencing', 'parallel pyrosequencing' and 'Illumina (Solexa) sequencing' are among the leading types.

So what does all this mean for patients and those of us interested in the mechanisms of disease? Let's look at four settings influenced by genes and their products:

1 Genetic effects on the embryo
2 Genetic effects on biochemistry
3 Genetic effects on the immune system
4 Molecular genetics of cancer.

These provide examples related to different actions of proteins:

- Structural
- Enzymes
- Receptors or ion channels
- Growth regulators.

GENETIC EFFECTS ON THE EMBRYO

The obvious place to start is in the embryo with body patterning, which is the term used to describe the regular features of our anatomy. If fertilisation is day 0 and implantation occurs at day 6, then the embryo is forming at 14 days and the major organs are formed by 42 days. The key embryological cell layers are ectoderm, endoderm and mesoderm, and the cells need to know what their position is in the embryo and how they should differentiate. This is achieved by creating gradients across and along the body to distinguish anterior from posterior, right from left, and dorsal from ventral. You wouldn't be surprised to learn that this is mediated through differential gene expression, but in a particularly interesting way because of the need to set up *linear* gradients so that the genes are arranged in a *co-linear* way.

The anteroposterior axis is dependent on the four *HOX* gene clusters and these gene clusters have up to 13 genes arranged in line (co-linearity), with the genes at the $3'$-end being activated earlier than those further down the line. Not surprisingly, this creates a linear gradient of gene products, but what are the gene products? They are proteins containing a domain that can bind specific DNA sequences in the nucleus to regulate other genes crucial for forming specific body structures.

Dorsoventral differentiation and right–left differentiation are more complex but one important clinical example of an abnormality in dorsoventral differentiation involves abnormalities of the *SHH* (sonic hedgehog) gene. This affects Shh protein expressed in the notochord and neural tube folding, potentially resulting in

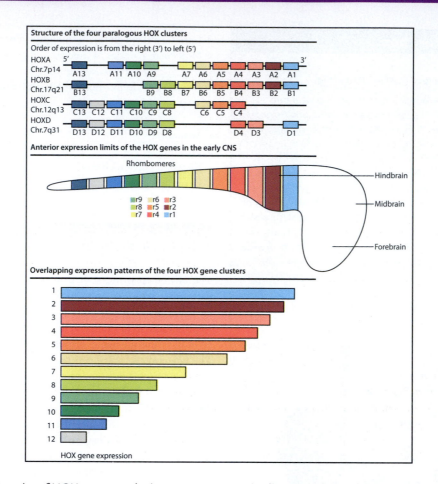

Figure 2.3 Linearity of HOX genes results in an antero-posterior linear gradient of gene products important for early body patterning. From *Medical Genetics at a Glance.* 3rd Edn, D J Pritchard, B R Korf.

Basic biology

- Homeobox: DNA sequence involved in morphogenesis
- Homeodomain: protein domain coded by homeobox DNA, which itself can bind and regulate nuclear DNA, generally by acting on promoter regions.
- Human genes are written in italic capitals (e.g. *SHH*) and non-human are in italic lower case (e.g. *Shh*).

non-division of the forebrain (holoprosencephaly) and cyclopia (single central eye). Shh protein is also one of the first signals in right–left differentiation as the heart tube loops. Other genes include *Lefty1* (*LEFTA*), *Lefty2* (*LEFTB*) and *Nodal* (*NODAL*). *Nodal* and *Lefty1* are activated by asymmetrical flow of fluid and this depends on having functioning cilia to waft the fluid.

Kartagener's syndrome is the combination of dysfunctional cilia, situs inversus (mirror imaging of heart, lungs, liver, spleen, stomach and small bowel), chronic sinusitis and bronchiectasis. It is caused by a variety of cilia-related genes but especially those coding for the dynein arms on the microtubules. In children and adults, the abnormal cilia fail to move bronchial mucus normally or create flow along the fallopian tubes and so they have increased chest infections and reduced fertility. They also have random situs (i.e. the right–left arrangements of the heart, lungs and abdominal organs can be the wrong way round), and this is all explained by the underlying problem with cilia.

Part 1: Disease, health and medicine

Mutations of *LEFTA*, *LEFTB* and *NODAL* can also cause situs inversus through a combination of actions, because Nodal and Lefty2 proteins affect genes specifically on the left side of the embryo and Lefty1 protein prevents signal transmission across the midline. Right–left symmetry is quite commonly abnormal in monozygotic twins, not because of a primary abnormality of genes, but because of diffusion of lateralising gene products from the left side so affecting the twin on the right.

The gradients extend right to the fingers and toes and *HOX* gene mutations can cause problems such as syndactyly (fusion of digits).

Before we leave the subject of embryogenesis, it is worth emphasising that the majority of congenital abnormalities are of unknown cause, 20% are genetic (chromosomal or monogenic), 20% are multifactorial and 10% are environmental, especially infection or alcohol related (Fig. 2.4). They can cause the following:

- A primary failure in development (*malformation*), e.g. cleft lip and palate
- Damage to a normally formed organ (*disruption*), e.g. loss of fingers or arm by an amniotic band, damage due to infections such as cytomegalovirus
- Mechanical distortion (*deformation*), e.g. dislocation of hips because of a lack of amniotic fluid (oligohydramnios)
- Abnormal tissue organisation (*dysplasia*), e.g. kidney cysts which are often monogenic in cause.

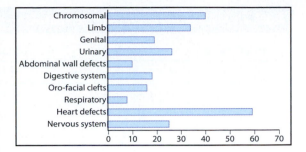

Figure 2.4 Estimated prevalence per 10,000 births of congenital anomalies in England and Wales 2010. Drawn from data in Springett A, Morris JK. Congenital Anomaly Statistics 2010: England and Wales. London: British Isles Network of Congenital Anomaly Registers, 2012.

being ingested, so DNA testing is likely to be used increasingly. After birth, around 40 disorders can be screened for using tandem mass spectrometry on heel-prick blood samples rather than the time-honoured approach of examining the urine of newborn babies (e.g. maple syrup urine disease, named because of its sweet smell).

CELL AND TISSUE DEPOSITS

In several of these disorders it is possible to actually see the accumulated substance with simple stains down a light microscope. Cells can accumulate substances that may cause damage or be harmless. In addition to inborn errors, the main mechanisms are as follows:

- Altered metabolism, e.g. fatty change in hepatocytes
- Inability to digest ingested substances, e.g. carbon, melanin and haemosiderin
- Misfolding of proteins, e.g. Creutzfeldt–Jakob disease (prion proteins), Alzheimer's disease (Aβ-peptides).

Material can also accumulate in tissues rather than cells and the most important of these is cholesterol in the intima of arteries in atheroma. This occurs through a combination of increased intake and reduced removal. For details see page 240 Fig. 8.5.

Proteopathies
Some genetic abnormalities can predispose individuals to conditions related to misfolding of proteins. Often, however, the misfolding seems to be caused by an

GENETIC EFFECTS ON BIOCHEMISTRY

'Inborn errors of metabolism' have been recognised for many decades, and over time the biochemical and genetic details have been elucidated. The most important are listed in Table 2.1.

There are some general principles worth appreciating. Most of these conditions involve loss of a specific enzyme, leading to accumulation of the enzyme substrate and lack of its normal product. Diagnosing these conditions can involve measuring the accumulated substrate, assaying the enzyme activity or looking specifically for DNA mutations. Prenatal diagnosis is often important at stages of development when the enzyme would not be functioning, the relevant tissue is not accessible or the dietary substrate is not yet

Chapter 2: What are the common mechanisms of disease?

Table 2.1 Main inherited defects of metabolism

Type of disorder	Disorder	Incidence	Genetics	Defective biomolecules
Amino acids	Phenylketonuria	1/10 000	AR	Phenylalanine hydroxylase
	Cystinuria	1/7000	AR	SLC3A1/SLC7A9
Carbohydrates	MODY	1/400	AD (>8 genes)	Various affecting beta cell function
	Diabetes type 1	1/400	Polygenic >40 genes	Various affecting autoimmunity
	Diabetes type 2	1/10	Polygenic around 30 genes	Various affecting insulin resistance
	Lactose intolerance	1/10 (white people)	AD	Lactase
Metal transport	Haemochromatosis	1/300 (white people)	AR – four genes	*HFE*, transferring receptor 2, ferroportin
Lipid metabolism	Familial hypercholesterolaemia	1/500	AD	Low-density lipoprotein receptor
	MCAD deficiency	1/20 000	AR	Medium chain acyl-CoA dehydrogenase
Porphyrias	For example, acute intermittent	All rare	Mostly AD	Defective enzymes
Purine metabolism	Severe combined immunodeficiency	1/70 000	AR	Adenosine deaminase affecting production of T/B cells
Sphingolipidoses	Gaucher's disease	1/900 (Ashkenazim)	AR	β-Glucosidase
Mucopolysaccharidoses	Hurler's disease (MPS1)	1/10 000	AR	α-L-Iduronidase
Peroxisomal	Adrenoleukodystrophy	1/20 000	XR	Very-long-chain fatty acid synthase
Hepatic glycogen storage	GSD1: von Gierke's disease	Rare	AR	Glucose 6-phosphatase
Muscular glycogen storage	GSD2: Pompe's disease	Rare	AR	Lysosomal α-1,4-glucosidase

GSD, glycogen storage disease; MCAD, medium-chain acyl-CoA dehydrogenase; MODY, maturity-onset diabetes of young people; MPS, mucopolysaccharidosis.

initiator (e.g. prion disorders), or because there are large amounts of identical proteins (e.g. light chain amyloid in myeloma or hormone-related amyloid). The proteins change their three-dimensional shape by increasing the amount of β-sheet secondary structures. This can cause disease through loss of the protein's

Table 2.2 Some important proteopathies

Condition	Protein
Alzheimer's disease	Amyloid β, Tau
Prion diseases	Prion proteins
Primary systemic amyloidosis	Monoclonal Ig light chains
Secondary systemic amyloidosis	Amyloid A (acute phase reactant)
Dialysis amyloidosis	
Cystic fibrosis	β_2-microglobulin, CFTR protein

Table 2.3 Risk factors for proteopathies

- Unstable primary amino acid sequence
- Introduction of an initiator
- Changes to temperature or pH
- Increased protein concentration through increased production or decreased clearance
- Post-translational modification, e.g. hyperphosphorylation

normal function (e.g. cystic fibrosis transmembrane conductance regulator protein [CFTR]) or interference with the tissue through accumulation (e.g. cardiac amyloid). The accumulation can be intracellular or extracellular.

Historically, the material was called amyloid because it could be seen down a microscope and stained in a similar way to starch (Latin for starch is *amylum*). More recently, there has been particular interest in proteopathies affecting the brain and causing dementia. It is appreciated that the classic amyloid plaques visible on Congo red light microscopical staining may be relatively benign, whereas invisible, non-fibrillary, misfolded oligomers are more potent causes of neuronal damage. A particularly important protein in neurodegenerative conditions is the tau protein which is the basis for the neurofibrillary tangles seen in Alzheimer's Disease and other conditions. The tau protein is associated with microtubules and aggregates after being hyperphosphorylated.

Case study: haemochromatosis

A 59-year-old man presents with a 1-month history of swelling of his ankles, urinary frequency and polydipsia. Further questioning elicits a history of reduced libido and shortness of breath on moderate exertion. On examination, the patient seems to have a greyish change to his skin, although he does not appear anaemic or cyanosed. There is mild pitting oedema of the ankles, an irregular pulse and bilateral basal crackles in the lungs. The testes show a mild degree of atrophy. Blood tests showed normal full blood count (FBC), normal urea and electrolytes (U&Es), abnormal liver function tests (LFTs) and a raised fasting glucose level.

Question: What endocrine abnormalities does this patient have?

Answer: The elevated fasting blood glucose indicates diabetes mellitus. This would also explain the polyuria and polydipsia caused by an osmotic diuresis induced by glycosuria. There is also likely to be an element of hypogonadism. There has been a change in the patient's libido and there seems to be some degree of testicular atrophy. Further investigation would reveal that the patient has a low blood follicle-stimulating hormone (FSH), luteinising hormone (LH) and testosterone.

Question: What is the diagnosis?

Answer: The patient has haemochromatosis. He has abnormal LFTs, cardiac disease, diabetes mellitus, pituitary disease (hypogonadism secondary to decreased levels of FSH and LH from the pituitary) and characteristic skin pigmentation changes.

(Continued)

(Continued)

Question: What is haemochromatosis?

Answer: Haemochromatosis is an autosomal recessive disease in which there is an abnormal increase in iron absorption from the gut.

Question: How could the diagnosis be confirmed?

Answer: Genetic studies can be performed and are the most useful, but measurement of iron transport and storage levels in the blood is also helpful and often quicker. Liver biopsy can reveal the presence of excess iron, as well as assessing other histology. The gene for haemochromatosis is located on chromosome 6p21.3, although alternative loci have been described for variants of the disease. There is linkage with HLA-A3 and -B14.

Read more in Pathology in Clinical Practice Case 46

Basic biology

Point mutation mechanism in somatic hypermutation

The latest experimental evidence suggests that these mutations are deliberately introduced by using the AID (activation-induced [cytidine] deaminase) enzyme and 'error-prone DNA polymerase'. The AID enzyme deaminates one or more cytosine bases in the DNA, so creating uracil bases, which are not normally found in DNA, and are detected by DNA mismatch-repair enzymes. The 'error-prone DNA polymerase' introduces the mutation as it is 'repairing' the DNA strand. This is why the mutations are predominantly single-base mutations rather than insertions or deletions.

GENETIC EFFECTS ON THE IMMUNE SYSTEM

The immune system and its role in defending the body against pathogens is covered in section X, its defence against cancer in section Y and its harmful effects in autoimmune disease in section Z. Here we will consider how the genes in B lymphocytes can produce billions of different antibody specificities (generation of diversity) and the genetic component of allergic reactions (atopy).

GENERATION OF DIVERSITY

See Figure 6.17 on page 196 for gene rearrangements in heavy chain production.

An immunoglobulin (Ig) molecule is composed of two identical heavy chains (H) and two identical light chains (L), each of which has constant (C), joining (J) and variable (V) regions. The heavy chains also have diversity (D) regions. The genes for light chain production are on chromosome 2 (κ chains) and chromosome 22 (λ chains), and there are around 40 V genes, 5 J genes and 1 C gene. For heavy chains, there are 9 C genes, 27 D genes, 44 V genes and 6 J genes on chromosome 14. This is the situation in the germline but, during B-cell maturation, there is rearrangement of the genes to bring together the required number of V, D, J and C genes to provide a specific code for that B cell. This code specifies the shape of the B-cell surface receptor and the antigen-specific antibody that it produces. You could multiply the number of V and J genes for light chains by the number of V, D and J genes for heavy chains to have an idea of the number of different B-cell specificities produced. However, this is only part of the story because there is *somatic hypermutation*, which is a programmed process of mutation that occurs in individual B cells after they have been stimulated by antigen to divide. The B-cell-receptor locus undergoes single-base substitutions at 'hotspots' called 'hypervariable regions' and the mutation rate is up to 1000 000 times higher than in normal dividing cells. Why? It is thought this allows the B cell with the closest fit to the new antigen to refine its progeny to be even better at responding to that antigen.

So the germline cell has a variety of genes that are rearranged to produce a diverse population of B cells; these then hypermutate when stimulated to proliferate by antigens. Of course, it is always possible that a new mutation could react with your own cells to cause an autoimmune (atopic) reaction or produce other cellular changes, leading to B-cell lymphoma.

Allergic reactions

Atopy is the triad of asthma, atopic eczema and allergic rhinitis, and it is both common and multifactorial. Some of the basic mechanisms are covered on page 80 (hypersensitivity reactions) and page 22 (gluten sensitivity in Chapter 1). As our knowledge of the genome increases, it is becoming clearer how many genes may be implicated, although it has long been appreciated that a family history of atopy is relevant. In particular there is a dominant autosomal allele carried by 25% of northern Europeans that can result in overproduction of IgE. The incidence is highest in offspring of affected mothers and it is thought that the gene may be imprinted (see page 37).

Read more about allergic reactions in Pathology in Clinical Practice Case 6

MOLECULAR GENETICS OF CANCER

This is covered in depth in Chapter 13.

It is worth mentioning here that the decades of work involved in looking at genetic changes in tumours is beginning to allow a personalised approach to therapy, increasing the likelihood of response and reducing the number of patients exposed to unpleasant side effects. In the UK, there is a trial for non-small-cell lung cancer involving 14 drugs, with patients having their tumour genetics assessed before being enrolled in the trial using the most appropriate drug. Around 21 genetic abnormalities have been identified in non-small-cell tumours but each occurs only in a small percentage of tumours, so it is important to know which patient's tumour has which genetic abnormality.

CELL DAMAGE AND CELL DEATH

Let us move from the genome to the cell but using a common clinical scenario, a patient who presents with sudden onset of neurological symptoms.

Case study: shortness of breath

A 23-year-old woman goes to see her general practitioner because she has had several episodes of shortness of breath in the past 2 months. The episodes last for up to 30 minutes and are sometimes associated with wheezing. She thinks that there is a connection with being exposed to dust because she had her worst episode when she was helping to clean a friend's loft and another when she was changing the bag on her vacuum cleaner. She also thinks that cats trigger her shortness of breath and has noticed that she sometimes has a dry cough, particularly at night. She has not had a productive cough, haemoptysis or chest pain. She is otherwise fit and well, takes regular exercise and her exercise tolerance remains unchanged.

Question: What is the likely diagnosis in this patient's case?

Answer: Asthma – the presentation in a young person of episodic dyspnoea that is associated with wheezing and a dry cough and related to precipitating factors is very suggestive. Extrinsic allergic alveolitis might also be considered, but the precipitating factors are somewhat diverse (assorted dusts and cats) and the cough is not associated with the exposure. The nocturnal exacerbation of the cough is also suggestive of asthma.

Question: What is asthma?

Answer: Asthma is a chronic condition in which there is bronchial hypersensitivity and hyperreactivity, leading to reversible episodes of bronchospasm that produce dyspnoea and/or wheezing.

Read more in Pathology in Clinical Practice Case 6

Chapter 2: What are the common mechanisms of disease?

CLINICAL SCENARIO: STROKE

A 76-year-old woman is brought into accident and emergency having had a sudden onset of weakness of her right arm and leg that occurred 90 minutes earlier and is associated with difficulty in speaking. The weakness was without warning or precipitating event. The patient's previous medical history is unremarkable.

On examination, the cardiovascular, respiratory and abdominal systems are unremarkable. Examination of the limbs reveals normal power in the left arm and lower limb but grade 0 movement in the right arm and lower limb. There is hyperreflexia of the biceps, brachioradialis and triceps, knee and ankle reflexes on the right, and the right plantar response is upgoing. Tone is increased. There are no abnormal movements, fasciculations or wasting. Coordination is not assessable in the right limbs due to the loss of power but is normal in the left arm and lower limb. Widespread sensory loss for all modalities is present in the right arm and lower limb.

Examination of the cranial nerves reveals a right homonymous hemianopia and weakness of the upper part of the right side of the face. The patient understands verbal instructions but has great difficulty in speaking, using only short, incomplete sentences. She seems to be frustrated by her inability to speak. She is able to sip a glass of water without problems.

She is treated with appropriate thrombolytic treatment but the symptoms persist. She has had a stroke and some cells in the brain had been deprived of oxygen (*hypoxia*) because of an interruption to the blood supply (*ischaemia*). The medical term for a stroke is a cerebrovascular accident (CVA) because the cause is most commonly an obstruction (occlusion) or rupture of a vessel supplying the brain.

The radiological images would show a sequence of changes from the initial insult through to limited repair. If the patient died, then there would be specific changes seen down the microscope that reflect cell damage and cell response.

🔗 Read more in Pathology in Clinical Practice Case 12

Microscopical appearances

The first phase is generalised oedema as the damaged cell membranes and altered permeability of the blood brain barrier affect both intracellular and extracellular fluid distribution. This produces acute cellular oedema or 'cloudy swelling'. At the light-microscopy level, this appears as enlargement of the cell and a pale granular look to the cytoplasm. Vesicles may also appear due to the distension of the endoplasmic reticulum. This picture of cellular oedema is also referred to as hydropic or vacuolar degeneration. Within 12 hours of irreversible injury, the microscopical changes in neurons involve shrinkage of the cell body with eosinophilia (red staining on the haematoxylin and eosin stain) of the cytoplasm, pyknosis and disintegration of the nuclei (karyorrhexis), disappearance of the nucleolus and finally complete dissolution of the nuclei (karyolysis), see fig 2.11 page 58. If the axon is damaged, there is a microscopically visible spheroid, caused by swelling and disruption of axonal transport systems. The brain has nerve cells (neurons) and supporting cells (glial cells). The neurons are most susceptible to ischaemia. The glial cells may be so badly damaged that they also die, or they may survive and work alongside other glial cells invading the damaged area, removing dead cells and attempting limited repair. This is called gliosis. The astrocytes increase in number (*hyperplasia*) and size (*hypertrophy*), and have some similarities to fibroblasts in other tissues but don't produce significant collagen. The microglia are phagocytic and cluster around damaged cell bodies and axons or areas of haemorrhage.

Our unfortunate patient has suffered an *infarction,* which is the term used when there is a localised area of irreversible ischaemic tissue damage, most commonly caused by a reduced blood supply. The brain does not have any glycogen or fat storage and so is dependent on a steady supply of glucose and oxygen via the blood. In the adult, the brain weighs 2% of the body weight but consumes 20% of the body's oxygen, and so is the most vulnerable organ to hypoxia. Neurons (nerve cells) are also so specialised, with their complex axonal and dendritic structures and myelin coatings, that they are unable to divide to repair damage. This is why strokes are

Radiology of a stroke

Series of images showing the evolution of a stroke are shown here in Fig. 2.5.

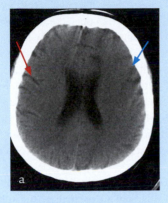
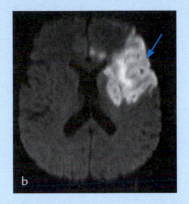
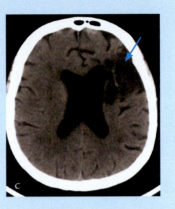
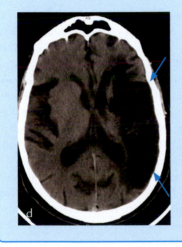

Figure 2.5 (a) The acute CT scan when the patient is brought into accident and emergency and is confused and has an impaired level of consciousness. The image shows that the patient is moving (hence the slight blurring). In the left hemisphere one notices a loss of the grey–white differentiation of the cerebral cortex and the subcortical white matter (abnormal side – blue arrow, normal side – red arrow). This indicates oedema as the blood–brain barrier breaks down and the cell membranes let fluid in. (b) MRI within the first week in hospital: this is a diffusion-weighted image (DWI) that shows the distribution of trapped water in the brain – the water filling the dead cells – cytotoxic oedema (blue arrow). The infarct is in the territory of the left middle cerebral artery. (c) The CT scan 3 months down the line. The infarct is on the way to becoming cystic and therefore blacker (blue arrow). There is also a loss of volume and the lateral ventricles dilate, filling the space previously taken up by brain tissue. (d) Another patient 5 years after his original stroke, which shows a large cystic cavity replacing much of the left hemisphere (blue arrows) after a large middle cerebral artery infarct.

so devastating. If symptoms persist beyond the first few hours, it indicates *irreversible* cell damage with no prospect of structural repair, but sometimes there is some compensatory functional recovery through neuronal plasticity and opening up of alternative neuronal pathways. So what is happening at a cellular level?

BIOCHEMICAL CHANGES IN THE CELLS

There are two important questions to consider:

1 What are the biochemical changes that occur in an injured cell?
2 What distinguishes reversible from irreversible injury?

Chapter 2: What are the common mechanisms of disease?

There are four sites within the cell that are of paramount importance in cell damage and death:

- Mitochondria
- Plasma membrane
- Lysosomes
- Cytoskeleton.

The key mechanisms are:

- depletion of ATP
- increased intracellular calcium
- increase in free radicals
- disruption of membranes
- DNA and protein damage.

Depletion of ATP

The first effect of ischaemia is to reduce the production of adenosine triphosphate (ATP) by the mitochondrial oxidative phosphorylation system (Fig. 2.6). If the production of energy slows down or stops, then the cells cannot function. In tissues that have glycogen, the ischaemic cells will switch over to anaerobic metabolism to produce ATP, but this has the unwanted effect of producing lactic acid so that the pH drops in the cell, potentially creating more damage by affecting cellular enzymes and clumping the nuclear chromatin, so-called *pyknosis*. Ischaemia also has profound effects on the plasma membranes and the ionic channels situated within the membranes. You will recall that these are vital in maintaining the normal ionic gradients across the cell membranes, with sodium and calcium at low concentrations inside the cells and potassium lower in the extracellular space. These concentrations are maintained by pumps that are energy dependent, so it is not difficult to see that the loss of oxidative phosphorylation and any direct damage to the membranes will disrupt the function of these pumps. So what is the effect?

First, the failure of the pumps will result in leakage of sodium into the cells and potassium out of the cells. Sodium has a larger hydration shell than potassium, so more water moves in association with sodium ions than exits with the potassium. Additional water enters because the acidosis and raised intracellular concentrations of high-molecular-mass phosphates will increase the osmotic pressure inside the cell. The result is acute swelling of the cell due to cellular oedema. The endoplasmic

reticulum (ER) also swells, the ribosomes detach from the ER, the mitochondria become swollen and blebs begin to appear on the cell surface. This last phenomenon is intriguing because the changes in cell shape and surface blebbing imply alterations in the cytoskeleton of the cell. The changes in the microfilaments of the cytoskeleton are believed to be due to the increased concentration of calcium, which also results from the failure of the membrane pumps. Calcium is a very important ion in cell death and we shall see why in a minute.

You might find it difficult to believe but all the changes described so far are reversible! If the oxygen supply is restored, the cells still have the capacity to return to the normal state and the neuron will transmit again. So what are the changes that finally tip the cell beyond the point of no return?

The morphological hallmarks are severe disruption of the membranes affecting, in particular, the mitochondria, plasma membranes and lysosomes. Calcium is thought to play a central role in this final progression to irreversible cell death.

Increased intracellular calcium

In the normal cell, the calcium concentration is tightly controlled by the calcium pump in the cell membrane. Ischaemia disrupts oxidative phosphorylation, so affecting the energy-dependent calcium pump, leading to a rapid influx of calcium and saturation of the calcium-regulating proteins. Damage to cell organelles also leads to release of intracellular calcium. The high levels of calcium are toxic to the cell, leading to changes in the cytoskeleton, cell surface blebbing, and damage to the mitochondria, lysosomal membranes and cell membranes. The release of enzymes from the ruptured lysosomes also contributes to the final destruction of the cellular components.

Increase in free radicals

In addition, there is another important pathway common in many types of cell damage that generates reactive oxygen species (ROS). This is called *oxidative stress* and involves free radicals derived from oxygen. As you will recall, free radicals have a single unpaired electron in their outer orbit, which makes them unstable and strongly reactive to cellular lipids, proteins and nucleic acids. ROS are produced normally in mitochondrial respiration but don't cause damage because they are removed through the action of superoxide dismutases,

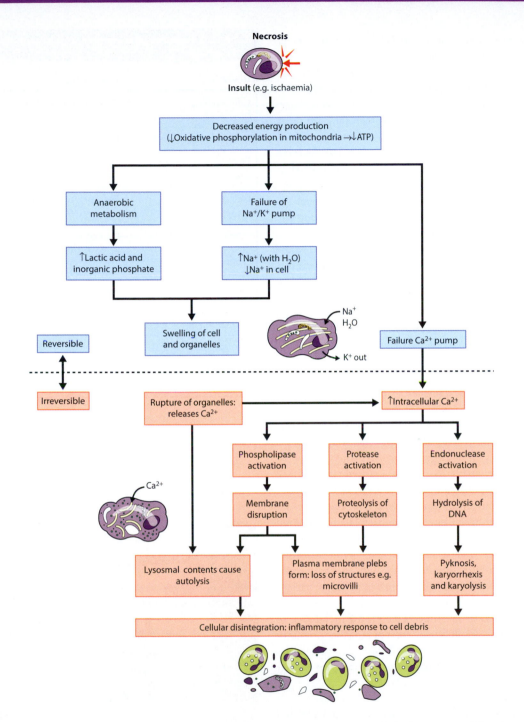

Figure 2.6 Necrosis is the culmination of a series of events, the first phases of which are reversible. The failure of the calcium pump marks the onset of irreversible change. The cell breaks down and its contents are exposed to the extracellular space, with leakage of lysosomal membranes and other cellular constituents. The cells appear swollen and degenerate under the microscope.

Chapter 2: What are the common mechanisms of disease?

glutathione peroxidases, catalases and antioxidants, such as vitamins A, C and E.

ROS are also essential, in low concentration, for many cell signalling pathways. They are produced in significant amounts by neutrophils and macrophages for killing harmful microbes, but then they are safely packaged in lysosomes that will fuse with phagosomes. (Phagosomes are membrane-bound ingested microbes or other pathogenic substances.) Free radical production is pathologically increased by ionising radiation, chemicals (e.g. carbon tetrachloride and paracetamol) and oxygen toxicity, and in cellular ageing and ischaemia–reperfusion damage.

ROS damage the DNA by producing single-strand breaks that contribute to cell death, ageing and cancer. They cause cross-linking or fragmentation of proteins so that the proteins cannot function normally. They attack polyunsaturated lipids in membranes, which not only damages the membrane but also generates more peroxides, so amplifying the process.

Disruption of membranes

Membranes can be damaged by physical and chemical agents, microbial toxins and components of the complement cascade (see page 137), as well as ischaemia and ROS. These generally act through some combination of increased degradation of membrane phospholipids, probably by calcium-induced activation of endogenous phospholipases, or decreased phospholipid synthesis because of a reduction in energy-dependent cell activities. The impact of phospholipid breakdown products can be devastating on the cell because they can insert into the membrane lipid bilayer, altering permeability and electrophysiological behaviour (obviously important in our patient with ischaemic nerve cells).

The cytoskeleton connects the plasma membrane to the cell interior and is important in signalling, motility and internal structure. If the skeleton is damaged by proteases then the membrane function deteriorates.

DNA and protein damage

The final item on our list of mechanisms is DNA and protein damage. These commonly cause a specific type of cell death called apoptosis, which we consider on page 62.

Table 2.4 Features of reversible and irreversible cell damage	
Reversible	Irreversible
Cell swelling	Release of lysosomal enzymes
Mitochondrial swelling	Protein digestion
Endoplasmic reticulum swelling	Loss of basophilia
Detachment of ribosomes	Membrane disruption
'Myelin' figures	Leakage of cell enzymes and proteins
Loss of microvilli	Nuclear changes: pyknosis, karyorrhexis, karyolysis
Surface blebs	
Clumping of nuclear chromatin	
Lipid deposition	

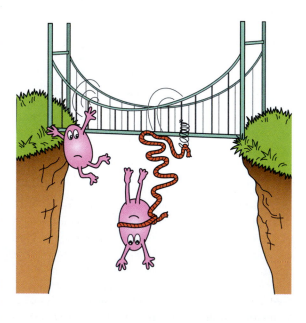

Figure 2.7 Reversible and irreversible changes.

CLINICAL RELEVANCE OF CELL CHANGES

So much for science! Do these biochemical and microscopical changes help us to understand any of the clinical manifestations of our patient with cerebral ischaemia?

The nerve cell's ability to conduct electrical signals is dependent on the integrity of its plasma membrane and the pumps that maintain the ionic gradients. For the transmission of signals from one nerve to another, it requires the production and controlled release of transmitter substances, followed by their reuptake or degradation. For the movement of substances from their site of synthesis in the cell body to the ends of the axon, it needs an intact cytoskeleton, especially the microtubules. You can easily appreciate that the changes that occur in an ischaemic cell totally disrupt these functions and nerve transmission ceases. The clinical impact depends on which area of the brain is affected because many nerve pathways are quite discrete and the blood supply to the brain has distinct territories, so the effects can be quite localised but still very disabling. Books on neuroanatomy and clinical neurology will provide the detail but, in the specific case of our patient, the CVA has occurred in the left cerebral hemisphere and the combination of neurological defects indicates damage to brain regions that are all located in the territory of the left middle cerebral artery. The left cerebral hemisphere deals with motor and sensory function for the right side of the body and also from the right visual field. The localisation of speech function is slightly more complex: >99% of right-handed people and around 50% of left-handed people have their speech centres in the left cerebral hemisphere.

Specialised cells are affected differently by ischaemia so it is worth looking at the clinical effects of damage on liver cells and cardiac muscle cells.

The hepatocyte is principally a chemical factory synthesising and degrading molecules through many complex biochemical pathways. This requires a range of enzymes and any severe damage to the cell that disrupts cell membranes will allow leakage of those enzymes. This provides the basis of some of our most commonly used laboratory blood tests for screening and assessing a patient. So-called LFTs look at the blood levels of key enzymes that are predominantly or exclusively found in hepatocytes. Their level in the blood will rise if there is acute or ongoing liver cell damage. The same is true of muscle cells that leak enzymes (such as creatine kinase [CK], aspartate aminotransferase [AST] and lactate dehydrogenase [LDH]) and troponin molecules, which can be used to help confirm a clinical diagnosis of myocardial infarction.

The brain and the heart are the two major organs affected by acute or chronic ischaemia. Just as in the brain, the heart is dependent on electrical conduction to keep it beating in a coordinated way. If the membrane potential is altered, it can result in cardiac arrhythmias, either because the pacemaker is affected or the conduction pathways are damaged, or because myocytes are not responding normally. This happens extremely quickly and can result in a sudden cardiac arrest and death. At that point, the cytoskeleton with all the mitochondria and myofilaments is still intact and so the damage is reversible. Later the proteins degrade, the membranes rupture and inflammatory cells, especially macrophages, infiltrate the damaged area to remove the debris and lay down a scar (see healing and repair, page 147). That section of the heart can no longer contract and so the ventricular pump is weaker and the patient may be breathless, have peripheral oedema or limited exercise tolerance – the features of chronic heart failure.

We have talked about reversible and irreversible cell injury; however, there is a situation where a cell that has suffered only minor damage can suffer a further fatal insult; this is ischaemia–reperfusion injury and it is clinically important in cerebral and cardiac ischaemia.

An increase in damage caused by ROS is thought to be a major mechanism and this could result from the following:

- Increased production of ROS by damaged mitochondria
- Increased action of oxidases in endothelial cells and infiltrating leukocytes
- Reduced cellular antioxidant mechanisms.

In addition, reperfusion will bring inflammatory cells and mediators, such as components of the complement system, which can cause damage.

Finally we must consider the clinical effects of alcohol on liver cells because it is such a common cause of disease. The earliest and most reversible damage is *fatty change*. This refers to an excess of intracellular lipid, which appears as vacuoles of varying size within the cytoplasm. Similar to cellular oedema, it is entirely reversible and is a non-specific reaction to a variety of insults. Sometimes it is present adjacent to tissues that are more severely damaged or show frank evidence of necrosis. Fatty change can occur in any organ but is most frequent in the liver, which is not surprising because the liver is the major site of lipid metabolism.

Figure 2.8 illustrates the main causes and effects of fatty change in the liver. Put simply, adipose tissue releases fat as free fatty acids which enter the hepatocytes, where they are converted to triacylglycerols and, to a smaller extent, cholesterol. Triacylglycerols are complexed with apoproteins to form lipoproteins, which are then secreted into the blood. Changes will lead to lipid accumulation within the hepatocytes. The liver appears enlarged and pale (Fig. 2.9) and fat globules are seen microscopically (Fig. 2.10).

Alcohol is a hepatotoxin that has wide-ranging effects on fatty acid metabolism. It increases peripheral tissue release of fatty acids so that more are delivered to the liver and, within the liver, it is implicated in increasing fatty acid synthesis, decreasing the utilisation of triacylglycerols, decreasing fatty acid oxidation and blocking lipoprotein excretion. Thus, it interferes with a variety of biochemical pathways.

Gross examination of the organs affected by fatty change will show that they are enlarged and yellow, and tend to be greasy to the touch. Microscopically, the characteristic finding is of vacuoles within the cytoplasm. These may begin as small vacuoles but, if the fatty accumulation continues, the vacuoles will coalesce to form larger vacuoles or 'fatty cysts'.

This type of change is entirely reversible if the insult is withdrawn. A binge in the medical school bar on a Friday night may produce fatty change but this will disappear if one is able to abstain for a few days afterwards! Chronic abuse of alcohol may produce sufficient fatty change to interfere with the normal function of the hepatocytes and, in the long term, excessive alcohol consumption may lead to cell death, scarring and cirrhosis, which is not reversible.

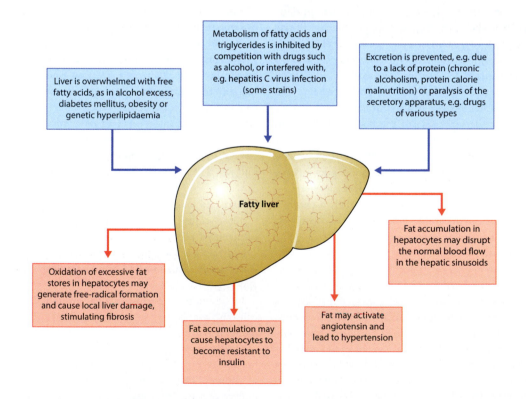

Figure 2.8 Fatty liver: main causes of fat accumulation and their deleterious effects.

Part 1: Disease, health and medicine

Figure 2.9 Normal and fatty liver (pale).

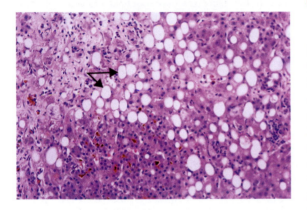

Figure 2.10 Photomicrograph of liver showing fatty change (vacuoles arrowed).

Key facts

Causes of fatty change in the liver

Diabetes, metabolic syndrome and obesity

Alcohol

Protein malnutrition

Acute fatty liver of pregnancy

Congestive cardiac failure

Ischaemia/anaemia

Drugs: steroids, methotrexate, intravenous tetracyclines

Carbon tetrachloride

The other two major causes of fatty change in the liver are malnutrition and diabetes. Malnutrition in particular affects two steps in fat metabolism. It increases the release of fatty acids from peripheral tissue and protein deficiency reduces the cell's ability to combine triacylglycerols with apoprotein. Disordered carbohydrate metabolism in uncontrolled diabetes leads to excessive peripheral release of fatty acids. Metabolic syndrome has many similarities to type 2 diabetes and can be considered a pre-diabetic state. It has similar disturbances of carbohydrate metabolism, insulin resistance and raised levels of triacylglycerols, so it is also a cause of fatty change in the liver.

Case study: fatty liver

Malcolm, a 48-year-old man, was hit by a car as he weaved his way home by bicycle (he lost his licence for drink driving a year ago). His injuries were mild, but he had hit his head so he was kept in for observation. After a day, the nurses noted that he was plucking agitatedly at his sheets and cringing from 'vultures' circling him. He was also sweating, with tachycardia and rapid breathing. The doctor diagnosed delirium tremens caused by acute alcohol withdrawal. Malcolm recovered over the next 2 days and a psychiatric referral was made, but Malcolm denied an alcohol problem and discharged himself.

One year later, Malcolm was rushed into hospital in a weak, jaundiced and semi-comatose state. His breath was foul smelling ('foetor hepaticus'). His sclerae were deep yellow and his liver was enlarged and tender. Investigations showed that he was in fulminant liver failure.

Question: Discuss the consequences of a fatty liver.

Answer: A fatty liver can impair blood flow in the hepatic sinusoids due to mechanical obstruction. There can be oxidation of fat causing free radical formation, which stimulates inflammation and fibrosis. Hepatocyte insulin resistance

(Continued)

Chapter 2: What are the common mechanisms of disease?

(Continued)

leads to type 2 diabetes mellitus. Fat can also activate angiotensin, leading to hypertension.

Question: Compare fatty liver disease, steato-hepatitis and acute liver failure.

Answer: *Fatty liver* shows fat with minimal inflammation and no fibrosis. This is reversible but about 50% of patients later develop fibrosis. *Steatohepatitis* has fatty change, inflammation and ballooning of hepatocytes (a feature of toxic damage), often with fibrosis. Patients are at risk of acute liver failure, advanced fibroses or cirrhosis. *Acute liver failure* has massive hepatocyte necrosis due to a toxic insult such as alcohol, paracetamol overdose or acute viral hepatitis. The mortality rate in acute liver failure due to steatohepatitis is around 40%.

Read more in Pathology in Clinical Practice Case 19

NECROSIS

There comes a stage at which reversible damage becomes irreversible and cell damage becomes cell death (Fig. 2.11 and see Fig. 2.6). The final events follow one of two distinct processes: 'necrosis' or 'apoptosis' (see page 62).

Necrosis is cell death due to lethal injury. Unlike apoptosis, cell death is not an energy-dependent active process but is a consequence of sudden changes in the microenvironment abolishing cell function. The changes seen in the tissues are a consequence of denaturation of proteins and release of digestive enzymes that destroy the tissue.

The principal types are:

- coagulative
- colliquative or liquefactive (Table 2.5)
- caseous.

COAGULATIVE NECROSIS

If you consider Fig. 2.12, (a) is of a normal kidney with normal glomeruli and tubules, whereas (b) is from a kidney that has suffered an ischaemic insult and is showing coagulative necrosis. Spot the difference?

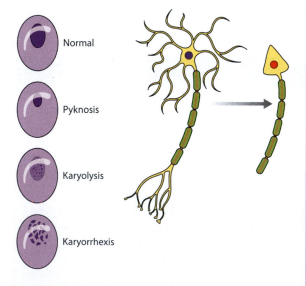

Figure 2.11 Cell death involves nuclear changes (left) and loss of dendritic and axonal structures in nerve cells (right).

Normal

Pyknosis

Karyolysis

Karyorrhexis

Table 2.5 Key facts about coagulative and liquefactive necrosis	
Coagulative	**Liquefactive**
Mechanism	
Severe ischaemia destroying proteolytic enzymes	Strong proteolytic enzyme action destroying tissue
Appearance	
Initial preservation of cell outlines and tissue architecture	Loss of cell outline Tissue becomes cystic or fluid
Occurrence	
Kidney, heart	Brain, bacterial infections

The second picture (Fig. 2.12b) is essentially the ghost outline of the first! The difference between the two is that the damaged kidney shows loss of nuclei from the cells and the cytoplasm stains a slightly darker pink. This pattern of necrosis is the most common type and occurs in many solid organs such as the heart and kidney (Fig. 2.13a).

The tissue, of course, doesn't remain in that state forever. Polymorphs start to move in within 24 hours of infarction, and release enzymes that digest the cellular components; the resulting debris will be removed by phagocytic macrophages. So the appearance of an area of coagulative necrosis will change with time and the final result will be an amorphous fibrous scar.

COLLIQUATIVE OR LIQUEFACTIVE NECROSIS

The hallmark of this type of necrosis is the release of powerful hydrolytic enzymes that degrade cellular components and extracellular material to produce a

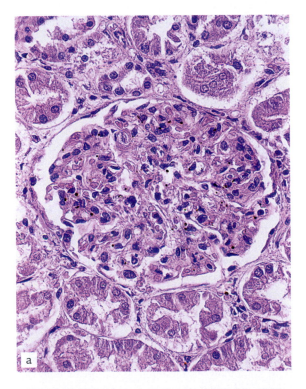

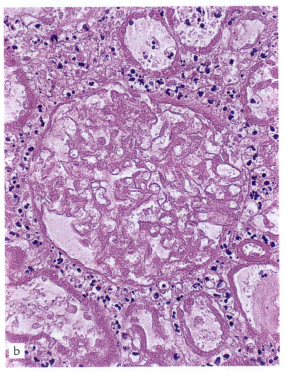

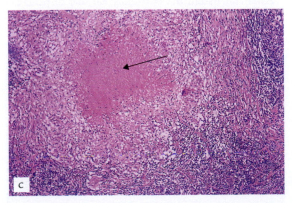

Figure 2.12 (a) Normal kidney; (b) coagulative necrosis; and (c) photomicrograph of lymph node with tuberculous granuloma and caseous necrosis (arrowed).

Chapter 2: What are the common mechanisms of disease?

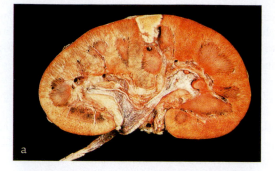

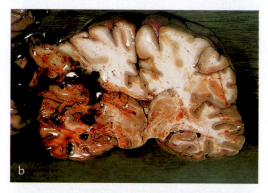

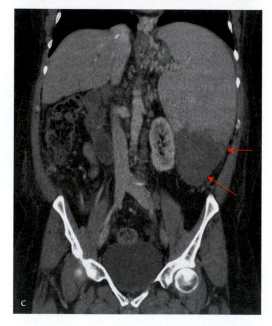

Figure 2.13 (a) Kidney, showing wedge-shaped cortical infarct; (b) brain with cerebral infarction; and (c) coronal CT after intravenous contrast showing massive splenomegaly with a non-enhancing infarct at the tip of the spleen (red arrows).

proteinaceous soup. Characteristically, it occurs in the brain where it produces a cystic cavity containing fluid and necrotic debris (Fig. 2.13b).

Liquefaction may also be encountered in tissues when there is a superadded bacterial infection. Then, enzymes are released from both the bacteria and the inflammatory cells that have been recruited to fight the infection.

CASEOUS NECROSIS

Caseous necrosis typically occurs in tuberculosis and is so called because of a resemblance to soft crumbly cheese! The necrotic area is not quite liquid but nor is the outline of the tissue retained as in coagulative necrosis. On microscopical sections stained with haematoxylin and eosin (H&E) (see Fig. 2.12c), the necrotic area appears homogeneously pink (eosinophilic) with a surrounding inflammatory response involving multinucleate giant cells, macrophages and lymphocytes (see granulomatous inflammation, page 165). It is believed that lipopolysaccharides in the capsules of the mycobacteria may be responsible for this peculiar reaction but the mechanism is unclear.

OTHER TYPES OF NECROSIS

Although these are the main types of necrosis, for completeness, we should briefly mention four others. These are fat necrosis, gangrene, fibrinoid necrosis and autolysis.

Fat necrosis
This type of necrosis is peculiar to fatty tissue and is most commonly encountered in the breast following trauma and within the peritoneal fat due to pancreatitis. Within the breast, trauma may lead to the rupture of adipocytes and the release of fatty acids. This will elicit an inflammatory response and the area will become firm due to scarring, forming a palpable lump. Clinically, the lump may be mistaken for a carcinoma and excision and microscopical examination may be required to determine the diagnosis.

In pancreatitis, damage to the pancreatic acini results in the release of proteolytic and lipolytic enzymes, which denature fat cells in the peritoneum and lead to an inflammatory reaction. Calcium is also deposited in the tissues in combination with fatty acids to form calcium soaps. This is a form of dystrophic calcification and we consider calcification again in the section on tissue response to necrosis.

Gangrene

This does not represent a distinctive type of necrosis but is a term used in clinical practice to describe black, dead tissue. It is most commonly seen in the lower limb in patients with severe atherosclerosis, which often causes irreversible ischaemic damage to the most peripheral tissues in the body. If the pattern of necrosis is mainly of the coagulative type, it is referred to as dry gangrene, whereas the presence of infection with Gram-negative bacteria converts it into a liquefactive type of necrosis, when it is called wet gangrene. It will be apparent from the preceding discussion that the type of necrosis encountered depends on a number of different factors, including the type of tissue involved and the nature of the offending agent. Gas gangrene is the disastrous complication that follows infection of tissue by the Gram-positive organism *Clostridium perfringens*, found in soil. The bacterium releases a toxin and also produces gas, which can be felt as crepitation when the affected area is pressed.

Fibrinoid necrosis or fibrinoid change refers to the microscopical appearance seen when an area loses its normal structure and resembles fibrin. It does not have any distinctive gross appearance.

Autolytic change is completely different from the others because it refers to cell death occurring after the person has died. Obviously, the heart stops pumping and all the tissues become irreversibly ischaemic. Enzymes leaking from the cells digest adjacent structures, but there is no inflammatory response because the inflammatory system is dead!

Calcification in necrotic tissue

Necrotic tissue may become calcified. When a cell undergoes necrosis, large amounts of calcium enter the cell and this combines with phosphates within the mitochondria to produce hydroxyapatite crystals. Extracellular calcification can also occur, the crystals forming in membrane-bound vesicles derived from degenerating cells. This is the initiation phase. There is then propagation of crystal formation, depending on the local concentration of calcium and phosphate and the amount of collagen present (this enhances calcification). Usually such *dystrophic* calcification is not a problem to the body, but if it affects an important site,

such as heart valves, it may affect function (Fig. 2.14). Dystrophic calcification may be useful, as in the microcalcification observed on mammograms; this can alert the radiologist to an early breast cancer (Figs 2.15 and 2.16). Dystrophic should be distinguished from *metastatic* calcification. This is linked to abnormally high serum calcium levels, due perhaps to hyperparathyroidism (increased parathyroid hormone secretion mobilises calcium from the bones), or excess vitamin D ingestion (increased calcium absorption from the gut). Metastatic carcinoma within bones may liberate

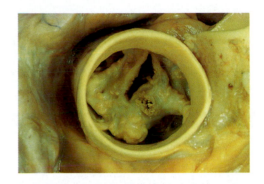

Figure 2.14 Dystrophic calcification in damaged aortic valve, producing stenosis.

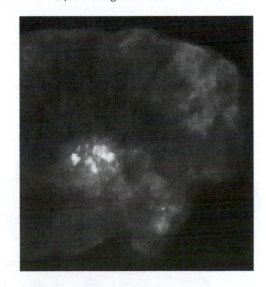

Figure 2.15 Radiograph of a resected breast specimen showing a circumscribed abnormality, which is radio-opaque because of dystrophic calcification.

Chapter 2: What are the common mechanisms of disease?

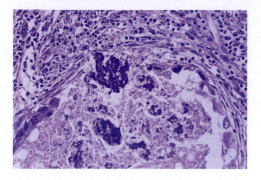

Figure 2.16 Dystrophic calcification in a necrotic tumour.

calcium and result in metastatic calcification, the term 'metastatic' referring to the widespread and scattered nature of the lesions encountered rather than inferring a similar mechanism.

APOPTOSIS AND AUTOPHAGY

Programmed cell death is a planned and coordinated mechanism to achieve the death of individual cells and is a process requiring energy input from the targeted cell. The packaged, membrane-bound bundles that are produced are quickly tidied away by nearby cells. On the other hand, necrosis is the result of accidental damage of various types, invariably involves groups of cells, and usually generates an inflammatory reaction, tissue damage and often scarring (Fig. 2.17).

The word 'apoptosis' is derived from Greek and was originally used to describe the falling of individual leaves from a tree. In pathology (Fig. 2.18), it is a specific type of cell death that involves single cells or small groups of cells in a tissue where the other cells are functioning normally.

Programmed cell death has an important role in all animals for controlling cell numbers, facilitating tissue modelling and removing damaged cells. It results in the death of the cell. Autophagy, literally self-eating, need not kill the cell but is an essential part of normal cell homeostasis that allows recycling of cell constituents and a source of energy when the cell is starving. Sometimes it can produce cell death and the main differences between autophagic and apoptotic cell death is that apoptosis requires the involvement of a phagocyte to tidy up the cell packages resulting from apoptosis: the 'come-and-eat-me' component. Autophagy is an adaptation for survival when there are limited nutrients, a means of removing damaged organelles or misfolded proteins, and an important route for destroying intracellular pathogens, such as *Mycobacterium tuberculosis*. Failure of autophagy is likely to lead to accumulation of damaged cells and the clinical effects of ageing. Autophagic processes are often identified in areas of apoptotic cell death, and it is not clear whether there is a form of autophagy that deliberately results in cell death or whether the autophagic processes are attempting to repair damaged cells and avoid death. Table 2.6 compares apoptosis with necrosis and autophagy.

Figure 2.17 Necrosis is a messy business, whereas programmed cell death (apoptosis) is an ordered event.

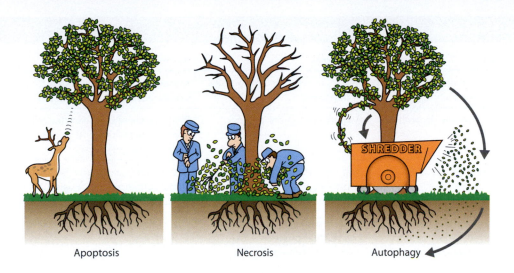

Figure 2.18 The term 'apoptosis' derives from the Greek for the falling of a single leaf from a tree. Extrapolating the analogy, in this diagram necrosis involves the death of many leaves at once, whereas in the process of autophagy there is death of leaves but their constituents are recycled.

CAN CELL DEATH BE USEFUL?

The importance of programmed cell death is evident from the earliest stages of the embryo through to the involutional changes of the menopause (Fig. 2.19). It is also a part of daily life because it is estimated that an adult loses more than 50 billion cells each day through apoptosis.

Let us consider the production of a limb with its five digits. To achieve this, tissue growth has to occur by cell division but it is also necessary to produce interdigital cell death. It is either that or ending up as a duck! This type of cell death is genetically controlled. Similarly, there are the stages of metamorphosis that take place to turn a tadpole into a frog. Metamorphosis requires not only mitotic activity and tissue growth, but also a large amount of programmed cell death. When a tadpole turns into a frog, the most obvious change is that limbs are formed and the tail is resorbed. During the process of resorption, there is an increase in thyroxine, which appears to lead to the activation of collagenases and, hence, destruction of the tail. Here we have an example of how programmed cell death may depend on the production of a hormone with activation of protein enzyme systems to assist the process.

The development of the nervous system is also dependent on programmed cell death. There is an excess of neurons and only those that produce the correct synaptic connections with their target cells survive. The rest, up to 50% of the neurons, die as a result of apoptosis. In gene knock-out mice unable to undergo apoptosis, the cells can be lost via autophagic cell death.

The endometrium is a hormone-dependent tissue that undergoes cyclical changes during the reproductive period as well as involutional changes after the menopause (Fig. 2.20). The oestrogens secreted by the ovary in the early part of the menstrual cycle induce endometrial proliferation and, if pregnancy does not occur, there is programmed cell destruction that results in menstrual shedding. If pregnancy occurs, then there is hyperplasia of the breast in preparation for lactation, which will be followed by physiological atrophy involving apoptosis after weaning. This atrophy not only is due to cell loss but also results from a reduction in cell size and loss of extracellular material. After the menopause, the withdrawal of the hormonal influence results in involution of the uterus and ovaries.

Apoptosis plays an important role in the immune system. It is necessary for the selection of specific subpopulations of both T and B lymphocytes, and is

Chapter 2: What are the common mechanisms of disease?

Table 2.6 Comparison of necrosis, apoptosis and autophagy

Necrosis	Apoptosis	Autophagy
Results in cell death once past the reversible point	Results in cell death from point of initiation	Mostly assists cell survival but sometimes results in cell death
Caused by energy deprivation	Requires energy for process	Requires energy but recycles cell constituents within individual cells so improving energy levels
Caused by injurious agent or event	Response to oxidative stress, hormones, growth factors and cytokines	Occurs normally in exercise, calorie restriction and defence against intracellular pathogens
Haphazard destruction of cell with release of contents including enzymes from ruptured lysosomes	Orderly packaging of organelles and nuclear fragments in membrane-bound vesicles	Cell integrity generally maintained
Cellular debris stimulates inflammatory response	New molecules expressed on membranes stimulate phagocytosis without inflammatory response	Cell integrity generally maintained
Clinical relevance		
Detrimental effect due to loss of cell function exacerbated by inflammatory response and fibrosis, e.g. myocardial and cerebral infarction	Normal process for cell turnover Depletion of CD4 cells in HIV Virally induced tissue damage Cancer pathobiology and treatment	Normal muscle homeostasis in exercise Response to calorie restriction Repair of internal cell structures by degradation of damaged organelles and proteins Elimination of intracellular *Mycobacterium tuberculosis* Cancer pathobiology and treatment

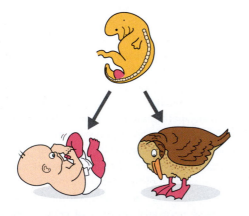

Figure 2.19 Programmed cell death (apoptosis) at the embryonic stage allows limbs to develop appropriately.

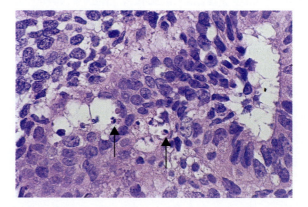

Figure 2.20 Programmed cell death in the endometrial glands (arrows show apoptotic debris).

Part 1: Disease, health and medicine

also important in the destruction of target cells by cytotoxic T cells.

Apoptosis, causes and mechanisms

You'll appreciate from the section above that apoptosis is crucial in normal physiological processes and will occur in normal healthy cells, for example:

- Normal cell turnover
- Normal tissue remodelling in embryogenesis or repair
- Normal hormone-dependent involution of tissues, e.g. endometrium
- Removal of self-reactive lymphocytes
- Removal of immune cells as the inflammatory response ends.

It is equally important in pathological circumstances where damage limitation is needed via the removal of affected cells without inducing inflammation, e.g.:

- Virally infected cells (either stimulated by the virus, e.g. HIV or the T-cell response, e.g. viral hepatitis)
- Radiation, cytotoxic drugs or oxidative stress inducing the DNA damage
- Misfolded proteins inducing ER stress.

The final common pathway for apoptosis involves enzymes called caspases, which are activated through an intrinsic (mitochondrial) or extrinsic (death receptor) pathway or through increases in cell calcium. The mitochondrial pathway operates by altering the balance of pro-apoptotic and anti-apoptotic members of the BCL-2 family, which affects the permeability of the mitochondrion and the release of factors that control the caspase cascade. This occurs particularly when cells are deprived of growth factors or have damaged DNA or accumulated proteins.

The death receptor pathway operates through members of the TNF (tumour necrosis factor) receptor family and the Fas (CD95) receptor. This is the most important mechanism for removing self-reactive lymphocytes and T-cell-mediated, virally infected cell death (Fig. 2.21).

It is increasingly recognised that accumulating misfolded proteins is an important mechanism in many diseases and ageing. It is called 'ER stress' because the endoplasmic reticulum is the normal site where newly synthesised proteins undergo folding with the assistance of chaperone molecules. The ER becomes overwhelmed in conditions where misfolding of proteins increases and this results in either apoptosis or cell adaptation (the 'unfolded protein response') so that protein production is reduced and chaperone synthesis increased. Conditions that increase misfolded proteins include:

- Genetic mutation in proteins or chaperones
- Viral infections
- Chemical insults
- Metabolic alteration that depletes energy stores.

You can actually see the changes of apoptosis down a light or electron microscope and, in contrast to necrotic cells which swell, apoptotic cells shrink. They also lose their contact with neighbouring cells early on. After 1–2 hours the nuclear chromatin condenses on the nuclear membrane and then the membrane 'packages' these small aggregates of nuclear material to give membrane-bound nuclear fragments. The cytoplasm shrinks and the cell's organelles also become parcelled into membrane-bound vesicles. These are called apoptotic bodies and they contain morphologically intact mitochondria, lysosomes, ribosomes, etc. Finally, these apoptotic bodies are phagocytosed by neighbouring cells or by macrophages and the contents degraded in secondary lysosomes. They are marked 'for disposal' by specific markers, such as phosphatidylserine or thrombospondin. Thus no messy acute inflammatory process is initiated. However, this process requires the expenditure of energy, as with any good garbage disposal system!

Defects in this apoptotic corpse clearance might be associated with the development of autoimmune and inflammatory conditions. The uptake of apoptotic fragments stimulates release of anti-inflammatory mediators and can inhibit secretion of pro-inflammatory mediators (i.e. the opposite to the effect of uptake of necrotic cells).

Autophagy

Autophagy is mostly concerned with cellular housekeeping. It takes damaged or redundant proteins, organelles, parts of organelles or areas of the nucleus, and digests them into their essential constituents through the action of lysosomal enzymes. Thus, proteins

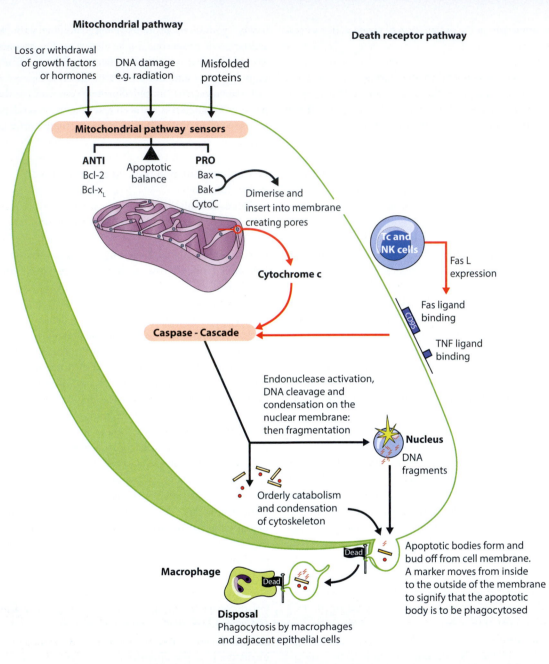

Figure 2.21 Programmed cell death (apoptosis) is an ordered and energy-requiring process that can be stimulated by diverse agents. Once initiated, the process is irreversible. There is no associated inflammation. It can be thought of in three stages: initiation, execution and disposal. Activation of the caspase cascade results in the breakdown of nuclear and cytoplasmic components which are packaged in membranes for disposal. Caspases are activated by the mitochondrial and death receptor pathways.

become amino acids and nucleic acids become nucleotides. All this occurs within a membrane-bound vesicle where the lysosomal enzymes mix with the substrate. How this vacuole forms is what distinguishes the three types of autophagy (Table 2.7). In macroautophagy, the substrates are first packaged in an autophagosome, which has an outer membrane that is not derived from a lysosome. This then joins with a lysosome and the dividing membranes break down, allowing the enzymes to reach the substrate. In microautophagy, the lysosome itself engulfs the substrate. In chaperone-mediated autophagy, receptors on the surface of the lysosome selectively bind specific substances and allow them to translocate into the lysosomal lumen. Autophagy (Atg) genes identified in yeast are conserved and function in many animals, including mammals.

Autophagy appears to be able both to protect a cell and to destroy it (Fig. 2.22). When nutrients are scarce, digestion of intracellular macromolecules can provide the energy to maintain minimal cell functioning. Pathogens and toxins may be segregated and degraded. Abnormal proteins or damaged organelles can be eliminated. All of these can play an important homeostatic role in the early stages of a disease process Why then is it also a mechanism for killing the cell? The theory is that it protects the cell by removing damaged cell components for as long as it can, but if it loses that battle it is best to remove the whole cell, i.e. to order cell death.

Apoptosis, autophagy and disease

It follows from the previous discussion that a lack of balance in initiating or suppressing programmed cell death can lead to problems.

Apoptosis is an important host defence mechanism against viral infection. When viruses infect cells, they attempt to take over the cell's replication machinery in order to proliferate and spread. Viral antigens are expressed on the host cell membrane and (in concert with $CD4^+$ lymphocytes) $CD8^+$ lymphocytes bind to this and secrete perforin, lysing the affected cell. Alternatively, natural killer (NK) cells recognise the viral antigen and initiate cell lysis. Viruses can sometimes get around this, with antiapoptotic mechanisms, which come into play when breaks occur in DNA strands as the viral genome incorporates itself into the cell's DNA. Many viruses code for proteins that block apoptosis. Examples include the inactivation of $p53$ by human papillomavirus type 16 (HPV16) and Epstein–Barr virus (EBV), which produce molecules that either simulate $bcl2$ or block molecules related to the TNF/CD40 pathway.

Too much apoptosis may also be seen if the effector mechanisms malfunction, as is seen in AIDS, when infection of T cells by HIV leads to deletion of the $CD4^+$ population of T cells, wreaking havoc in the immune system due to the loss of its most crucial regulatory cell. HIV expresses $gp120$, which activates the fas ligand on $CD4^+$ cells and leads to apoptosis. $CD4^+$ cells are essential for the generation of memory to intercurrent and opportunistic infections.

It has become clear that apoptosis (or the lack of it) has a role in the development of tumours. Inactivation of the cell cycle regulatory genes (which act as 'quality control officers' on the integrity of the DNA) will permit mutations to be passed to daughter cells by allowing cell replication to take place; usually such cells would be commanded to undergo apoptosis.

Table 2.7 Comparison of the types of autophagy			
Characteristic	Macroautophagy	Microautophagy	Chaperone-mediated autophagy
Occurs in	Stress	Normal physiology	Stress
Sequestering membrane	Non-lysosomal	Lysosomal	–
Receptor mediated	No	No	Yes
Engulfs organelles	Yes	Yes	No
Digests soluble cytosolic proteins	Yes	Yes	Only KFERQ-tagged

KFERQ, the one-letter code for the amino acid sequence Lys-Phe-Glu-Arg-Gln.

Chapter 2: What are the common mechanisms of disease?

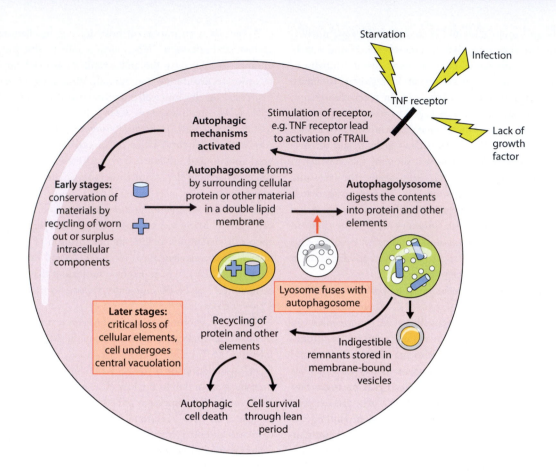

Figure 2.22 Autophagy: its role in cellular conservation and in cell death. In autophagic cell death, which is often a response to chronic adverse environmental factors such as starvation, the process of lysosomal degradation of cellular components within membrane-bound vesicles has already occurred before cell breakdown. The cells often appear vacuolated under the microscope as they undergo this process.

Table 2.8 Possible effects of autophagy in various conditions	
Disease	**Change in autophagy**
Cancer – early stage	Inactivation favours tumour growth
Cancer – late stage	Activation allows survival of cells in centre of tumour
Vacuolar myopathies	Inactivation leads to accumulation of vacuoles that weaken muscles
Neurodegeneration – early	Activation helps remove protein aggregates (helpful)
Neurodegeneration – late	Damaged neurons undergo autophagic cell death
Axonal injury	Inactivation prevents removal of damaged organelles and transmitter vacuoles so that neurotransmitters are released in cells and induce apoptosis
Infectious disease	Inhibition facilitates viral infection of cells and allows survival of bacteria

DEGENERATIVE DISEASES AND AGEING

This is a pathology book about the mechanisms of disease and this is a chapter on cell damage and cell death – so you won't be surprised that we will concentrate on the tissue, cellular and molecular changes in ageing. However, that is the reductionist approach, and thinking more holistically about patients and how and why they age would lead us to psychological and sociological theories. Just as we debated 'what is a disease?', we can similarly ask 'what is ageing?'. Is it just the passage of time (*chronological ageing*), changes that will happen to us all if we live long enough (*universal ageing*) (see box) or specific age-linked diseases that affect only some people (*probabilistic ageing*)? Many societies treat people differently as they become older and retire, and these changes in expectations can affect health, especially mental health. Some people benefit from keeping active whereas others may be content to disengage; personality, past events and current circumstances will all have an influence.

To return to the cell, you could deduce that the factors which cause damage to DNA, cell membranes and cell organelles are likely to play a part in accelerating ageing and factors important in repair, such as autophagy and poly(ADP-ribose) polymerase, slow down cellular ageing. Starting with DNA, it is important to distinguish between DNA *damage* and DNA *mutations*. *Damage* occurs normally every day as single- and double-strand breaks occur and are detected and repaired by enzymes. In the mouse, this is estimated to be thousands of lesions per hour in each cell! *Mutations* are changes in base sequences in DNA and these are not detected or repaired. You won't be surprised to learn that there are a variety of human disorders called 'DNA repair deficiency disorders' and these cause accelerated ageing and increased risk of cancer. Generally, if the tissue is composed of frequently dividing cells (e.g. gut), then there is an increased risk of cancer, whereas cells which don't divide (e.g. heart and brain) have increased ageing. Hereditary non-polyposis colorectal cancer is a common tumour and sufferers have mutations in one of the mismatch repair genes (see page 346).

An alternative mechanism is that transcription declines with age and that affects function. It is known that transcription of certain genes in brain tissue responsible for synaptic plasticity, vesicular transport

Physiological changes of ageing

Cardiovascular system
↓ Vessel elasticity
↓ Number of heart muscle fibres
↑ Size of muscle fibres
↓ Stroke volume

Respiratory system
↓ Chest wall compliance
↓ Alveolar ventilation
↓ Lung volume

Gastrointestinal system
↓ Bowel motility
↓ Enzyme, acid and intrinsic factor production
↓ Hepatic function

Urinary system
↓ GFR
↓ Concentrating ability

Nervous system
Degeneration and atrophy of 25–45% of neurons
↓ Neurotransmitters and conduction rate

Musculoskeletal system
↓ Muscle mass
↑ Bone demineralisation
↑ Joint degeneration

Immune system
↓ Inflammatory response

Skin
↓ Subcutaneous fat and elastin
↓ Sweat glands
↓ Temperature regulation through arterioles

and mitochondrial functioning declines between the age of 40 and 100 years.

Replicative senescence is the term describing a cell's inability to divide. Cells do appear to have a finite number of divisions and this may be linked to telomere shortening. Telomeres are at the linear ends of chromosomes and are important for ensuring their complete replication. With each replication, the telomere becomes shorter until the chromosome ends are damaged and this stops

Key facts

Structural and biochemical changes in cellular ageing

- Reduced oxidative phosphorylation in mitochondria
- Reduced synthesis of nucleic acids, transcription factors, proteins and cell receptors
- Decreased uptake of nutrients
- Reduced repair mechanisms
- Irregular, abnormally lobed nuclei
- Pleomorphic mitochondria
- Decreased endoplasmic reticulum
- Distorted Golgi apparatus
- Accumulation of lipofuscin
- Accumulation of abnormally folded proteins

Figure 2.23 Ageing is normal!

Case study: osteoporosis

A 70-year-old woman, Mrs Moore, comes to accident and emergency with severe back pain after stretching to open her bedroom window. On examination she is tender over her thoracolumbar junction. She has a lateral radiograph of her thoracolumbar spine. which shows osteoporosis.

Question: What is osteoporosis?

Answer: Osteoporosis occurs when bone becomes demineralised faster than the body can replace the minerals. Bones become fragile and weak and may break with normal stresses, such as climbing stairs and opening windows. Sites most likely to have osteoporotic fractures include the distal radius, pelvis, femoral neck and vertebral column. The condition is very common in elderly people, affecting almost 80% of postmenopausal women. Osteoporotic patients gradually lose height over time as they develop crush fractures in several vertebrae. The back also becomes more curved (kyphotic) as the vertebrae become wedged.

Question: What are the risks for osteoporosis?

Answer: Menopause – lack of oestrogen leads to rapid demineralisation of bone
Decreased calcium and vitamin intake
Inactive lifestyle
Alcohol
Smoking

Question: What do the three pictures in Fig. 2.24 show?

Answer: (a) An isotope bone scan showing activity in the L1 vertebral fracture indicating that there is an osteoblastic response at the fracture site, so it is a relatively recent injury.

(b) Sagittal short TI inversion recovery (STIR) MRI showing bone marrow oedema in the L1 crush fracture. The bright signal reflects increased water content at the fracture site and tells the radiologist that this is an acute and symptomatic injury that may benefit from vertebroplasty.

(c) Typical appearance of a pathological wedge fracture of a vertebra. In this example the spine contains metastatic tumour.

Part 1: Disease, health and medicine

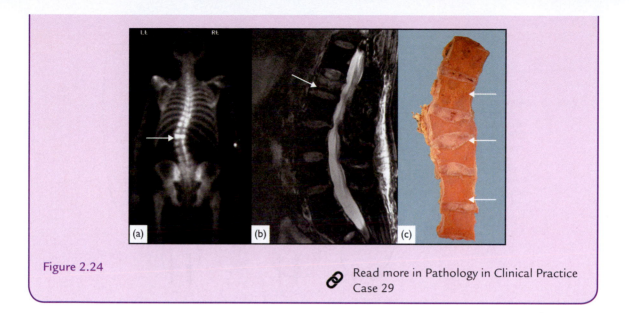

Figure 2.24

Read more in Pathology in Clinical Practice
Case 29

the cell cycle. In cells, such as stem cells or germ cells, that need to continue to divide, there is an enzyme called telomerase that can lengthen the chromosomes. This might also be important in some cancers.

Ageing is not just a random process but involves specific genes, receptors and signals. One of the mechanisms involves insulin growth factor 1 (IGF-1) pathways and another involves mTOR, which you will recall affects autophagy. Several of these seem to be affected by calorie restriction diets and this has been shown to delay ageing in the nematode *C. elegans*, yeast, flies and mice. However, the effect decreases in higher order animals. Around 800 genes are currently being investigated for their role in ageing and are collated on the GenAge Database.

MECHANISMS IN INFECTIOUS DISEASES

Two common mechanisms of bacterial disease are toxin production (endotoxin and exotoxins) and direct cell damage.

BACTERIAL ENDOTOXIN

Endotoxin is not secreted by living bacteria but is a cell wall component that is shed when the bacterium dies. The component is called lipid A, which is part of the lipopolysaccharide (LPS) in the outer cell wall of Gram-negative bacteria. It causes fever, and macrophage and B-cell activation by inducing host cytokines. Only Gram-negative bacteria have endotoxin, with the one exception being the Gram-positive *Listeria monocytogenes*.

Endotoxin is a potent stimulator of a wide range of immune responses. To the immune system, the recognition of LPS spells danger and warrants an immediate and dramatic response, which is often detrimental to the host itself. Clinically, this manifests as fever and vascular collapse or shock. Macrophages are stimulated by LPS to produce TNF and interleukin 1 (IL-1) which have many effects, including acting directly on the hypothalamus to produce fever (see page 139). LPS also stimulates, directly or indirectly, the complement and clotting pathways and platelets to produce DIC (disseminated intravascular coagulation, see page 236), thrombosis and shock (see page 259, Fig. 9.1). Shock results from increased vascular permeability produced by mediators from mast cells and platelets, combined with the TNF and LPS affecting endothelial cells. LPS also stimulates the liver to produce acute-phase proteins (see page 139) and hypoglycaemia. In Gram-positive organisms, lipotechoic acid within the bacterial wall causes problems similar to LPS in Gram-negative bacteria (Fig. 2.25).

BACTERIAL EXOTOXINS

Exotoxins are proteins released by living bacteria and there are a wide variety with different actions. Neurotoxins act on nerves or endplates to produce

Chapter 2: What are the common mechanisms of disease?

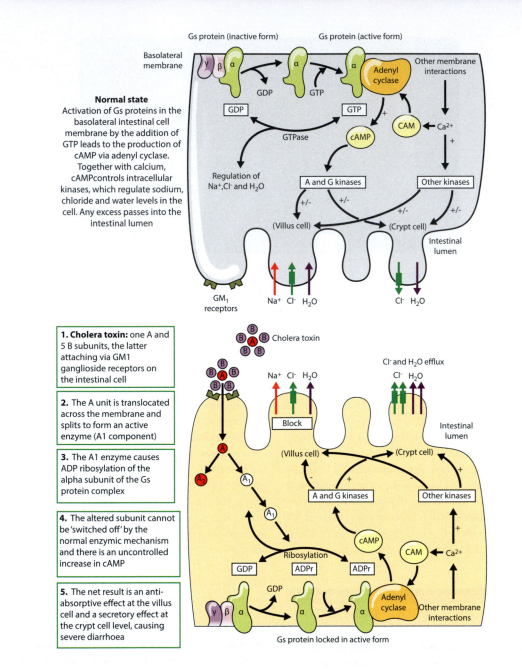

Normal state
Activation of Gs proteins in the basolateral intestinal cell membrane by the addition of GTP leads to the production of cAMP via adenyl cyclase. Together with calcium, cAMP controls intracellular kinases, which regulate sodium, chloride and water levels in the cell. Any excess passes into the intestinal lumen

1. Cholera toxin: one A and 5 B subunits, the latter attaching via GM1 ganglioside receptors on the intestinal cell

2. The A unit is translocated across the membrane and splits to form an active enzyme (A1 component)

3. The A1 enzyme causes ADP ribosylation of the alpha subunit of the Gs protein complex

4. The altered subunit cannot be 'switched off' by the normal enzymic mechanism and there is an uncontrolled increase in cAMP

5. The net result is an anti-absorptive effect at the villus cell and a secretory effect at the crypt cell level, causing severe diarrhoea

Figure 2.25 Cholera toxin permanently activates a Gs protein in the intestinal cell wall. The diagram shows an intestinal cell in the normal state (top) and affected by cholera toxin (below); the diagrams are mirror images with the intestinal lumen in the middle. Cells in the intestinal villus: Na^+ and Cl^- absorption is blocked by cholera toxin. Crypt cells: this is the site most affected by cholera toxin. Here Cl^- secretion is stimulated. Water passively follows the active ion transport. Dehydration occurs. Although the main Na^+ absorption path (Na^+/H^+ exchange) is blocked by cholera toxin, Na^+ can be co-transported with glucose or amino acids, which is why oral rehydration solutions using glucose are effective in cholera. (Courtesy of Professor Phil Butcher, St George's, University of London.)

Part 1: Disease, health and medicine

paralysis, cytotoxins damage a variety of cells, tissue-invasive toxins are often enzymes capable of digesting host tissues, and pyrogenic toxins stimulate cytokine release and cause rashes, fever and toxic shock syndrome (TSS). A very important group is the enterotoxins which act on the gastrointestinal tract to produce diarrhoea by inhibiting salt absorption, stimulating salt excretion or killing intestinal cells. There are two broad categories: infectious diarrhoea and food poisoning with preformed toxin. In infectious diarrhoea, the bacteria proliferate in the gut, continuously releasing enterotoxin. The symptoms do not occur immediately after ingestion but require a day or two for the bugs to become established in sufficient numbers. In food poisoning, the bacteria grow in the food releasing their exotoxin. This acts very quickly after ingestion to produce diarrhoea, abdominal pain and vomiting, but the symptoms last for only 24 hours because no new toxin is created.

Bacterial exotoxins may be classified by their site of action:

- Extracellular, e.g. epidermolytic toxin *Staphylococcus aureus* which causes scalded skin syndrome
- At cell membrane level (not transported into cell, but cause changes in intracellular cGMP), e.g. *Escherichia coli* heat-stable enterotoxin (ST), causing travellers' diarrhoea, and *Staphylococcus aureus* TSST 1 toxin leading to toxic shock syndrome
- On the cell membrane, causing pore formation or disruption of lipid by enzymatic activity, e.g. phospholipase C activity toxin *Clostridium perfringens*, spore-forming toxins (thiol-activating haemolysins) such as streptolysin (*Streptoccus pyogenes*),

History

John Snow and the Broad Street pump

It would be quite possible to go through the entire medical curriculum without ever hearing the name of John Snow (1813–58). He was a man of simple habits and seemed to lack the charisma that is vital in attracting attention on the world stage. Yet his contributions to medicine were certainly 'world class'. If you think that epidemiological studies might be dull, Snow's biography is well worth a read. It demonstrates eloquently how they can have a truly dramatic impact on public health. Snow is perhaps best remembered for having anaesthetised Queen Victoria in 1853 and 1857, giving credibility to the use of pain relief during childbirth! But it is his contribution to the understanding of cholera that is relevant here. Snow's link with cholera evolved over a number of years. His first encounter with the disease was in Newcastle upon Tyne during the epidemic of 1831–2, when he had just started his medical training. It was during the next cholera epidemic of 1848–9 that he made his seminal contribution. By now Snow was in London and here he began to unravel the mode of transmission of the disease. Snow's work was a masterpiece in epidemiological investigation. He meticulously mapped the houses in which new cases of cholera were being diagnosed and observed a marked difference in the incidence and mortality of cholera in the south of London (8 deaths per 1000 inhabitants) compared with other areas (1–4 deaths per 1000 inhabitants). This led him to hypothesise that cholera was spread by water. He identified the public pumps from which the families living in the affected and unaffected areas drew their water. He noticed that there were surprising sites of sparing within otherwise heavily affected areas. His suspicion that the infection was in the water supply grew when he found that those living in the spared areas worked at a local brewery and received free beer, which was made with water from a source away from that supplying their homes. Even before the identification of bacteria, he postulated that the transmission was the result of a *living* organism that had the ability to multiply. Snow is of course remembered for urging that the handle be removed from the pump that supplied contaminated water in Broad Street in London during the epidemic of 1854 (Fig. 2.26). This led to a dramatic decrease in new cases in the area. Snow postulated that social conditions and hygiene were of paramount importance in the spread of infection. Although there was no medical treatment for cholera, it became apparent that the way to stop an epidemic was through good sanitation and good hygiene. It follows from this that our first line of defence against infection is the prevention of the multiplication and spread of organisms. Open sewers, overcrowded living conditions, contaminated drinking water, poor food storage and preparation, inadequate personal hygiene and unprotected sexual

(Continued)

Chapter 2: What are the common mechanisms of disease?

(Continued)

contact are a recipe for disaster. This is not a text on public health medicine or politics, so we will not dwell on these points. But remember that, each year in Asia, Africa and Latin America, roughly 4–6 million people die from diarrhoea and 1–2 million die from malaria. One-third of the world's population is subclinically infected with tuberculosis and 3 million die from TB each year; 12 million people are infected with HIV worldwide. All these problems are more likely to be solved by engineers and politicians than by the latest advances in molecular biology!

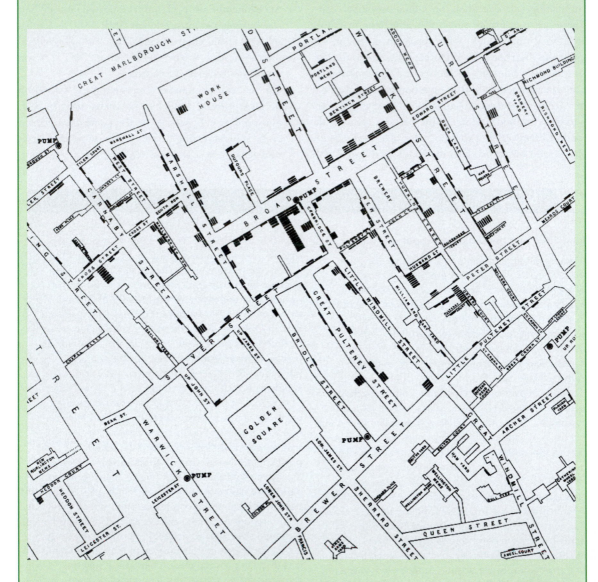

Figure 2.26 Deaths from cholera (–) in Broad Street, Golden Square, London, and the neighbourhood, 19 August to 30 September 1854. Water pumps are shown. John Snow realised that deaths were clustered around the Broad Street pump. Families working for the local brewery received free beer in preference to water and did not catch cholera! (Reproduced with permission from Wellcome Library, London.)

pneumolysin (*Strep. pneumoniae*), listeriolysin (*Listeria monocytogenes*), perfringolysin (*C. perfringens*; gas gangrene), cerolysin (*Bacillus cereus*; food poisoning)

- Type III toxins, which act intracellularly by translocating an enzymatic component across the membrane (A subunit of domain) which modifies an acceptor molecule in the cytoplasm. These can be grouped by type of enzymatic activity:

  - ADP ribosylation (cholera, diphtheria, pertussis)
  - *N*-glycosidases (shiga toxin)
  - glucosyl transferases (*C. difficile* toxin A and B)
  - Zn^{2+}-requiring endopeptidases (tetanus and botulism toxins).

GTP-binding proteins are often the target for ADP ribosylation by type III bacterial exotoxins. These proteins are involved in signal transduction and regulation of cellular function either by cAMP levels or kinase cascades leading to transcription modification, for example:

- G proteins (stimulatory or inhibitory of adenylate cyclase): cholera, pertussis, *E. coli* LT toxin
- Elongation factor 2 (translational control): diphtheria toxin
- Rho proteins (small G proteins; regulate actin cytoskeleton): inactivated by *C. difficile* A and B toxins, and can also be activated by deamination by pertussis necrotising toxin.

The structure of these toxins consists of either A:B5 (enzyme active A subunit and 5 × B subunits required for binding) or A:B type with A (enzyme active) and B (binding) domains on a single polypeptide chain. A:B5 types are seen in cholera and pertussis (whooping cough), and *E. coli* LT1 and LT2. A:B types are encountered in diphtheria, botulism and tetanus.

It is fascinating to compare the contrasting actions between two structurally similar neurotoxins produced by members of the same bacterial family, *C. tetani* and *C. botulinum*, which cause tetanus and botulism respectively. These diseases are purely due to toxin-mediated action following infection and are quite different in pathology, yet the molecular action of the two toxins is identical. They are both endopeptidases specific for synaptobrevin, a protein found in the cytoplasm of synaptic vesicles. However, the binding (B) domains of the toxins show different specificities for cell receptors. Tetanus toxin binds to the gangliosides of the neuronal membrane, is internalised and moves by retroaxonal transport from peripheral nerves to the CNS, where it is released from the post-synaptic dendrites and localises in presynaptic nerve terminals. This blocks the release of inhibitory neurotransmitter, γ-aminobutyric acid to cause unopposed, continuous, excitatory synaptic activity, leading to spastic paralysis. Botulism toxin binds ganglioside receptors of cholinergic synapses and prevents release of acetylcholine at the neuromuscular junctions, causing flaccid paralysis (Fig. 2.27).

ANTIMICROBIAL RESISTANCE

In the battle between humans and microbes, antibiotics are produced and bacteria respond by developing resistance. How does that occur and what can be done to minimise it?

The rapid rate of reproduction aligned with horizontal gene transfer by conjugation, transduction and transformation means that bacteria can readily acquire new characteristics. It is thus inevitable that, after introduction of any new antibacterial, resistance will develop at some stage. Overusage or inappropriate usage will accelerate this problem. Resistance mechanisms can spread from one bacterium to another (and from different species to different species) and the resistant bacteria can of course spread from human to human, or in some increasingly recognised situations from animals to humans and vice versa. The acquisition of antibacterial resistance may sometimes come at a fitness cost to a bacterium – in other words, resistant bacteria are not always as virulent as the wild-type strains. This is because there is a finite capacity, spatially and energetically, for genetic material within a bacterial cell and acquisition of resistance genes may result in loss of virulence genes. However, sometimes resistance factors and virulence factors may be co-acquired, e.g. on a plasmid, resulting in a virulent and difficult-to-treat organism. Over the years there have been waves of different resistant bacteria and the problem organisms seem to shift from Gram-negative bacteria to Gram-positive bacteria and back to Gram-negative bacteria.

Chapter 2: What are the common mechanisms of disease?

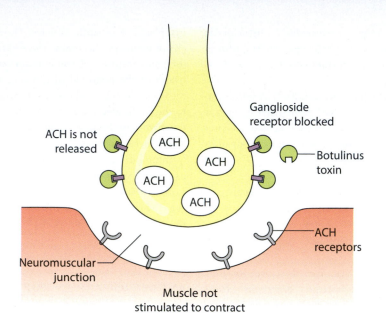

Figure 2.27 Botulinum toxin binds ganglioside receptors of cholinergic synapses and prevents release of acetylcholine at the neuromuscular junctions, causing flaccid paralysis.

It is not entirely clear why this may be, but changes in the use of different classes of antibacterials must play a big part. The mid-1990s saw the rise of problems with multiply resistant Gram-positive bacteria such as MRSA (meticillin-resistant *Staphylococcus aureus*) and GRE (glycopeptide-resistant enterococci) and *C. difficile*. Concerted efforts using well-defined infection prevention and control interventions (screening, isolation, decolonisation, improved environmental decontamination and – perhaps the most important – hand washing by healthcare workers), as well as improved antimicrobial stewardship (use of the right drug at the right time by the right route for the right duration), has helped to reduce the problems with these organisms. But despite this there are still major problems currently and ahead of us. Most worryingly there has been a rise in the incidence of multiply resistant Gram-negative bacteria, especially the Enterobacteriaceae such as *Klebsiella* sp. which can cause urinary tract, respiratory, abdominal and surgical site infections. Some of these multiply resistant bacteria have new resistance mechanisms such as carbapenemase enzymes, which means that the effectiveness

of carbapenems such as meropenem is curtailed by lysis of the drug. Often these isolates are resistant to many other classes of antibiotics including aminoglycosides and quinolones, as well as the remaining β-lactams. With virtually no new antibacterials likely to be appearing soon, it means that older drugs have been rejuvenated because, interestingly, these bacteria often remain sensitive to them, probably because there has been little selective pressure to limited recent usage. A good example is the polymyxin antibiotic colistin. However, once usage of these old drugs picks up it is likely that resistance will start to be seen against these agents as well.

DIRECT CELL DAMAGE BY VIRUSES

Microorganisms can also damage cells directly and this is particularly true of viruses. Viruses are obligate intracellular organisms requiring the host cell's machinery for replication (Figs 2.28–2.30 and see Table 1.4 in Chapter 1). Viruses use three main methods for entering cells:

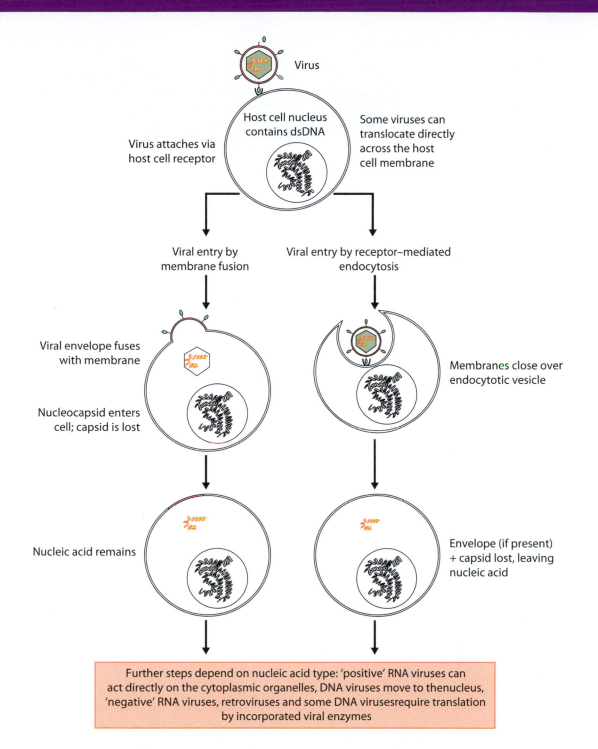

Virus

Host cell nucleus contains dsDNA

Virus attaches via host cell receptor

Some viruses can translocate directly across the host cell membrane

Viral entry by membrane fusion

Viral entry by receptor–mediated endocytosis

Viral envelope fuses with membrane

Nucleocapsid enters cell; capsid is lost

Membranes close over endocytotic vesicle

Nucleic acid remains

Envelope (if present) + capsid lost, leaving nucleic acid

Further steps depend on nucleic acid type: 'positive' RNA viruses can act directly on the cytoplasmic organelles, DNA viruses move to thenucleus, 'negative' RNA viruses, retroviruses and some DNA virusesrequire translation by incorporated viral enzymes

Figure 2.28 After attaching to the host cell via a receptor, the viral particle enters the cell by fusion of the viral and host cell membranes or by receptor-mediated endocytosis. It then uncoats and releases its nucleic acid. Replication occurs either in the cytoplasm alone, as with most RNA viruses and rare DNA viruses (the pox viruses), or involves both cytoplasmic and nuclear steps. DNA viruses can integrate directly with the host DNA but the only RNA viruses that can achieve nuclear integration are the retroviruses.

Chapter 2: What are the common mechanisms of disease?

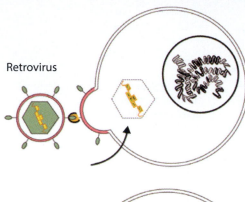

Retrovirus

Retrovirus binds to a human surface receptor (e.g. HIV-1virus binds using its gp120 antigen to human CD4 onT lymphocytes). The viral RNA contains the followinggenes:pol (encodes reverse transcriptase and integrase), gag(encodes viral structural proteins), env (encodes envelope proteins, including gp120), reg (encodes regulatory genes, e.g. tat,rev)

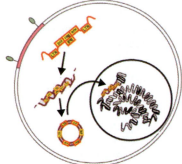

The viral and host cell membranes fuse, the viral capsid disintegrates and viral RNA enters the human cell. Viral reverse transcriptase catalyses the production of double-stranded DNA, which enters the host cell nucleus and integrates into host DNA under the influence of integrase

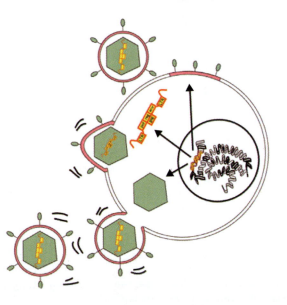

The host cell is induced to manufacture and a ssemble new virions by long terminal repeat sequences and genestatandrev.The new virions peel off an envelope of host cell membrane as they exit

Figure 2.29 Retroviruses, similar to DNA viruses, can insert their genes into the host cell DNA, but require an additional step. Their RNA is translated into DNA by reverse transcriptase, supplied by the virus.

1 Translocation of the entire virus through the cell membrane

2 Fusion of the viral envelope with the cell membrane

3 Receptor-mediated endocytosis of the virus, followed by fusion with the endosome membrane.

What happens once the virus is inside the cell? First the particle must uncoat and separate its genome from its structural components. It then uses specific enzymes of its own or present in the host cell to synthesise viral genome, enzyme and capsid proteins. These must be

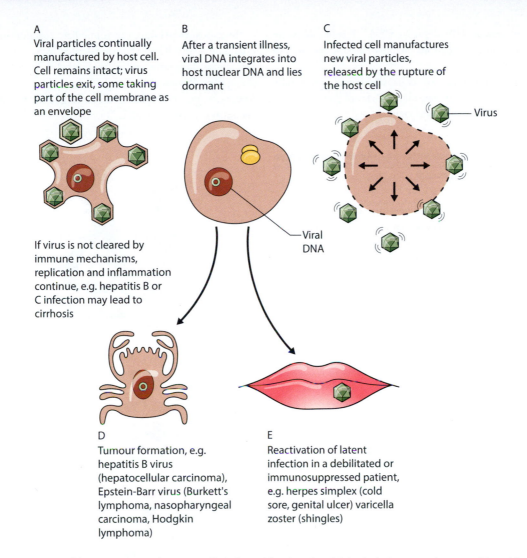

A
Viral particles continually manufactured by host cell. Cell remains intact; virus particles exit, some taking part of the cell membrane as an envelope

If virus is not cleared by immune mechanisms, replication and inflammation continue, e.g. hepatitis B or C infection may lead to cirrhosis

B
After a transient illness, viral DNA integrates into host nuclear DNA and lies dormant

Viral DNA

C
Infected cell manufactures new viral particles, released by the rupture of the host cell

Virus

D
Tumour formation, e.g. hepatitis B virus (hepatocellular carcinoma), Epstein-Barr virus (Burkett's lymphoma, nasopharyngeal carcinoma, Hodgkin lymphoma)

E
Reactivation of latent infection in a debilitated or immunosuppressed patient, e.g. herpes simplex (cold sore, genital ulcer) varicella zoster (shingles)

Figure 2.30 Possible outcomes in human cells infected by virus in which viral clearance is not achieved by the host's immune mechanisms.

assembled and released, either directly (unencapsulated viruses) or by budding through the host cell's membrane (encapsulated viruses).

Viruses can damage the host cell directly in a variety of ways:

- Interference with host cell synthesis of DNA, RNA or proteins (e.g. polio virus modifying ribosomes so that they no longer recognise host mRNA)
- Lysis of host cells (e.g. polio virus lyses neurons)
- Inserting proteins into host cell membrane, so provoking an immune attack by host cytotoxic lymphocytes (e.g. hepatitis B and liver cells)
- Inserting proteins into host cell membrane to cause direct damage or promote cell fusion (e.g. herpes, measles, HIV)
- Transforming host cells into malignant tumours (e.g. EBV, HPV, HTLV-1).

Alternatively, the host cell may suffer secondary damage due to viral infection. This may be due to the following:

- Increased susceptibility to infection due to damaged host defences (e.g. influenza viral damage to respiratory epithelium facilitates bacterial pneumonia with

Chapter 2: What are the common mechanisms of disease?

Staph. aureus, HIV depletes CD4$^+$ T cells, allowing opportunistic infections such as *Pneumocystis jirovecii* pneumonia)
- Death or atrophy of cells dependent on virally damaged cell (e.g. muscle cell atrophy after motor neuron damage by polio virus).

HYPERSENSITIVITY REACTIONS

The phenomenon of damage caused by the immune system while trying to combat an insult is referred to as hypersensitivity. As a cause, it is partially 'intrinsic' and partially 'extrinsic'. The precipitating factor may be an external agent but, rather than the immune system coping and restoring the body's homeostasis, the immune response is altered and becomes the 'internal' cause of the disease.

Four main types of hypersensitivity reaction were described by Gell and Coombs, although several types may operate together. Types I–III involve antibodies and type IV involves cells.

The following are the mechanisms involved:

- Release of allergic mediators in type I (anaphylactic) hypersensitivity
- Binding of self-reactive antibodies to cells in type II (antibody-dependent cytotoxic) hypersensitivity
- Damage to blood vessel walls and tissues by circulating immune complexes in type III (immune complex-mediated) hypersensitivity
- Delayed tissue damage due to interactions between sensitised T lymphocytes and other inflammatory cells in type IV (cell-mediated) hypersensitivity.

TYPE I: ANAPHYLACTIC HYPERSENSITIVITY

This is the mechanism behind atopic allergies such as asthma, eczema, hay fever and reactions to certain food. Patients who have this problem have been previously exposed to the allergen and have generated IgE antibodies against it. The IgE antibodies attach to mast cells via cell surface receptors. A common feature of many allergens is that they have repetition of the same antigenic determinant on their surface. This means that cross-linking of IgE molecules attached to the same mast cell is likely. It is cross-linkage that activates the mast cell.

Re-exposure produces an immediate reaction to the offending agent! The extrinsic allergen (e.g. grass pollen, house dust mite faeces, seafood) binds to IgE on the surface of mast cells in the mucosa of the bronchial tree, nose, gut or conjunctivae, leading to the release of chemical mediators (Fig. 2.31).

First exposure: nasal and bronchial mucosa exposed to pollen, dust etc.

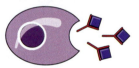

Soluble antigen stimulates production of IgE antibodies

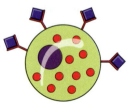

Fc of IgE attaches to receptor on mucosal mast cell

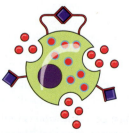

Second exposure to pollen; cross-linking of IgE molecules on mast cell stimulates release of primary and secondary inflammatory mediators

Figure 2.31 Type I hypersensitivity.

Some mediators, such as histamine, are already formed within mast cell granules and the cross-linking of IgE molecules attached to the mast cell surface stimulates the release of the granule contents into the tissues (Fig. 2.32). This is why the effect of antigen exposure can be seen within 5 minutes; the granule contents elicit an inflammatory response that lasts up to an hour.

In the meantime, the cross-linked IgE molecules stimulate the membrane of the mast cell to generate arachidonic acid and its metabolites within its cell membrane. Release of leukotrienes, prostaglandins and platelet-activating factor by this mechanism sparks a response that starts 8–12 hours later and may last from 24 hours to 36 hours.

Generally, the mediators released by these processes act locally but they can also produce life-threatening systemic effects. Effects include constriction of smooth muscle in bronchi and bronchioles, causing wheezing, dilatation and increased permeability of capillaries, and resulting in localised tissue oedema, increased nasal and bronchial secretions, red watery eyes, skin rashes and diarrhoea. If laryngeal oedema or bronchospasm is severe, death may result from respiratory obstruction.

What can be done about this problem? Antihistamines antagonise the action of histamine and can relieve many of the symptoms. Steroids, which inhibit the leukotriene pathway, can prevent or alleviate the longer-term symptoms.

If an 'atopic' patient knows that he is likely to encounter an antigen on a particular occasion (e.g. pollen in a hay fever sufferer) he may be able to take drugs that stabilise the mast cell membrane, reducing or preventing its activation.

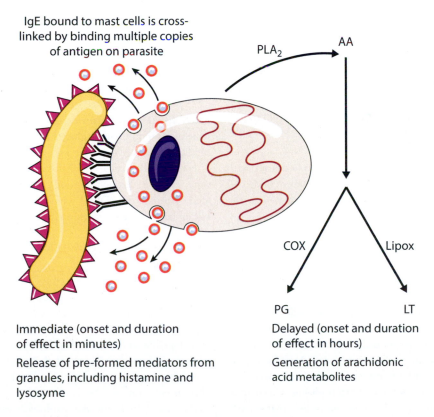

IgE bound to mast cells is cross-linked by binding multiple copies of antigen on parasite

PLA_2 AA

COX   Lipox

PG   LT

Immediate (onset and duration of effect in minutes)

Release of pre-formed mediators from granules, including histamine and lysozyme

Delayed (onset and duration of effect in hours)

Generation of arachidonic acid metabolites

Figure 2.32 Mast cell activation is typically stimulated by IgE antibody cross-linkage as illustrated here, but this is not always necessary. Mast cells can also be directly stimulated by some substances, e.g. mellitin from bee stings or C3a5a complement fragments. AA, arachidonic acid; COX, cyclooxygenase; Lipox, lipoxygenase; LT, leukotrienes; PLA_2, phospholipase A_2; PG, prostaglandins.

Chapter 2: What are the common mechanisms of disease?

A course of injections of a very dilute solution of the offending allergen can eventually lead to IgG instead of IgE being generated; because IgG does not stick to mast cells, the problem is avoided. But care must be taken with this approach not to generate anaphylactic shock, a life-threatening complication. In this condition, the patient must be injected immediately, subcutaneously, with adrenaline, which reverses the actions of the mediators causing bronchospasm and oedema. Patients known to be susceptible to this type of response (often to bee stings or peanut protein) carry an EpiPen with them, which can deliver a single dose of adrenaline. The action of adrenaline lasts only a finite time, and patients should also be given antihistamine or steroids as back-up.

Why should some people produce the IgE molecules that mediate this reaction? Genes play a part, and also a particularly strong reaction by T_H2 cells (helper T lymphocytes) to the allergen, resulting in release of three crucial cytokines: IL-4 that makes B cells switch to producing and secreting IgE; IL-5 that attracts eosinophils which contribute to the response through the release of their granules; and IL-13 that stimulates epithelial cell mucus production.

TYPE II: ANTIBODY-DEPENDENT CYTOTOXIC HYPERSENSITIVITY

In this type of hypersensitivity, antibodies react with antigens *fixed to the surfaces* of various types of body cell to cause damage. The effects can be mediated in the following ways:

- Activation of the complement system
- Targeting the cell for phagocytosis
- Affecting the cell's function.

In the complement-fixing type, antibodies bind to an antigen, fix complement, and so cause destruction of the target cell or trigger inflammatory cells.

When circulating cells are coated with antibody, the Fc part of the antibody can bind the cell to phagocytic cells, such as neutrophils and macrophages. Circulating blood cells and blood components are particularly vulnerable and so there is autoimmune haemolytic anaemia (destruction of red blood cells), autoimmune thrombocytopenic purpura (platelet destruction) and drug-related agranulocytosis (granulocyte destruction).

You will recall that IgG crosses the placenta and so blood group incompatibility between the mother and the baby can cause problems. This is illustrated in Fig. 2.33 where rhesus antibodies from a rhesus-negative (Rh-) mother cross the placenta to damage the red cells of a Rh+ baby.

For routine blood transfusions, it is essential that any blood which is transfused into a patient is first cross-matched to ensure that the recipient does not possess antibodies to antigens on the donor red blood cells. This is because the donor red cells will become coated with antibody and/or complement which may promote phagocytosis due to opsonisation, via the Fc or C3b, or fix complement to produce cell membrane damage through C8 and C9 membrane attack complex.

In the functional type of type II hypersensitivity, antibody binding interferes with cell function; often by interfering with hormone receptors, but also with cell growth and differentiation or cell motility. Included in this group are antibodies that block cell function (as in some forms of Addison's disease where autoantibodies develop to adrenocortical proteins) or antibodies which stimulate receptor function (as in Graves' disease, causing thyrotoxicosis).

Type II reactions are also involved in some drug reactions (e.g. chlorpromazine-induced haemolytic anaemia and quinidine-induced agranulocytosis) where binding of a drug may alter a normal self-antigen.

Natural killer cells, which are not restricted by HLA type, can kill through antibody-dependent cell-mediated cytotoxicity (ADCC) which may be important for killing large parasites or tumour cells.

TYPE III: IMMUNE COMPLEX-MEDIATED HYPERSENSITIVITY

Similar to type II, this is an antibody-driven process, but the difference is that antibodies react with free antigen. Whether the complexes are soluble or insoluble depends on the ratio of antibody to antigen. These immune complexes can circulate in the blood, giving rise to 'serum sickness' or they can be deposited locally (Arthus reaction).

Both the soluble and the insoluble types of complex can activate macrophages, aggregate platelets and initiate the complement cascade. Circulating immune complexes can lodge in the small vessels of many organs to cause a vasculitis (inflammation of the blood vessel

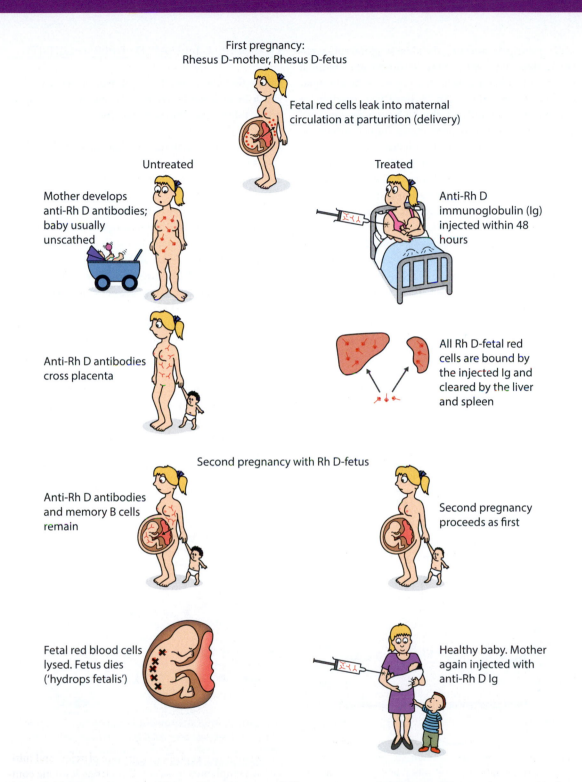

First pregnancy:
Rhesus D-mother, Rhesus D-fetus

Fetal red cells leak into maternal circulation at parturition (delivery)

Untreated

Treated

Mother develops anti-Rh D antibodies; baby usually unscathed

Anti-Rh D immunoglobulin (Ig) injected within 48 hours

Anti-Rh D antibodies cross placenta

All Rh D-fetal red cells are bound by the injected Ig and cleared by the liver and spleen

Second pregnancy with Rh D-fetus

Anti-Rh D antibodies and memory B cells remain

Second pregnancy proceeds as first

Fetal red blood cells lysed. Fetus dies ('hydrops fetalis')

Healthy baby. Mother again injected with anti-Rh D Ig

Figure 2.33 Type II hypersensitivity: rhesus incompatibility: an untreated rhesus-negative mother may develop antibodies if she has a Rh+ baby. In the next pregnancy her antibodies cross the placenta, attach to Rh antigens on the surface of red blood cells and lead to their lysis.

Chapter 2: What are the common mechanisms of disease?

wall), principally affecting the kidney (glomerulonephritis), skin and joints. Nucleoproteins are common sources of *endogenous* antigens, e.g. in systemic lupus erythematosus (SLE). An Arthus-type reaction is encountered clinically in the lung after an *exogenous* antigen is inhaled and precipitates locally within the alveolar walls. The inhaled antigens are usually animal or plant proteins and often associated with specific occupations. The resulting damage to lung alveoli, with repeated episodes of inflammation and scarring, leads to a restrictive form of lung disease called extrinsic allergic alveolitis (Fig. 2.34).

Of course, antibody binding to antigen is a normal defence mechanism, so why does it also cause disease? Pathogenic immune complexes are more likely to form when there is antigen excess and the complexes are small and harder to remove from the circulation by phagocytosis. It is also affected by the valency of the antigen, avidity of the antibody and charge of the complex.

TYPE IV: CELL-MEDIATED HYPERSENSITIVITY

Unlike the other three forms of hypersensitivity, all of which involve antibody, type IV involves T lymphocytes (Table 2.9). This takes two forms:

1 CD4$^+$ T-cell-related, cytokine-mediated inflammation
2 CD8$^+$ T-cell-related, direct-cell cytotoxicity.

The details of how CD4 (helper) cells and CD8 (suppressor/cytotoxic) cells recognise and respond to antigen is covered on page 197.

Naive CD4 cells can differentiate into T-helper 1 (Th1) cells and Th17 cells. Th1 cells secrete interferon γ (IFN-γ) which is a strong activator of macrophages. Th17 cells secrete IL-17 and other cytokines that recruit and stimulate neutrophils. Typically the response (delayed-type hypersensitivity) occurs over several

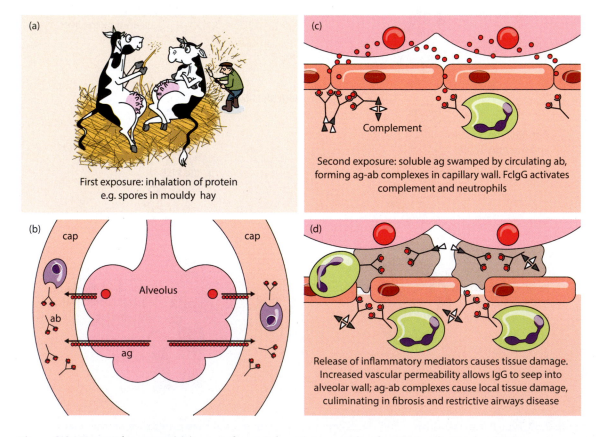

(a) First exposure: inhalation of protein e.g. spores in mouldy hay

(b) cap · cap · Alveolus · ab · ag

(c) Complement
Second exposure: soluble ag swamped by circulating ab, forming ag-ab complexes in capillary wall. FcIgG activates complement and neutrophils

(d) Release of inflammatory mediators causes tissue damage. Increased vascular permeability allows IgG to seep into alveolar wall; ag-ab complexes cause local tissue damage, culminating in fibrosis and restrictive airways disease

Figure 2.34 Type III hypersensitivity, e.g. farmer's lung (an example of extrinsic allergic alveolitis).

Table 2.9 T-cell-mediated type IV reactions	
CD4 T cell	Rheumatoid arthritis
	Multiple sclerosis
	Inflammatory bowel disease
	Contact sensitivity
CD8 T cell	Viral infection
	Transplant rejection
	Tumour cell killing

hours and days, recruits and activates other T cells and macrophages, and produces local tissue damage and granuloma formation. It is the basis of the Mantoux test (tuberculin test, Fig. 2.35), used to see whether a person has some T-cell immunity to tuberculosis. It involves injecting a small amount of a (non-infectious) purified protein derivative of *M. tuberculosis* into the skin and observing whether a localised red induration occurs over the next 48 hours; T cells and macrophages mount a type IV reaction in a sensitised person. This test is not 100% sure, because some patients with overwhelming tuberculous infection are anergic and show no reaction.

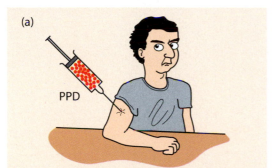

(a) Soution of tuberculoprotein purified protein deri vative (PPD) injected into skin of previously sensitized patient

(c) 12–24 hours later: erythema at injection site

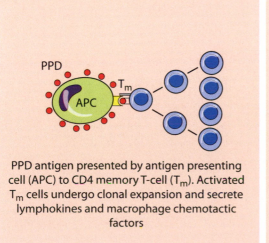

(b) PPD antigen presented by antigen presenting cell (APC) to CD4 memory T-cell (T$_m$). Activated T$_m$ cells undergo clonal expansion and secrete lymphokines and macrophage chemotactic factors

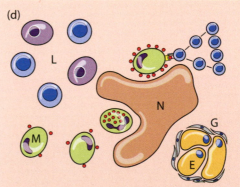

(d) Cellular responce includes recruited lymphocytes (L), and macrophages (M). Tissue damage and necrosis (N) occurs due to release of inflamatory mediators. Macrophages may become epithelioid cells (E) which aggregate to form granulomata (G)

Figure 2.35 Type IV hypersensitivity: Mantoux test reaction.

Chapter 2: What are the common mechanisms of disease?

Cytotoxic (CD8 cell-mediated) reactions involve direct killing of target cells through the perforin granzyme system, which are granules contained in cytotoxic cells that are released on activation. Perforin binds to the target cell membrane, allowing the entry of granzymes that activate cellular caspases. This induces apoptotic death of the target cells (see page 62).

In the first part of this book, we considered the interrelationships of disease, health and medicine, and then looked at the different categories of causes of disease and the major mechanisms that operate. We have finished this section with the topics of infection and inflammation, which leads us nicely to the second part of the book, which explores the ways in which we defend against disease and how the body can repair itself.

PART 2

DEFENCE AGAINST DISEASE

INTRODUCTION
THE ROLE OF EPIDEMIOLOGY IN DISEASE

In Part 1 we looked at the causes and mechanisms of disease. Part 2 examines how the body deals with disease and the way in which the body responds to noxious agents, be they from outside the body (exogenous), such as infectious organisms, toxins or traumatic injury, or arising within the body (endogenous), as in ischaemic damage, autoinflammatory disorders, metabolic stress or mechanical stress.

The body's response may be appropriate to the degree of tissue damage sustained, in which case restoration of normal function is the expected outcome. Sometimes there is a problem with the normal processes – an inadequate inflammatory response may lead to overwhelming sepsis, or an exaggerated reaction may cause extensive tissue damage, scarring, contractures, critical illness syndromes or death. Even cancers can be linked to inflammation, either directly due to organisms that modify and mutate DNA, or indirectly through mechanisms that provoke the release of reactive oxygen species by inflammatory cells, causing DNA damage. There is so much that could go wrong with the finely balanced mechanisms that regulate the innate and acquired immune responses to infection that it is a wonder that problems are not encountered more frequently.

In the coming chapters we discuss the details of the interactions of the inflammatory cells, host cells and the communication role played by mediators and cell receptors and ligand molecules. Inherited defects in neutrophil killing mechanisms, acquired deficiencies of key regulator cells such as CD4 T-helper lymphocytes in HIV/AIDS, or the body turning on itself to produce autoimmune diseases are just some of the potential problems that may be encountered in immune function.

First, as an introduction, consider the fascinating story of AIDS. It demonstrates a complex interplay between an infectious agent and the body's normal defence mechanisms. Its origins, epidemiology and modes of transmission needed unravelling to reduce its spread. Investigation of its biology has resulted in major improvements for treatment and survival.

HIV is a recently discovered infective disease that exemplifies the role of CD4+ cells in the defence against disease.

HIV/AIDS, first recognised in the early 1980s, is now the largest infective cause of death in the world. This virus infects and destroys the CD4+ cells that are at the heart of the body's adaptive immune system. CD4+ helper T lymphocytes (Th cells) recognise microbial antigens presented to them by antigen-presenting cells (APCs). They also stimulate effector cells, the CD8+ T-cytotoxic lymphocytes (Tc cells) and B cells, which produce antibodies. Both the CD8+ Tc cells and the antibodies are specifically targeted to attack the same antigen recognised by the CD4+ Th cell. Not only this, but the generation of memory Tc and B cells is also stimulated by Th cells. By studying how HIV/AIDS evolves, and how the body responds, we can understand much of the way the immune system works. CD4 receptors are found on cells other than Th cells.

> ## 🔑 Key facts
>
> Cells with CD4 receptors are vulnerable to infection by HIV.
>
> CD4 is a glycoprotein receptor found on Th cells, dendritic cells, macrophages and monocytes (including microglial cells, the macrophages of the central nervous system).
>
> However, the virus must bind a co-receptor, such as CCR5, to gain entry to the cell.

WHERE AND WHEN DID AIDS BEGIN?

Almost certainly, HIV originated from a mutated simian virus (SIV) found in certain monkeys and chimpanzees, which at some point made the mutational leap to become infective to humans. When and in what parts of the world this occurred has been debated – retrospective studies show that it may have been present in humans in the 1950s or even earlier. The earliest confirmed HIV infection was in preserved blood from a patient who died in the Congo in 1959.

First reports in the early 1980s of clusters of *Pneumocystis carinii* (now reclassified as *Pneumocystis jirovecii*) pneumonia (PCP or PJP) in gay men in California and elsewhere in the USA were followed closely by a number of outbreaks of aggressive Kaposi's sarcoma in several parts of the world, and curious immunodeficiency diseases were described in many parts of Africa, the USA, Haiti and elsewhere. Patients developed chronic diarrhoea, lymphadenopathy, tuberculosis and candidiasis. 'Slim disease' was reported in Ugandan men and women, who became mysteriously wasted (now attributed to HIV lipodystrophy), and the terms 'gay bowel syndrome' and then GRID (gay-related immunodeficiency disease) were coined in the western world. The disease became a scourge in both sexes in Africa and the East, whereas in western Europe the disease was for many years largely restricted to promiscuous MSM (men who have sex with men). Patients with haemophilia who received pooled blood products were early victims.

A combination of promiscuity, unprotected sexual intercourse, intravenous drug use, prostitution and transmission in jails has meant that the disease now constitutes a global pandemic in both sexes. Currently the disease is a major problem in parts of eastern Europe.

The term 'autoimmune deficiency syndrome' (AIDS) was coined in 1981. Luc Montagnier, in France in 1983, and Robert Gallo, in the USA in 1984, almost simultaneously isolated what we now know as the human immunodeficiency virus (HIV). In 1984, Dalgleish *et al.* demonstrated that the virus entered cells after attaching to the CD4 receptor.

Quite early on in the investigation into these new immunodeficiency diseases, later collectively recognised as AIDS, a link was made between genital herpes simplex virus (HSV type II) and an increased risk of infection with the virus later shown to be HIV. HSV-induced ulceration deprives the mucosa of its protective

Key fact

The 2008 Nobel Prize for medicine was awarded jointly to Dr Luc Montagnier and Francoise Barre-Sinoussi, a mere 25 years after discovering HIV, and to Harald zur Hausen who discovered in 1983–4 that human papillomavirus (HPV) types 16 and 18 caused cervical cancer and in 2006 developed an HPV vaccine.

epithelial barrier, increasing the risk of transmission of HIV. Added to that is the herpes simplex stimulus to CD4+ cell proliferation, which increases HIV viral synthesis by the infected CD4+ cells. The link with homosexual activity reflects the reduced barrier to infection offered by glandular surface epithelium in the

Key fact

Initial concerns that, because mosquitoes can spread malaria, HIV might also be transmitted in this way have been allayed. Malarial parasites utilise the mosquito as part of their life cycle. They migrate to the mosquito's salivary glands and multiply, to be squirted into the next victim, because mosquito saliva contains antithrombotic substances, allowing free blood flow. Any HIV inadvertently sucked out of a positive host dies in the mosquito's gut. Insufficient contaminated blood can be transferred by mosquito mouthparts to infect by the equivalent of a mozzie needlestick injury!

PLEASE USE SHARPS BIN

Key fact

Needlestick injury to health workers

The risk of transmission of HIV by needlestick injury involving minimal amounts of blood is extremely low – only around 1/300 people exposed to the very highest-risk patients contract HIV if no subsequent protective measures are taken.

The wearing of protective gloves at the time of injury drops the risk by 80% and, if the wound site is subjected to 10 minutes of prolonged washing with soap and running water, the risk is virtually nil.

Postexposure prophylaxis with antiretroviral drugs, instigated immediately, also brings the risk of infection to virtually zero.

rectum, which is only one-cell thick, compared with the stratified squamous epithelium lining the vagina and ectocervix.

HIV can be transmitted sexually, via infected blood (usually from shared needles) and in some body secretions, such as breast milk, but *not* by saliva, sweat or tears. HIV does not usually cross the placenta. However, the risk of transplacental infection is increased in patients with either recently acquired or very advanced HIV with high blood viral levels, or if there is an intrauterine infection or the mother is malnourished. If the mother has malaria, other sexually transmitted infection, urinary tract infection or respiratory infection, such as tuberculosis (TB), the risk is also increased. Antiretroviral treatment given to the mother greatly reduces this risk, as does the correction of any of the infectious or nutritional problems mentioned above. Infection of an infant more usually occurs at the time of birth or via breast milk (perinatal vertical transmission), if not protected by antiretroviral drug therapy. Around 25% of breast-fed babies are likely to catch HIV in this way, whereas if antiretroviral therapy is instigated the risk of transmission is less than 2%.

HOW DOES HIV CAUSE DISEASE?

Enveloped viruses, such as HIV or herpes simplex, can enter cells by binding a receptor on the host cell (Fig. 1). (Non-enveloped viruses are either taken up by receptor-mediated endocytosis or injected into the host cell by phages, see page 101.) In the case of HIV, its gp120 surface antigen binds the CD4 receptor on Th cells and some macrophages, but the virus also needs to bind a co-receptor in order to achieve its goal. This is usually the CCR5 receptor, which is attached to a G protein within the cell. It is a multifunctional receptor that normally recognises chemokines and is central to several inflammatory responses.

When HIV binds, it melds and fuses its envelope with the host cell's phospholipid bilayer membrane and uncoats its viral RNA, which enters the cytosol. The viral proteins and genome, once released, are able to activate the transcription of new viral particles. HIV, a retrovirus, can 'reverse transcribe' its RNA into DNA, which enters the host cell nucleus and is then replicated, forming new viral particles, which bud out of the cells.

After the initial infection, which may be silent or produce a rash and flu-like symptoms that last approximately 2 weeks, the virus may be unobtrusive and cause few if any problems for many years. During this period, lymph node enlargement may occur at many sites.

The virus gradually causes the death of infected T cells, thought to be largely achieved by the T cell's own pattern recognition receptors (PRRs), which assemble into an inflammasome, activating caspases and causing the death of the host T cell (see page 114). It may take up to 8 years before the number of circulating CD4 T cells in the blood falls below the critical level of $500/\mu L$. At this point, the body finds itself without the key coordinating cells of the adaptive immune response (see page 188) and starts to be prone to unusual infections.

CD4+ Th cells are central to adaptive immunity (see page 189). They both receive signals from APCs, which alert the immune system to the presence of invading microbes, and give signals for the activation and proliferation of effector CD8+ Tc cells and B cells – the latter become plasma cells and secrete antibodies. These can cause the destruction of free microbes or infected host cells displaying parts of the microbes on their surfaces. Without CD4+ Th cells, there is no generation of memory T and B cells, and so no memory of past infection.

Without CD4+ cells, the body is left to rely on the innate immune system. This is geared to recognise common bacterial and fungal pathogens because they bear 'PAMPs' (pathogen-associated molecular patterns, so called because these were first found in pathogenic organisms). PAMPs are actually found on most bacteria and many fungi, not all of which are pathogenic. Certain stress-related chemicals are

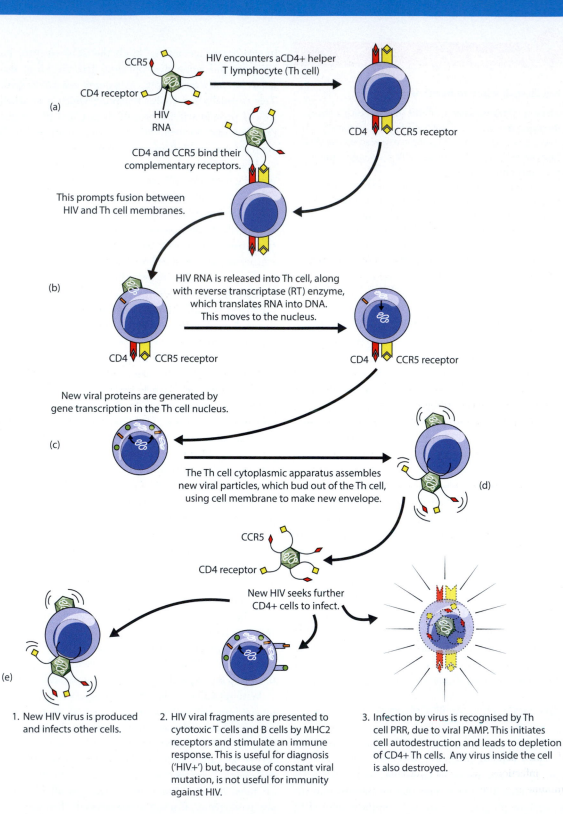

(a) CCR5

CD4 receptor

HIV encounters aCD4+ helper
T lymphocyte (Th cell)

HIV
RNA

CD4   CCR5 receptor

CD4 and CCR5 bind their
complementary receptors.

This prompts fusion between
HIV and Th cell membranes.

(b) HIV RNA is released into Th cell, along
with reverse transcriptase (RT) enzyme,
which translates RNA into DNA.
This moves to the nucleus.

CD4   CCR5 receptor

CD4   CCR5 receptor

New viral proteins are generated by
gene transcription in the Th cell nucleus.

(c)

The Th cell cytoplasmic apparatus assembles
new viral particles, which bud out of the Th cell,
using cell membrane to make new envelope.

(d)

CCR5

CD4 receptor

New HIV seeks further
CD4+ cells to infect.

(e)

1. New HIV virus is produced
and infects other cells.

2. HIV viral fragments are presented to
cytotoxic T cells and B cells by MHC2
receptors and stimulate an immune
response. This is useful for diagnosis
('HIV+') but, because of constant viral
mutation, is not useful for immunity
against HIV.

3. Infection by virus is recognised by Th
cell PRR, due to viral PAMP. This initiates
cell autodestruction and leads to depletion
of CD4+ Th cells. Any virus inside the cell
is also destroyed.

Figure 1 HIV entry to cell: binding by the cell receptors, receptor-mediated membrane fusion and endocytosis:

a. HIV gp120 receptor binds CD4 and co-receptor CCR5 receptor on the T-cell surface, fuses its envelope with the cell membrane, and viral RNA and proteins enter the cytosol.

b. Viral reverse transcriptase engineers the synthesis of DNA from viral RNA.

c. DNA is incorporated into the host genome – synthesis of viral particles starts.

d. Virus buds from cell membrane, which forms the new viral envelope.

e. Three options are shown: 1. infection of further CD4+ cells, 2. stimulation of HIV antibodies, 3. destruction of infected CD4+ cells, leading to depletion.

Below the critical level of 200 CD4+Th cells/ μL of blood, the adaptive immune system is bereft of its coordinator. There can be little or no response to viral and other infections requiring CD4+ Th-cell input. The innate immune system is intact, although dendritic cells and macrophages, both of which carry CD4, may also be lost, and the response may be suboptimal. The patient falls prey to opportunistic infections but also develops unusual malignancies, such as intracerebral lymphomas and aggressive Kaposi's sarcoma – almost unique to HIV, both related to human herpesvirus 8 (HHV8). This demonstrates the role of immunosurveillance and of viruses in neoplasia.

also PAMPs – see Chapter 4, Table 4.1). The innate immune response is generally effective against all except encapsulated microbes, but of little help against viruses and more complex organisms.

AIDS

Thus it is that HIV infection becomes the acquired immune deficiency syndrome (AIDS), characterised by either a CD4 count <200/μL or the development of 'AIDS-defining' illnesses. These include 'opportunistic' infections, which take advantage of the deficient immune response, such as pneumonia due to *P. jirovecii* or cytomegalovirus (CMV), oesophageal candidiasis, small bowel cryptosporidiosis or atypical mycobacteriosis (*Mycobacterium avium-intracellulare*), or cryptococcal

meningitis (*Cryptococcus neoformans*). These organisms are widely prevalent in the air, soil and water, and do not generally trouble people with a working immune system.

Fascinatingly there are also AIDS-defining malignancies, such as intracerebral lymphoma and Kaposi's sarcoma (both secondary to infection with HHV8), which demonstrate both the role immunosurveillance plays in guarding against tumour formation and spread, and the fact that some tumours are linked to viruses.

NATURAL IMMUNITY TO HIV/AIDS

Curiously, some people have natural immunity against HIV, those with Δ32 mutations in their CCR5 receptor genes (Fig. 2). This mutation is seen in around 16% of people in northern Europe, but is rare in sub-Saharan Africa, Asia, and North and South America. A Δ32 mutation is effective only against strains of HIV that use CCR5 as a co-receptor to initiate entry to the CD4 cell. However, most HIV infections do use CCR5. Δ32 prevents the HIV gp120 antigen from binding the mutated CCR5 receptor on the host surface molecule.

Berlin has become synonymous with good news for two HIV patients! The first was an anonymous man, infected with HIV in 1995, who in 1998 was successfully treated with a combination of antiretroviral drugs and chemotherapeutic agents, and is still well at the time of writing (2015). Timothy Ray Brown, the 'second Berlin patient', was also found to be infected with HIV in 1995. To compound his luck he developed acute myeloid leukaemia in

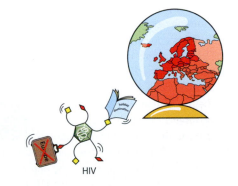

HIV

Figure 2 The Δ32 mutation in the *CCR5* gene confers protection against HIV infection and is most common in northern Europe, where 10% of people carry the mutation. It has been speculated that the mutation evolved around 700 years ago to select a population with decreased susceptibility to the then highly prevalent smallpox.

2006. In 2007 and 2008 he was given two stem cell transplants from a donor with homogeneous Δ32 gene mutations. Timothy Ray Brown took no further antiretroviral medication after the initial transplantation, but showed immediate suppression of viral replication, with a rise in CD4+ cells. Since 3 months after transplantation he has shown no trace of the virus and is alive at the time of writing, and the Timothy Ray Brown Research Foundation has launched the Cure for AIDS Coalition.

WHAT HAS BEEN ACHIEVED IN TERMS OF TREATING HIV/AIDS?

Early on in the course of our understanding of the disease, it was recognised that the use of condoms would decrease the risk of HIV sexual transmission by around 80% (WHO estimate). Unfortunately many men refuse to wear condoms, or use them incorrectly.

Microbicidal vaginal lubricant gels have undergone trials, with varying success. Some of the gels actually increased the risk of HIV infection, but one, containing the retroviral agent tenofovir, reduced the risk of HIV in the woman by 39–54%. Condoms will protect against sexual transmission – but also against pregnancy. How can an HIV-negative woman safely become pregnant with an HIV-positive man (*in vitro* fertilisation excluded)? A new approach currently in early clinical trials in the UK is the inexpensive production of antibodies to HIV in genetically modified tobacco plants. The antibodies are administered in a vaginal gel, with the object of coating the HIV with antibody, blocking its ability to infect and targeting it for destruction by the innate immune system.

The development of highly active antiretroviral therapy (HAART), with its combinations of drugs that interfere with the viral life cycle and its reproductive mechanisms, has been life saving for those who have access to the medication. For those who start this medication at the moment the CD4 count drops to <500/μL, life expectancy is 70–100% of that of uninfected people. Once the CD4+ T-cell count has been restored to >500/μL, there is no increased risk of opportunistic infection. This comes at a cost, e.g. in the UK the average cost of HAART is £15,000 per patient per year. Magnificent philanthropic funding, together with initiatives by the WHO and others, have brought antiretroviral treatment to parts of the world that would not otherwise have been able to fund it. However there are many people who are not yet able to access treatment and are dying of TB, pneumonia or other diseases because of HIV.

A multinational trial run by the US National Institutes of Health into starting antiretroviral therapy immediately on diagnosis of HIV, rather than waiting for the CD4 count to drop, or AIDS-defining illness to develop, was started in 2011. The intention was to run the trial until 2016, but this was abandoned early in 2015 because of the strikingly beneficial effects that the early treatment conferred. Not only was viral replication suppressed effectively, but also the reduced viral levels in the blood of infected patients led to reduced disease dissemination.

There is as yet no proven effective anti-HIV vaccine ready for use in humans. One problem has been the fact that the virus mutates constantly and, if (as is frequently the case) a patient has been infected by more than one strain of the virus, parts of each virus may join to make novel viral proteins. However, exciting developments, which may circumvent the need for a vaccine, are currently in trial in rhesus macaque monkeys. Monkey muscle cells (*in situ*), genetically modified by being transfected with a gene encoding an 'immunoadhesin', produce an antibody that binds avidly to the HIV viral envelope protein. This has been shown to prevent HIV1, HIV2 and SIV infection when the virus is subsequently introduced into the monkeys. The antibodies can be continually produced by the transfected cells, seemingly for several years, and may be suitable for passive immunisation.

CAN HIV EVER BE A GOOD THING?

HIV, in a modified, non-pathogenic form, has recently been used as a vector for gene therapy. It has an almost unique ability to transfer a segment of RNA into a host CD4+ cell and then reverse transcribe this into DNA, which is then inserted into the host genome and translated into a working gene. Bone marrow stem cells can be transfected with working genes using modified HIV (HIV with its pathogenic components removed). So far, there appears to be promise from trials using modified HIV as a vector in the genetic modification of the *WAS* gene in patients' own bone marrow stem cells in six boys with the X-linked Wiskott–Aldrich syndrome. This syndrome is characterised by weeping eczematous skin lesions, low serum IgM levels and thrombocytopenia (low platelet numbers), and often causes premature death due to infection or bleeding, or the development of haematological malignancies.

THE BODY'S RESPONSE TO INFECTION

MAJOR ROUTES FOR THE TRANSMISSION OF INFECTION

Infections start from a reservoir of some kind, which acts as a source of pathogens. The most common source of human infections is infected humans, although some diseases involve animal reservoirs or soil organisms.

It is not difficult to work out the main routes of spread from one infected human to another. Organisms in one person's respiratory tract are coughed out as droplets and aerosols, become airborne and are inhaled by other people. A simple bus journey is laden with opportunities for bacteria and viruses to visit new human hosts. Direct mucosal contact is important in sexually transmitted infections or STIs (HIV, herpes simplex, hepatitis B, human papillomavirus [HPV], *Chlamydia* sp., gonorrhoea and syphilis), and viruses infecting salivary glands (herpes and mumps).

Gut pathogens are excreted in the faeces and can re-enter the same or another gut via faecal contamination of water or food (faeco-oral route). Contamination of food is generally due to either lack of human handwashing or failure to cover food to protect it from flies.

There is a much more sinister group of arthropods than the simple fly with dirty feet. These are insects that are themselves infected with an organism that they transfer by injection into humans (e.g. malaria and yellow fever). This is one means of blood-borne spread. An insect sucks up infected blood from a human or animal and transfers it to another human or animal when it next bites. In some instances (e.g. malaria) the insect is an essential component because part of the infective agent's life cycle takes place within its body. Humans traditionally fear rats, possibly because of their historical links with bubonic plague. But, similar to flies, they are relatively innocent, being mere vehicles for the fleas that live on their bodies and carry the bacterium *Yersinisa pestis* to humans.

Comparable methods of spread are via hypodermic needles shared by drug addicts (e.g. hepatitis B) or through use of contaminated blood products (e.g. HIV and hepatitis C in people with haemophilia).

Before we consider how the body can respond to disease we should consider its natural defences. For practical purposes the defences of the body can be separated into three components: structural barriers, innate immunity and adaptive immunity. This is a rather artificial separation. As we shall see, these three components are intimately linked and act together in many instances to protect the body. Let us start with the structural barriers.

Chapter 3: The body's response to infection

THE BODY'S STRUCTURAL DEFENCES AGAINST INFECTION

The body is covered by epithelium, inside and out. The skin, with its stratified squamous epithelium topped with a waterproof layer of keratin, covers the outside. It contains some large holes, leading to areas covered by more permeable epithelium. From the top down these are the eyes, nasal cavity, mouth, anus, urethra and vagina, and each one has its own defence system (Fig. 3.1).

The eye drops its portcullis and floods the moat at the slightest provocation, because it is determined to repel invaders before they can produce any tissue damage. The main parts of the eye must be transparent and able to transmit light without distortion. A scarred cornea can render the eye useless and anything worse than a little inflammation of the conjunctiva can be devastating. So the eye's defences include a nerve reflex to close the eyelids as danger approaches, and a lacrimal gland to secrete tears to wash away particles,

chemicals and bugs in a matter of seconds. In addition, tears contain lysozyme, an enzyme capable of degrading bacterial cell walls. If you have ever had even the smallest scratch to the surface of your eye, you will know that the flow of tears and the desire to keep the eyelids closed keeps going until the epithelial covering is re-established. Fortunately, this generally takes less than 24 hours.

The nose can be considered with the respiratory tract, because both are covered predominantly by respiratory epithelium. This is characterised by the presence of mucus-secreting cells and cilia. Cilia are tiny hair-like structures that beat in a synchronous fashion to move particles up the respiratory tract. A layer of mucus lines the respiratory tract, providing a barrier against infectious agents and trapping inhaled particles. The mucous layer acts as a conveyor belt propelled by the underlying cilia. This is the mucociliary clearance mechanism and it is severely damaged by smoking. The airways also have, in coughs and sneezes, nervous mechanisms of defence that forcefully expel unwanted material. This is good for the individual but potentially hazardous for those in the vicinity, because it is a super method of spread for airborne bugs. If microbes get past the mucociliary defence, there is a second line of defence involving the macrophages in the alveoli – sometimes termed the 'sentries of the lung'. These highly skilled phagocytic cells can recognise invading microbes using pattern recognition receptors (PRRs), and engulf and destroy them or recruit help from other inflammatory cells.

The mouth, oesophagus and anus form the ends of the gastrointestinal tract and are protected by non-keratinising, stratified, squamous epithelium and a layer of mucus from local glands. The mouth has teeth and an enzyme (amylase), both of which, it could be argued, have a defensive role, but their main purpose is clearly related to digestion. There are similar dual-purpose roles for the hydrochloric acid and enzymes in the stomach or the secretions of the small intestine; their main role is digestive but they may also be harmful to many microorganisms. The gut, similar to the respiratory tract, has a layer of protective mucus. Immunoglobulins are specifically made for the lining of the gut; they disable many bacteria and other organisms ingested with food. The proximal small intestine attempts, usually successfully, to maintain the near sterility of the upper gastrointestinal tract, utilising not only secreted

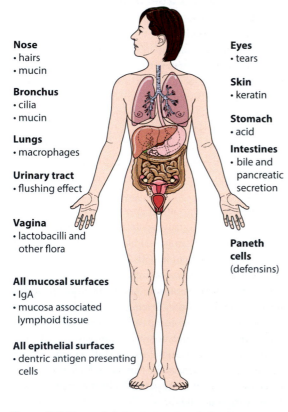

Nose
• hairs
• mucin

Bronchus
• cilia
• mucin

Lungs
• macrophages

Urinary tract
• flushing effect

Vagina
• lactobacilli and other flora

All mucosal surfaces
• IgA
• mucosa associated lymphoid tissue

All epithelial surfaces
• dentric antigen presenting cells

Eyes
• tears

Skin
• keratin

Stomach
• acid

Intestines
• bile and pancreatic secretion

Paneth cells
(defensins)

Figure 3.1 Natural defences against infection.

IgA immunoglobulin and mucus, but the granules of the Paneth cells. These secrete a number of microbicidal agents in response to the recognition of Gram-positive and Gram-negative bacteria – chief among which are lysozyme and the defensins. It is believed that Paneth cells recognise PAMPs (pathogen-associated molecular patterns) via their PRRs (see Chapter 4, page 114) and empty their granule contents into the lumen of the small intestinal crypts. Here they are best placed to protect the precious intestinal stem cells, which constantly replenish the supply of villous and crypt epithelial cells from harmful bacteria. (Fungi and protozoa do not cause Paneth cells to degranulate.)

The distal small intestine and the large intestine, of course, are not sterile, but are colonised by a range of around 400 different species of bacteria, both aerobic and anaerobic, which often live in peaceful coexistence as 'commensals' unless something upsets the balance. This commonly occurs when a course of antibiotics destroys one type of bacterium, so allowing another to over-proliferate and cause problems, so-called antibiotic-associated or pseudomembranous colitis. *Clostridium difficile* is a bacterium that proliferates under these circumstances and secretes two or more toxins to produce abdominal pain and profuse diarrhoea, said to have the odour of horse manure. In some familial forms of Crohn's disease, the PRR NOD2 may be mutated, with a deranged response to bacterial flora.

The urinary tract is normally protected by a one-way flow of urine from the kidney to the urethra, along tubes lined by a multilayered transitional epithelium. There is no mucus barrier here but it should not be needed, because the urine formed in the kidney is sterile. Its defences are the high volume of urine flushing the system, the physical barrier of the empty urethra and the natural variation of the pH of the urine. Women have a short urethra and are more liable than men to urinary tract infections (UTIs), which often occur after sexual activity. This can cause perineal bacteria to be massaged up the urethra into the bladder ('honeymoon cystitis'). The urine may have an acidic or alkaline pH, favouring the growth of some bacterial strains over others. Hopefully, natural variations in the urinary pH will prevent a particular bacterium from becoming established. People with diabetes are susceptible to UTIs if the urine contains glucose, which will nourish bacterial growth. In elderly people, UTIs should be suspected even in the absence of the usual symptoms of dysuria or cloudy or blood-stained urine, because symptoms may be vague, such as increasing confusion, or related to septicaemia.

The genital tract of the female starts its defences with the vagina, which is lined by non-keratinising stratified squamous epithelium and mucus, and friendly colonies of Döderlein bacillus, a lactobacillus that metabolises glycogen to lactic acid to produce an environment with a pH of 5 which inhibits colonisation by most other bacteria. Unfortunately, glycogen is present only when the vaginal epithelium is stimulated by oestrogens, between puberty and the menopause. At other ages, the vagina is alkaline and liable to infection with pathogenic staphylococci and streptococci. The main defence is, however, the action of mucus and lysozyme in the normal cervix. The uterus and tubes are specialised for their reproductive role, but it is possible that the monthly shedding of the endometrium may be a useful defence against infection. The stereocilia in the fallopian tubes, while wafting the ovum down, produce currents that may help to prevent bacteria from ascending. Theoretically there is access to the peritoneal cavity via the fallopian tubes, although in practice this does not appear to be a problem.

Should the first line of defences fail, the body has a regular guard of armed and dangerous cells and molecules, the brief of which is to resist attack (innate immunity). This army comes swiftly to the defence within seconds and minutes of tissue damage or microbial invasion occurring.

The innate immunity guard includes neutrophil polymorphs and tissue- and monocyte-derived macrophages, which immobilise and engulf damaging agents, complement proteins, which attract inflammatory cells to the site of damage and prepare them to be engulfed, molecules related to blood clotting, and factors released from a variety of cells which assist in the inflammatory response. We hear more of these when we come to discuss acute inflammation in Chapter 4.

As microorganisms sometimes manage to break through the first- and second-line defences, we need another line of defence involving the inflammatory response and the immune system, so-called adaptive immunity. This involves specific defences, which take time to develop, but, as they are targeted specifically at the causative agent, they generally cause less in the way of 'collateral damage' because they exert their effect on the innate defence mechanisms. T and B cells work to

Chapter 3: The body's response to infection

produce agents precisely manufactured to attack particular intruders.

Sometimes there are problems that are related to under- or over-activity of the immune system, or aberrant activity. Among these problems are diseases related to the absence of important inflammatory cells or agents, such as complement molecules, and those related to the systems that downregulate the activity of the immune system after an acute event. There may be hyperreactivity, as in the hypersensitivity diseases, or a loss of self-tolerance in the autoimmune diseases.

THE IMMUNE DEFENCE AGAINST INFECTION

Apart from the physical barriers, which make up the first line of defences, there is also an army of cells and inflammatory mediators lying in wait for the unwary microbe.

Acute inflammation is covered in Chapter 4. It occurs as an immediate response to injury or to invasion by pathogens and involves in particular dendritic cells, macrophages, mast cells and platelets, and numerous inflammatory mediators, all of which result in an influx of neutrophils, the characteristic cell of acute inflammation. When immunologists refer to the innate immune response they are considering the body's standard response to tissue damage or invasion by microbial pathogens. PAMPs and PRRs are key to the instigation of acute inflammation. Innate immunity is a bit like setting a minefield just in case someone tries to invade your country: if a mine is stepped on, an immense explosion takes place, usually killing whatever set off the mine and causing a considerable amount of local damage in the process (see Fig. 6.1, page 180). It does not really matter who steps on the mine, the damage is the same.

Healing and repair are natural consequences of acute inflammation. Normal healing processes and examples of problems related to healing and repair are discussed in Chapter 5.

The adaptive immune response is a precisely targeted response to a pathogen, which might be likened to a skilled sniper. The body has been alerted by the sentries or scouts on duty at the border, usually the dendritic cells but commonly also the tissue macrophages and less commonly the B cells – these three cell types are the antigen-presenting cells (APCs). In addition, the cytokines released as a consequence of innate immune mechanisms, i.e. acute inflammation, macrophages, mast cells, natural killer (NK) cells and inflammatory mediators of the innate immune system, may upregulate the APCs. If the agent is a microbe, ingested or intracytoplasmic fragments are broken up and presented on the surface of the cell or APC, to generate an immune response by activating T and B cells. It is as if a photograph of the microbe has been circulated to the highly trained sniper, who recognises and assassinates the target cleanly, with little 'collateral damage'. The cells involved are once again the macrophages and dendritic cells, but also the T and B lymphocytes (these crop up many times, so for ease of discussion we term them T cells and B cells). These are also the cells of chronic inflammation, which when encountered clinically is always an abnormal pathological phenomenon.

Students can become confused when pathologists talk about acute and chronic inflammation but immunologists speak of innate and adaptive (or acquired) immunity. Clinically apparent inflammation is a pathological manifestation of usual physiological mechanisms that have either taken place inappropriately, or failed to control and eliminate the causative agent. In Chapters 4–6 we discuss not only the normal patterns of acute inflammation, healing and repair, and chronic inflammation, but also some of the ways in which these very precisely organised and choreographed processes may go wrong.

HOW DO MICROORGANISMS EVADE OUR FIRST-LINE DEFENCES?

SKIN

Entry through intact skin is most easily achieved via a wound or using an animal vector designed for the purpose. Many microorganisms associate with biting insects that pierce human skin and so provide a route into the human body, e.g. mosquito-borne malaria. The rabies virus relies on larger animals such as infected dogs or bats biting humans, and then uses its acetylcholine-binding receptor to enter a neuronal cell. HIV, once in the blood (e.g. introduced by a contaminated needle), binds helper T cells via the CD4 receptor, which binds to gp120 antigen on the viral surface. In these examples, the microorganism itself does not have special characteristics for skin penetration but some, such as the helminth larvae of hookworm or *Schistosoma* sp., have lytic proteases to allow them to digest and burrow through intact skin.

Some viruses start out with a 'key to the door', in that they have receptors that bind specific molecules on the host cell surface, which allow them to gain entry. A good example of this is the rhinovirus, which binds intercellular adhesion molecule 1 (ICAM-1) on mucosal cells.

> 🔗 Read more about what can go wrong in Pathology in Clinical Practice Case 43

GASTROINTESTINAL TRACT

Helicobacter pylori is the only pathogen that actively survives in the gastric acid. It does this by producing a urease that converts urea to ammonia and, thereby, changes the pH of its microenvironment. Protective coverings protect bacterial spores, protozoan cysts and thick-walled helminth eggs, which can evade the digestive tract's acid and enzymes (Fig. 3.2). Many of the non-enveloped viruses (hepatitis A, rotavirus, reovirus and Norwalk [now more commonly known as Norovirus] agents) are resistant to digestive juices. Bacteria can 'hide' in insufficiently chewed food to

avoid the acidity. Some parasites' eggs (e.g. *Giardia* sp.) actually need gastric acid as a stimulus to cause them to hatch into the trophozoites that infect the intestine.

RESPIRATORY TRACT

The normal mucociliary defence system may be impaired by the host's own actions, such as cigarette smoking, or the aspiration of gastric acid. Neuraminidase-producing microorganisms may degrade the mucin layer and toxins produced by *Haemophilus* and *Bordetella* spp. can paralyse mucosal cilia. Even if the mucociliary mechanisms are working well, some viruses can still avoid being expelled by having specific methods for adhesion to the epithelial surface, such as haemagglutinins on influenza virus. Once the defences are damaged, e.g. by a bad cold or influenza, bacterial pneumonias are common and may overwhelm, causing severe bronchopneumonia with a high mortality if untreated (Fig. 3.3). It is this that may explain the fact that the Spanish flu epidemic of 1918–19 killed more people than the whole of the First World War!

Many microorganisms avoid our first line of defence by not even attempting to pass through the epithelial surface, but being content to grow on the top. Skin fungi (dermatophytes, e.g. tinea pedis [athletes' foot] and skin viruses [papilloma viruses causing warts]) live in the superficial layers. Gut pathogens,

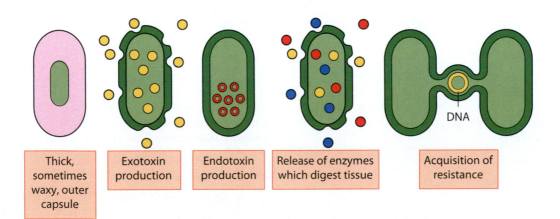

| Thick, sometimes waxy, outer capsule | Exotoxin production | Endotoxin production | Release of enzymes which digest tissue | Acquisition of resistance |

DNA

Figure 3.2 Bacterial protection against attack.

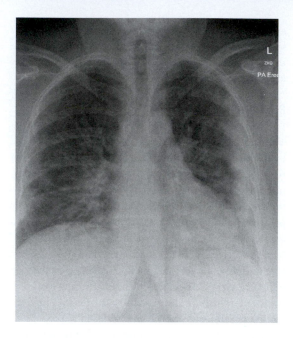

Figure 3.3 Influenza pneumonia in a 40-year-old man who presented with fever and cough and generally feeling unwell. The chest radiograph shows widespread 'fluffy' or alveolar shadowing in both lungs. Within 24 hours in hospital his condition deteriorated to the extent that he had to be intubated and ventilated in ICU.

such as *Vibrio cholerae*, multiply in the mucus layer, releasing exotoxins that cause watery diarrhoea. Some bacteria cause us harm without even entering the body if they produce exotoxins that contaminate food (e.g. staphylococci, see page 73).

So the first line of defence has been breached but the immune system and inflammatory mechanisms should still be able to defend us, shouldn't they? Thankfully, they normally do but the microbes have some clever tricks to avoid them.

First, the immune system must spot the invader, so the invaders change their surface antigens by shedding them, changing the antigens during the course of an infection or having numerous antigenic variants so that infective episodes do not produce useful immune memory.

The next phase is killing the microorganisms by phagocytosis, complement-mediated lysis, antibody-mediated mechanisms or neutrophil activity. Yes, you've guessed. Somewhere there is a bug that will have evolved a way round each of these and a few examples are given in Table 3.1.

By now you will have realised that, in any war, a lot of innocent bystanders get hurt. The bugs need to survive, replicate and be shed to find new hosts. The normal flora or commensals manage this without causing damage but the pathogens are greedy, have more

Table 3.1 Mechanisms used by microbes to evade the immune system

Mechanism	Example
Shedding of antigens	*Schistosoma mansoni*
Changing antigens during infection	*Neisseria* sp., altering pilin locus
	African trypanosomiasis
Many antigenic variants so no protective cross-immunity	Rhinovirus
	Influenza virus
Resistance to phagocytosis	Carbohydrate capsules of pneumococci, meningococci and *Haemophilus* sp.
Interference with antibodies	Protein A molecules of staphylococci blocking Fc portion of immunoglobulin or complement
	Digestion by proteases of *Neisseria*, *Streptococcus* and *Haemophilus* spp.
Resistance to complement-mediated lysis	K antigens on some *Escherichia coli*

Mechanism	Example
Resistance to macrophage killing	*Legionella*, *Mycobacterium* and *Toxoplasma* spp. inhibit acidification
Inaccessible to immune system	Gut luminal proliferation of *Clostridium difficile*
	Papillomavirus or fungi in superficial layer of skin
	Direct viral transfer between adjacent cells
Specific damage to immune cells	*Pseudomonas* sp. secretes a leukotoxin to kill neutrophils
Mimicry of host antigens	Group A streptococci and myocardium
Interference with MHC presentation of antigens	Herpes simplex inhibits peptide transporter
	Cytomegalovirus blocks MHC-I presentation and expresses an MHC-I mimic
Bind to and inhibit cytokines	Vaccinia virus produces soluble interferon receptor
Immunosuppression	HIV, Epstein–Barr virus

weapons, and provoke a conflict with the inflammatory and immune cells. Damage to the host cells occurs because of competition for nutrients, release of inflammatory mediators and toxic substances by the host's inflammatory cells, production of substances by the microbes that damage the host's tissues and direct cellular damage by microorganisms.

ARE SOME PEOPLE MORE SUSCEPTIBLE TO INFECTION THAN OTHERS?

Since the antiseptic treatment has been brought into the full operation, and wounds and abscesses no longer poison the atmosphere with putrid exhalations, my wards, though in other respects under precisely the same circumstances as before, have completely changed their character: so that during the last nine months not a single instance of pyaemia, hospital gangrene or erysipelas has occurred in them. As there appears to be no doubt regarding the cause of this change, the importance of the fact can hardly be exaggerated.

Lord Joseph Lister (1827–1912), British surgeon

Some people are undoubtedly more susceptible than most to infection. People with inherited syndromes such as severe combined immunodeficiency (SCID), complement deficiencies and HIV/AIDS, are all at increased risk.

Let us consider a normal healthy individual: this person is not malnourished, not on immunosuppressive drugs nor just recovering from an operation. From a microbe's point of view, what are the possible routes into the body? The surface of the body is covered by skin which, as discussed, has holes in it. Most of the holes lead down into sweat ducts, hair shafts and other skin appendages that still have an epithelial lining, be it a more delicate one. Microorganisms can live down these holes without causing particular problems unless something disturbs the delicate balance between host and bug. This occurs, for example, in scarring acne when the composition of the sebaceous gland secretion is altered, leading to blockage of the neck of a hair follicle, proliferation of bacteria behind the blockage, and the destruction of the follicle's epithelial line of defence to provoke inflammation in the surrounding skin.

In fact, skin is built to take knocks and we are all liable to develop small cuts and grazes on a frequent

Chapter 3: The body's response to infection

basis. We know that when this defence is down we must be extra-vigilant and keep the area clean and dry, and resist picking at the delicate protective layer of scab that forms at the site of damage. This becomes even more important when the skin contains a large wound, as may follow surgery. This is the time when bacteria have a really good chance to successfully invade a human body, because they have a band of helpers. These are the doctors and medical students who move rapidly from one patient to the next in the postoperative surgical wards, generously ensuring that all patients have the opportunity to acquire each other's skin flora.

Insistence on handwashing before and after examining a patient has undoubtedly reduced the risk of wound infection and cross-infection in British hospitals over recent years. Less well evidenced is the policy adopted in the UK of not wearing white coats or ties, and a 'bare below the arms' rule (this is not what it sounds, but simply means that there is no clothing on the lower arms). Stethoscopes must be regularly cleaned and can no longer be worn casually around the neck. Regardless of which has been the most effective measure, the results have been gratifying: infection rates from MRSA (methicillin-resistant *Staphylococcus aureus)* and pseudo-membranous colitis caused by *Clostridium difficile* have plummeted since the new Department of Health guidelines were instituted in the UK in 2007. The incidence of MRSA decreased by 57% and of *C. difficile* by 63% in England over the first 2 years after implementation of the new measures.

The combination of overcrowding (i.e. many sick people living in a closed environment), difficulty with personal hygiene (just try having a bed bath!) and unprotected (non-sexual!) contact with a large number of strangers describes both a postoperative surgical ward and a bacterium's idea of heaven. Add to this the likelihood that a bacterium that has been circulating in hospital for a while has learned a few tricks in terms of antibiotic resistance (see below), and it is understandable that a patient with compromised resistance, including the convalescing post-surgical patient, can be highly susceptible to infection. In fact 'nosocomial' (hospital-acquired) infection is a big risk to recovering patients, whose immune systems may have just been instructed to 'stand down', having dealt with the trauma of, for instance, surgery or a severe infective episode. Organisms prevalent in hospital may be resistant to multiple antibiotics. This is one reason for discharging patients from hospital as soon as is practicable.

People with diabetes are particularly prone to skin infections because, if their blood glucose levels are hard to control, all their body fluids can be high in glucose, and this provides an excellent culture medium for bacteria. Often people with diabetes have poor circulation due to vascular disease, so bacteria can flourish relatively undisturbed by an immune response. The problem may be compounded by traumatic damage to the skin if patients with diabetes also have a loss of sensation due to peripheral nerve damage, meaning that they might not feel, for example, the early stages of blisters or small abrasions.

ANTIBIOTIC RESISTANCE

Since the discovery of penicillin and sulfonamides in the first half of the twentieth century, we have been able to treat many bacterial infections. However, superbugs have begun to emerge, resistant to many antibiotics.

MRSA, which is resistant to numerous antibiotics, is a problem that has plagued hospitals. The introduction of an intensive handwashing policy, which was instituted to combat this and *C. difficile* colitis, has had considerable success.

Now multi-drug-resistant tuberculosis (MDR-TB, resistant to the two most powerful first-line anti-tuberculous drugs, isoniazid and rifampicin) and extensively drug-resistant TB (XDR-TB, TB resistant to both first- and second-line anti-tuberculous therapies) has become one of the major microbiological problems of our age (see page 166).

We know that the same genes that encode the various resistance factors can be found in several unrelated classes of bacteria. How do the bugs transmit resistance to each other?

Resistance involves the transfer of genes, which can be achieved in four main ways:

1 Transformation
2 Transduction
3 Conjugation
4 Transposon insertion.

TRANSFORMATION

Naked DNA fragments are released from a bacterium as it is lysed. These fragments then bind to the cell wall of another bacterium, are taken in and become incorporated into the recipient's DNA (Fig. 3.4). This sounds simple but generally it occurs only between closely related bacteria with similar DNA (i.e. extensive homology) and suitable binding sites on the cell wall.

TRANSDUCTION

Bacteria themselves can be infected by viruses, called bacteriophages (Fig. 3.5). These viruses bind to specific

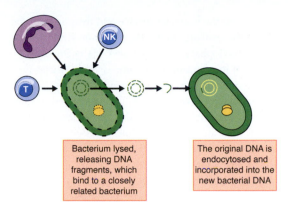

Bacterium lysed, releasing DNA fragments, which bind to a closely related bacterium

The original DNA is endocytosed and incorporated into the new bacterial DNA

Figure 3.4 Transformation.

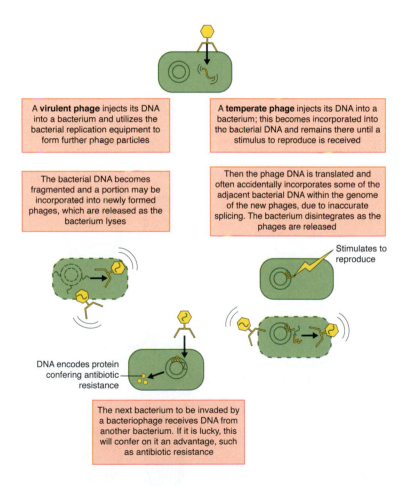

A **virulent phage** injects its DNA into a bacterium and utilizes the bacterial replication equipment to form further phage particles

A **temperate phage** injects its DNA into a bacterium; this becomes incorporated into the bacterial DNA and remains there until a stimulus to reproduce is received

The bacterial DNA becomes fragmented and a portion may be incorporated into newly formed phages, which are released as the bacterium lyses

Then the phage DNA is translated and often accidentally incorporates some of the adjacent bacterial DNA within the genome of the new phages, due to inaccurate splicing. The bacterium disintegrates as the phages are released

Stimulates to reproduce

DNA encodes protein confering antibiotic resistance

The next bacterium to be invaded by a bacteriophage receives DNA from another bacterium. If it is lucky, this will confer on it an advantage, such as antibiotic resistance

Figure 3.5 Transduction: bacterial DNA is transmitted via bacteriophages, of which there are two main types – virulent and temperate phages.

Chapter 3: The body's response to infection

receptors on the bacterial surface and then push a tube through the wall and inject the viral nucleic acid. The virus (bacteriophage) can take over the bacterial cell's replication mechanism in the same way as viruses replicate in human cells. New capsid proteins, nucleic acid and enzymes are produced and assembled so that, when the bacterial cell lyses, the phages are released. This is what occurs with virulent phages but there are also temperate phages that can insert their DNA into the host's DNA and lie dormant for long periods. Eventually, these little time bombs go off, and replicate and lyse the host cell in a similar fashion to virulent phages, but the fact that they have inserted into the bacterium's DNA is relevant to our discussion of transfer of resistance.

Generalised transduction involves virulent phages taking over the bacterial cell and accidentally packaging some bacterial DNA fragments rather than viral nucleic acid into one of the daughter phages. The mutant daughter phage will inject this bacterial DNA into the next cell that it tries to infect and the bacterial DNA may incorporate into the host cell DNA (comparable to the changes in transformation). The host cell survives happily because no viral nucleic acid is injected and it may gain either useful resistance or virulence factors (or both), depending on the source of the fragment of bacterial DNA.

Specialised transduction utilises temperate phages. When the phage DNA in the host genome (prophage) is reactivated, it is spliced out of the bacterial chromosome, replicated, translated and packaged into a capsid. The splicing may include taking some bacterial DNA immediately adjacent to the prophage, and this will be incorporated into daughter phages. If the bacterial DNA confers resistance or virulence, this is a powerful method for sharing the information and is called 'lysogenic conversion'. The genes for exotoxin production by *Corynebacterium diphtheriae*, *Vibrio cholerae* and *Streptococcus pyogenes* (scarlet fever) can spread in this way.

CONJUGATION

Conjugation is the major mechanism for transfer of antibiotic resistance and involves the exchange of plasmids. Plasmids are pieces of circular double-stranded DNA, separate from the chromosome, that carry a variety of genes, including some for drug resistance and

some coding for the enzymes and proteins needed for conjugation. These are called self-transmissible or F plasmids and bacteria can be F+ or F- depending on whether they contain the plasmid.

Bacterial sex involves a specialised sex pilus protruding from the surface of an F+ bacterium. This binds to and penetrates the cell membrane of an F- bacterium and a single strand of the F-plasmid DNA passes from one cell to the other (Fig. 3.6).

The former F- cell is now F+ and has the information for resistance and conjugation with other cells. The plasmid DNA is not totally fixed but can acquire additional bacterial genes if it integrates into the bacterial chromosome in a similar fashion to a temperate bacteriophage, i.e. when it is spliced out of the host chromosome it includes some of the adjacent bacterial DNA and is called an F prime (F′) plasmid. An alternative is that the bacterial chromosome with

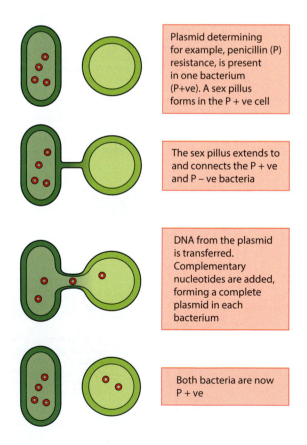

Plasmid determining for example, penicillin (P) resistance, is present in one bacterium (P+ve). A sex pillus forms in the P + ve cell

The sex pillus extends to and connects the P + ve and P – ve bacteria

DNA from the plasmid is transferred. Complementary nucleotides are added, forming a complete plasmid in each bacterium

Both bacteria are now P + ve

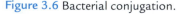

Figure 3.6 Bacterial conjugation.

the incorporated plasmid is all transferred to the F- cell during conjugation.

Plasmids encode medically important enzymes, such as penicillinase, and virulence factors, such as exotoxin and fimbriae.

TRANSPOSONS

Transposons are also DNA fragments that can carry genes for resistance and virulence, but they differ in that they can insert into host DNA which is dissimilar (i.e. lacks homology) and so can spread resistance across bacterial genera. They can insert into phages, plasmids and bacterial DNA (Fig. 3.7). Transposons (named because they are 'transposable elements') have been termed 'jumping genes'.

HOW CAN WE PREVENT OR OVERCOME BACTERIAL RESISTANCE?

Life is a struggle for survival and the animal, plant or microorganism that is best adapted to a particular environment is most likely to flourish. Humans can have an enormous impact on the environment and we must use this ability with care. In this context, we are concerned that we might alter the microbes' environment so as to give the resistant bacteria a survival advantage. If everyone were taking a particular antibiotic, only microbes resistant to that drug would survive and so that specific environmental niche would become populated with resistant bugs and the antibiotic would be useless. This is the fear with tuberculosis as the incidence of dormant multiple-drug-resistant tuberculosis increases. Of course, we don't have everyone on the same antibiotic and we have to hope that the non-resistant bugs will predominate when the antibiotic is stopped.

Problems occur when antibiotics are widely used, particularly in a hospital setting, and the resistant bug gains a clear advantage. We consider MRSA as an important example of how humans and microbes each develop strategies to defy the other. *Staph. aureus* is a Gram-positive coccus that was originally sensitive to penicillin. By the 1950s, this Gram-positive coccus had developed enzymes (β-lactamases) to destroy penicillin. Scientists then produced a penicillin analogue called meticillin, with a side chain that reduced lactamase binding to the enzyme. Further drug development led to the widely used flucloxacillin, so humans had the upper hand until the late 1970s and early 1980s, when *Staph. aureus* strains resistant to multiple antibiotics, including meticillin and gentamicin, emerged. These organisms are an enormous problem in hospitals, particularly on surgical wards where wound infections are a major concern.

MDR-TB strains now represent more than 20% of cases in parts of New York and are being found in mainland Europe and the UK. XDR-TB is resistant to all known antibiotics. Gentamicin-resistant enterococci (GRE) and, recently, the emergence of multiple-drug-resistant Gram-negative bacteria have caused great concern.

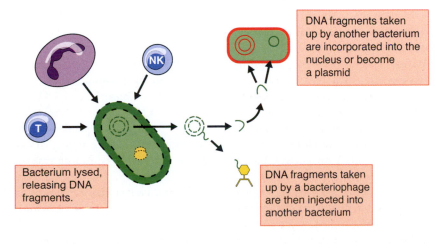

DNA fragments taken up by another bacterium are incorporated into the nucleus or become a plasmid

Bacterium lysed, releasing DNA fragments.

DNA fragments taken up by a bacteriophage are then injected into another bacterium

Figure 3.7 Transposon insertion.

Chapter 3: The body's response to infection

There are three main strategies for trying to avoid or overcome bacterial resistance:

1 Control of antibiotic use
2 Modification of existing antibiotics or development of new antibiotics capable of bypassing the bacterium's method of resistance
3 Use of combinations of antibiotics employing different mechanisms.

ANYTHING NEW?

The best approach is preventive. If antibiotics are used less and only when appropriate, then resistant strains are less likely to evolve. This means that there should be great caution about using antibiotics as growth promoters in animal feeds for fear that this will result in resistant bugs, which may be transferred to humans. In humans, antibiotics should not be used unnecessarily and care should be taken to avoid incomplete treatment. Incomplete treatment is a problem in tuberculosis where lengthy treatment with multiple drugs may be necessary to eradicate the organisms but patient compliance may be poor – new supervision programmes that oversee the taking of antibiotics to completion of the course have been found to improve TB cure rates. Incomplete treatment may allow the most resistant bug to survive, regrow and spread to other susceptible individuals. Common sense also demands that, particularly in hospitals, transfer of bugs from one patient to another must be minimised; this requires special attention to disinfection of endoscopes, bronchoscopes and other instruments that cannot be autoclaved.

The second strategy involves developing new drugs or modifying existing ones. The idea is that, once the mechanism for bacterial resistance is understood, it should be possible to redesign the molecule to get around the problem. We have already mentioned the example of meticillin with its side chain to prevent it from binding to the enzyme β-lactamase. Another example involves the tetracyclines, which have little effect on bugs that have developed a highly efficient mechanism for excreting the drug from the bacterial cell through specific 'efflux proteins'. A new group of tetracyclines called glycylcyclines have an altered side chain that prevents their excretion via efflux proteins.

The third approach is to use combinations of antibiotics. Sometimes the combinations have been developed empirically, but some combinations have been specifically designed to overcome resistance. This is the case with clavulanic acid, which is only a weak antibiotic but binds irreversibly to many β-lactamases, and so can protect β-lactam antibiotics from destruction. This is used clinically as Augmentin (co-amoxiclav) which is a combination of amoxicillin and clavulanic acid.

A fourth approach is to study the history books or visit remote regions of the world to examine previously undocumented medicinal plants. A recent UK report described success against MRSA using an ancient Anglo-Saxon remedy for a stye (infection of the follicle of an eyelash), which utilised 'cropleek and garlic, of both equal quantities, pound them well together … take wine and bullocks gall, mix with the leek … let it stand nine days in the brass vessel'. This encourages us to realise that people were sometimes successfully treated for infection in the pre-antibiotic era.

HOW DO VIRUSES DEVELOP RESISTANCE TO ANTIVIRAL AGENTS?

Viruses also become resistant to antiviral agents through the selection of naturally occurring mutants, which have amino acid changes at the active site or binding site of the antiviral agent. This is normally seen only in immunocompromised hosts in which the number of viral particles (viral load) is high, thus increasing the chances of a natural mutant occurring. Viruses have small genomes (HIV = 9 kb, herpes = 500 kb, bacteria = 3 Mb, human = 4000 Mb), so the chance of a mutant occurring is dependent on the number of virions and the size of the genome. HIV is a good example of this, in terms of both drug resistance and immune evasion.

The first part of HIV replication is the conversion of its RNA genome into DNA before integration into the host cell chromosome as proviral DNA. This reaction is unique to retroviruses and is mediated by the virally encoded, RNA-dependent DNA polymerase: reverse transcriptase (RT). Combined anti-HIV chemotherapy (highly active antiretroviral therapy or HAART) is now used, which includes several anti-RT compounds and a protease inhibitor because it is much less likely that any one virus will acquire mutations in two or three separate sites to affect the binding and activity of all the inhibitors. HIV also uses this natural mutation rate to its advantage to change the amino acid sequences of its exterior glycoprotein (gp120), which binds the cell CD4 receptor. The V3 region of gp120 shows epitope changes in isolates from the same patient infected with HIV, and these have been shown to permit escape from a cytotoxic T-cell immune response. Thus, the RT error rate provides a mechanism for generating immune escape mutants and drug resistance.

It is hardly surprising that infection is still a challenge – in fact globally, in less wealthy countries, infection kills more people than any other form of disease, including cancer and cardiovascular disease – and

 Read more about viruses in Pathology in Clinical Practice Case 44

Chapter 4 is devoted to inflammation, which includes conditions of infective and immune aetiology.

The innate and acquired immune system, the second and third lines of defence, are covered in detail in the next three chapters, but there follows a quick overview to set the scene.

Acute and chronic inflammation, fibrosis and scarring occur when the body's responses to infective agents or other traumatic insults go wrong, i.e. the physiological response, which is good, turns to the pathological response, which is bad.

The innate immune response (the acute inflammatory response) is the body's first line of protection against damage. This depends on a mixture of inflammatory cells, such as neutrophil polymorphs, tissue macrophages and NK cells, in addition to molecules

known as cytokines. Cytokines are molecules that are produced by one cell to act on another (or on the original cell).

If the clearance mechanisms of the innate immune system are successful, any invading pathogens are eliminated and damaged tissue is repaired and the inflammatory response subsides. Overwhelming of the innate immune response is usually recognised clinically as acute inflammation – we have probably all encountered an infected skin wound after a scratch or graze; another example of acute inflammation is acute appendicitis. Had the defences been successful, we would have remained unaware of these infections.

Healing following the acute inflammatory response to damage is a process known as organisation. This usually results in a small amount of fibrous tissue forming a discreet scar (repair), but ideally returns the tissue to complete normality (resolution). There are various circumstances in which unsightly or damaging fibrous scar tissue forms and occasionally the process of fibrosis is more diffuse and can be life threatening, e.g. pulmonary fibrosis, cirrhosis or glomerular sclerosis.

Under some circumstances healing and repair are not possible, e.g. an organism that can resist destruction, or if indigestible foreign material such as silica has entered the wound.

Should the first line of defence fail, the body's second line is the acquired immune response. This is a more targeted response requiring cells that can link to specific receptors on infective agents or cells within the body and eliminate them. The cells involved are the B and T cells and macrophages (both tissue derived and those derived from monocytes) and specialised APCs. B cells differentiate into plasma cells, which secrete immunoglobulins. Another range of cytokines is essential in this response.

Inflammatory cytokines are protein molecules secreted by cells that recruit and activate the immune cells and are vital to the generation of an effective adaptive immune response. Examples of cytokines are the tumour necrosis factors (TNF-α, -β, -γ) or interferons (IFN-α, -β, -γ), for instance. Interleukins are cytokines secreted by leukocytes (white blood cells), which act on other leukocytes. Growth factors act specifically to stimulate growth of particular cell types, e.g. epidermal

Chapter 3: The body's response to infection

growth factor (EGF), fibroblast growth factor (FGF), platelet-derived growth factor (PDGF) or vascular endothelial growth factor (VEGF).

If the initiating stimulus cannot be cleared, acute inflammation gives way to chronic inflammation. Sometimes, due to the nature of the stimulus (e.g.

viruses, autoimmune disease), the first response uses the acquired immune system and generates chronic inflammation.

CHAPTER 4
THE ACUTE INFLAMMATORY RESPONSE

No natural phenomenon can be adequately studied in itself alone, but to be understood must be considered as it stands connected with all nature.

Sir Francis Bacon (1561–1626)

WHAT IS INFLAMMATION?

John Hunter, surgeon to St George's Hospital from 1768 to 1793, was one of the first to observe that inflammation was not a disease but a response to tissue injury.

Inflammation is a mechanism by which the body deals with an infection, injury or insult. Causes of tissue damage have been discussed in Chapter 2 – infection is just one such cause. Some infective agents are discussed in Chapters 2 and 3, and the body's mechanical barriers to infection are also discussed. To understand inflammation and the innate and adaptive immune response, healing and repair is to get to the heart of understanding disease of all types.

When living tissue is damaged, a series of transient, coordinated local processes are initiated in order to contain the offensive agent, to neutralise its effect, to limit spread and hopefully to eradicate it. As part and parcel of this process, there is initiation of healing and repair of the injured tissue. Inflammation, healing and repair are like the black-and-white stripes of the

History

John Hunter (1728–1793)

John Hunter, surgeon to St George's Hospital from 1768 to 1793, was interested in inflammation and repair (Fig. 4.1). He was an incredible man whose aim was the total understanding of mankind! Hunter was born in Scotland on 14 February 1728, the last of 10 children. He spent the first 20 years there and his childhood has been described as 'wasted and idle'. This is because he 'wanted to know all about the clouds and the grasses, and why the leaves changed colour in the autumn'. Hunter's inquisitiveness and fascination with nature stood him in good stead when he began to unravel the mysteries of the human body. His book, *A Treatise on the Blood, Inflammation, and Gunshot Wounds*, is a monument to his thoroughness and powers of observation in delineating the processes of disease.

Figure 4.1 John Hunter (1728–93). © CORBIS.

Chapter 4: The acute inflammatory response

109

zebra: in order truly to understand the zebra, one cannot study the stripes in isolation. In the same way, the processes of healing and repair have to be addressed in their relationship to the process of inflammation.

The circulatory system is of fundamental importance in the inflammatory response. In general terms, the offending agent, whatever it may be, causes a change in the microvasculature of the injured area, leading to a massive outpouring of cells and fluid. This collection of cells and fluid is known as the inflammatory exudate and, within this exudate, we find ingredients that are needed to combat the offending agent and to begin the process of healing and repair.

Immediately after the injury of a piece of tissue, there is initiation of the acute inflammatory response and, usually, clotting as blood from damaged blood vessels coagulates. Platelets from the blood clot secrete cytokines important in many phases of the inflammatory response, but particularly in initiating and driving the repair process. Instant responses come from mast cells, which degranulate on pressure, signalling to the blood vessels to become permeable to plasma, and the complement system and native immunoglobulin within the tissues, which bind to and inactivate microbes if present. Almost instantly, the sentry cells such as dendritic cells and tissue macrophages recognise alarm signals ('stressors') and engulf microbes or secrete chemical messengers ('cytokines') to alert other inflammatory cells to come and assist, and messengers are sent to the bone marrow to generate extra acute inflammatory cells.

The initiation phase of acute inflammation is followed within around 6 hours by the recruitment phase, in which neutrophil polymorphs and macrophages are attracted to the site of inflammation. Signalling molecules are upregulated on the blood vessels so that the inflammatory cells know where to enter the tissues, and a chemical gradient is set up to guide them from the bloodstream to the site of damage or infection.

Almost at once the healing and repair processes are also stimulated: fibroblasts and myofibroblasts are activated to secrete collagen and extracellular matrix as a scaffold for the repair processes. The early healing phase is the proliferative phase, during which a

Key facts

Sequence of events in acute illness

Many events in acute inflammation take place simultaneously. There are roughly three phases:

1 **Inflammatory phase**

Initiated by microbial invasion, tissue damage, other exogenous or endogenous stressors.

In some instances, pathogen-associated molecular patterns (PAMPs) are recognised by pattern recognition receptors (PRRs), usually by dendritic cells and tissue macrophages. Inflammasomes form and release cytokines, which attract inflammatory cells and influence other cell types.

Damaged blood vessels/bleeding into tissues initiates coagulation.

Mediators released by cells and tissues attract neutrophils, tissue macrophages and other inflammatory cells.

Cytokines, interferons and growth factors are secreted by the antigen-presenting cells (APCs) and inflammatory cells.

2 **Proliferative phase**

Epithelium proliferates to reunite breach, if present.

Endothelial proliferation from existing vessels generates and secretes collagen.

Fibroblasts mature into myofibroblasts, proliferate and secrete collagen. Initial phase of wound contraction reduces the size of the injured site.

3 **Healing phase**

Neutrophil polymorphs and tissue macrophages undergo apoptosis when replete with phagocytosed material; no further recruitment. Cellular content diminishes.

New type III collagen is replaced with type I as wound remodelling and further wound contraction occurs. Type I collagen cross-linking stabilises wound.

'Restoration' macrophages help to restore the inflammatory cell types and numbers in the tissues to normal levels.

temporary matrix forms, made of type III collagen, hyaluronic acid and fibronectin. Capillaries are stimulated to form buds, which grow into the matrix, supplying vital oxygen and nutrients – this is granulation tissue. Simultaneously, the epithelial cells from the edges of the damaged tissue proliferate and migrate to close the defect. There is wound contraction as the collagen fibrils are aligned and shortened, and the wound is later remodelled. This phase can take days, weeks or many months, and later the type III collagen is resorbed and replaced with stronger type IV collagen, and the loose hyaluronic acid matrix with laminin.

It is clear, then, that the acute inflammatory response (the innate immune response) is closely tied up with repair processes, but there is so much to discuss regarding the mechanisms of acute inflammation that we will delay discussing healing and repair until Chapter 5. The same processes that alert the body to danger and trigger the innate immune system to respond with acute inflammation also alert the adaptive immune system, via cytokines. Adaptive immunity alerts T and B lymphocytes to the need to develop a highly targeted response, particularly to microbes, and these cells participate, together with antigen-presenting cells (APCs), to generate antibodies and memory cells in chronic inflammation, should it complicate the picture of healing and repair (see Chapter 5). We discuss adaptive immunity in a bit more detail in Chapter 6.

However, this is only half the story. It is romantic to imagine an army being sent to deal with an invading force to restore peace and tranquillity to the area. Life is not quite so simple; there is a price to be paid for war! The ugly side of it ranges from cosmetic problems, such as keloid scars, to life-threatening illnesses, such as autoimmune diseases.

Anyone who has had a boil or any other skin infection will be familiar with the four cardinal signs of inflammation: rubor (redness), calor (heat), tumor (swelling) and dolor (pain). A Roman physician, Cornelius Celsus, first described these in the first century AD. To this a fifth sign was later added, *laesio functae* (loss of function); however, this is not a necessary accompaniment to the inflammatory process.

Not all the features are immediately demonstrable if the focus of acute inflammation is deep within the body, as in acute appendicitis (Fig. 4.2). This typically presents with abdominal pain. Radiologically, the appendix may be swollen. The surgeon removing the inflamed appendix will note that it is swollen and reddened, and has prominent, congested blood vessels over its surface. If the appendix has perforated it will be coated in a thick fibrinopurulent exudate. The pathologist examining it will see sheets of neutrophils as part of a necro-inflammatory process, and acutely inflamed surface exudate ('serositis' or 'peritonitis') (Fig. 4.3).

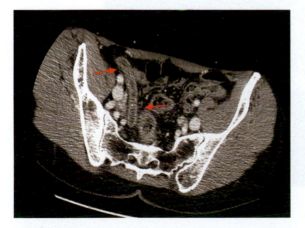

Figure 4.2 A 65-year-old man with pain in the right iliac fossa. CT shows a tubular, thick-walled dilated blind ending structure – acute appendicitis.

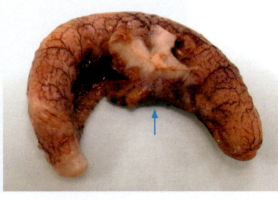

Figure 4.3 Acute appendicitis. The mid-appendix is encrusted with whitish-yellow fibrinopurulent exudate (arrow). The rest of the serosal surface is covered by normal, shiny peritoneum, with congested capillaries.

Chapter 4: The acute inflammatory response

WHAT IS THE PATHOPHYSIOLOGY BEHIND THE CLINICAL SIGNS OF INFLAMMATION?

MICROVASCULATURE

The microvasculature plays a central role in inflammation and the redness is caused by vasodilatation. This is important for increasing the flow of blood to the affected area and, hence, delivering the cells and plasma-derived substances needed for combat. By increasing the amount of blood in the region, vasodilatation also produces heat. The swelling results from increased permeability of vessel walls, leading to the outpouring of fluid and cells into the tissues, the inflammatory exudate (Fig. 4.4).

In some circumstances, the swelling may cushion the affected part and lead to immobilisation (*loss of function*). The sign that is the most difficult to explain is *pain*. This arises from the combination of tissue stretching by exudate and the action of some of the chemical mediators involved in inflammation.

But what are the underlying mechanisms of this process? In broad terms, there are three aspects to consider:

1 How is inflammation initiated, and why does it cease?
2 Which are the cells of inflammation, and where do they come from?
3 What are the chemical mediators that coordinate the inflammatory response, and how do they do it?

THE INITIATION OF ACUTE INFLAMMATION

How is inflammation initiated?

There are several different possible starting points for acute inflammation – to some extent they reflect the types of stimuli that may cause inflammation. Crushing of tissue by blunt injury may cause histamine to be released from mast cells, whereas cell damage resulting from trauma, ischaemia (loss of the blood supply) or tissue invasion by microorganisms may first be detected

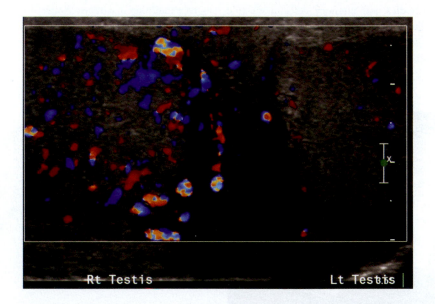

Rt Testis   **Lt Testis**

Figure 4.4 Acute inflammation: ultrasound scan of the scrotum showing an asymmetrical right testicle that is heterogeneous in texture and hypervascular – the blips of colour denote flow towards (red) and away (blue) from the ultrasound probe. The testicle is hypervascular and oedematous. The patient, a 40-year-old man, presented with a swollen red painful scrotum – **acute epididymo-orchitis.**

by the tissue macrophages. If the external defences are attacked, a population of dendritic cells (DCs) is resident in the epithelium of all surfaces exposed to the outside world (such as the skin, gut and respiratory tract). Mast cells and macrophages are resident within the tissues. The natural killer (NK) cell, a type of lymphocyte, also resides within the tissues and can respond directly to microorganisms.

DCs, macrophages, NK cells and neutrophil polymorphs (see below) have pattern recognition receptors (PRRs) for molecular structures known as PAMPs (pathogen-related molecular patterns), which are present on bacteria, viruses, fungi and parasites, not only marking them out as pathogens but also essential to their pathogenicity. They are not present on host cells. PRRs can also recognise non-infective 'stressors', such as ATP, chemicals and crystals. This is how many infections or damaging events are first recognised, e.g.

in the case of lobar pneumonia discussed in Chapter 1, the alveolar macrophage would have initiated the acute inflammatory response.

There are also protein products (particularly immunoglobulin and complement) that lie within the interstitium and that, once attached to microorganisms, can stimulate the innate immune system (see Chemical mediators below). The lectin-binding component of complement is a 'soluble PRR', which can set off an acute inflammatory response (Table 4.1).

Types of PRRs include the extracellular, membrane-linked, transmembrane 'toll-like receptors' (TLRs). Within the cytosol are two further types of PRRs, the NOD-like receptors (NLRs) (NOD stands for nucleotide-binding oligomerisation domain) and RIG-like receptors (RLR) (RIG stands for retinoic acid-inducible gene 1-like receptors). Lectin-binding PRRs are mentioned below.

Table 4.1 Some examples of pathogen-/damage-associated molecular patterns (PAMPs/DAMPs) and pattern recognition receptors (PRRs)

Site of PRR expression	Type of PRR	Subtype of PRR	Type of PAMP recognised	Typical effect of PRR activation
Surface	TLR	TLR3	Double-stranded viral DNA	Proinflammatory cytokine release
		TLR4	LPS in Gram-negative bacterial wall and some viruses	Proinflammatory cytokine release
		TLR5	Flagellin in organisms with flagella	Proinflammatory cytokine release
	CLR	Dectin 1	β-Glucan	Fungal cell wall
Secreted into the extracellular compartment	MBP		Mannose residues in bacterial and fungal cell walls	Complement activation
Intracellular: cytosol or attached to membranes	NLR	NOD2	Peptidoglycan in many bacteria	Inflammasome assembly, proinflammatory cytokines
	RLR	RIG1	Double-stranded viral RNA	Proinflammatory cytokine release

CLR, C-type lectin receptor; LPS, lipopolysaccharide; MBP, mannose-binding protein; NLR, NOD-like receptor; RIG, retinoic acid-inducible gene 1-like receptor; RLR, RIG-like receptor; TLR, toll-like receptor.

Once PRRs have been stimulated, the immune process becomes activated and proinflammatory cytokines are produced via one of two main routes. The best characterised 'transcription regulator' is nuclear factor κB (NF-κB) pathway, activation of which results in the production of interleukins IL-1, -6, -12 and -17, tumour necrosis factor α (TNF-α) and interferon γ (IFN-γ). However, there is another route that may be taken, which is the formation of a temporary structure, an inflammasome, which forms on demand within the stimulated cell. Inflammasomes are composed

of NLR-type PRRs, adaptor proteins and caspase precursor proteases. By cleaving IL-1, these caspases produce IL-1β and other IL-1 derivatives, such as IL-18, which play an important role in initiating healing and repair. Usually, caspase activation will also drive apoptosis (see Chapter 2) of the cell but recently an alternative, 'pyroptosis', has been described. Pyroptosis utilises many of the same mechanisms as apoptosis, but the resultant fragmented, packaged-cell fragments stimulate an inflammatory response, unlike apoptosis (Fig. 4.5).

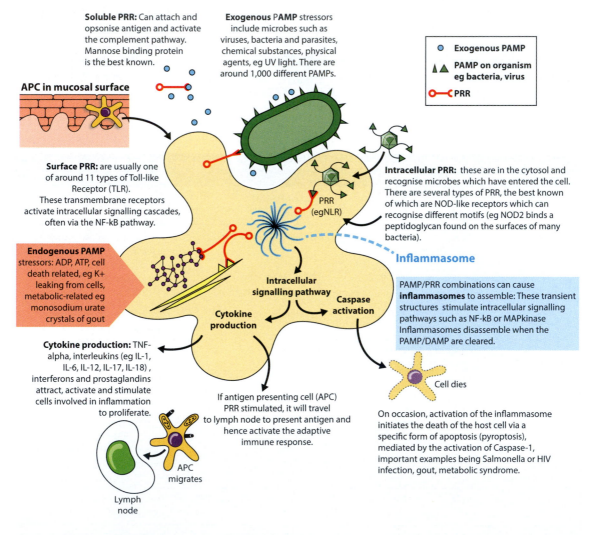

Soluble PRR: Can attach and opsonise antigen and activate the complement pathway. Mannose binding protein is the best known.

Exogenous PAMP stressors include microbes such as viruses, bacteria and parasites, chemical substances, physical agents, eg UV light. There are around 1,000 different PAMPs.

- ○ **Exogenous PAMP**
- ▲▲ **PAMP on organism eg bacteria, virus**
- ○━⊂ **PRR**

APC in mucosal surface

Surface PRR: are usually one of around 11 types of Toll-like Receptor (TLR). These transmembrane receptors activate intracellular signalling cascades, often via the NF-kB pathway.

PRR (egNLR)

Intracellular PRR: these are in the cytosol and recognise microbes which have entered the cell. There are several types of PRR, the best known of which are NOD-like receptors which can recognise different motifs (eg NOD2 binds a peptidoglycan found on the surfaces of many bacteria).

Endogenous PAMP stressors: ADP, ATP, cell death related, eg K+ leaking from cells, metabolic-related eg monosodium urate crystals of gout

Inflammasome

Intracellular signalling pathway

Caspase activation

Cytokine production

PAMP/PRR combinations can cause **inflammasomes** to assemble: These transient structures stimulate intracellular signalling pathways such as NF-kB or MAPkinase Inflammasomes disassemble when the PAMP/DAMP are cleared.

Cytokine production: TNF-alpha, interleukins (eg IL-1, IL-6, IL-12, IL-17, IL-18), interferons and prostaglandins attract, activate and stimulate cells involved in inflammation to proliferate.

If antigen presenting cell (APC) PRR stimulated, it will travel to lymph node to present antigen and hence activate the adaptive immune response.

Cell dies

On occasion, activation of the inflammasome initiates the death of the host cell via a specific form of apoptosis (pyroptosis), mediated by the activation of Caspase-1, important examples being Salmonella or HIV infection, gout, metabolic syndrome.

APC migrates

Lymph node

Figure 4.5 Pattern recognition receptors (PRRs) are fundamental to the innate immune response. Binding of pathogen-/damage-associated molecular patterns (PAMPs) initiates acute inflammation locally. Most cells carry PRRs. If PRR activation is in an antigen-presenting cell (APC), such as a dendritic APC, it will travel to the local lymph node and present antigen to T cells, stimulating the adaptive immune system.

Part 2: Defence against disease

THE CELLS INVOLVED IN ACUTE INFLAMMATION

We have mentioned the roles played by the cells already resident in the tissues – DCs, macrophages (sometimes called histiocytes), mast cells and NK cells. However, the cell that is *most* characteristic of acute inflammation is the polymorphonuclear leukocyte, or neutrophil polymorph (we refer to these as neutrophils), which is not resident in the tissues.

Inflammatory cells descend on a focus of tissue damage in 'waves'. The first cell type recruited is the neutrophil polymorph (the 'classic' acute inflammatory cell). Also recruited to the scene from the blood are monocytes, which on entering the tissues develop into new macrophages. Activated macrophages attract lymphocytes, which arrive a little later and are part of the adaptive immune response. DCs, macrophages, lymphocytes and plasma cells are the cells of chronic inflammation, and generate the adaptive immune response (Table 4.2). You will already note that the first two are major participants in the innate immune response, so there is overlap.

The neutrophil is the main type of granulocyte and is made in the bone marrow (the other granulocytes are the eosinophils and basophils, mast cells). Figure 4.6 illustrates the origins of haematopoietic and lymphoid cells and where they tend to localise to in the body.

Table 4.2 Cells involved in inflammation

Cell category	Cell type	Origin	Percentage white cells	Major function
Circulating cells				
Granulocytes	Neutrophils (polymorpho-nuclear leukocytes)	Bone marrow	75	Acute inflammatory cell involved in bacterial killing and phagocytosis
				Granule contents for increasing vascular permeability, chemotaxis, killing organisms and digesting extracellular matrix
	Eosinophils	Bone marrow	1	Acute inflammatory cell particularly common in allergic and parasitic conditions. Granules include major basic protein
	Basophils	Bone marrow	<1	Circulating cells that home to sites of inflammation. Granules include histamine
	Mast cells	Bone marrow	0	Found in tissue in "firstline defence" locations. Granules include histamine. Bind IgE
Lymphocytes	T cells	Lymphoid organs and thymus}	20	Various subtypes involved in antigen recognition and presentation, cell killing and regulation of immune responses (e.g. helper, suppressor and NK cells)
	B cells	Lymphoid organs and bone marrow}		On antigen stimulation, proliferate and give rise to **plasma cells**, which synthesise immunoglobulins
Macrophage system	Monocytes	Bone marrow	4	Migrate into tissues to be **macrophages** capable of phagocytosis, cytokine production and antigen processing and presentation
	Macrophages	Bone marrow derived		Macrophages move within tissues and can circulate to the draining lymph nodes.

(*Continued*)

Chapter 4: The acute inflammatory response

Table 4.2 Cells involved in inflammation (*continued*)

Cell category	Cell type	Origin	Percentage white cells	Major function
Circulating cells				
Dendritic cells				Mainly marrow derived, populate epithelium e.g skin or respiratory tract and become motile on activation by recognition of stimulus by PRRs. Migrate to local lymph nodes to activate adaptive immune response but are also active in the innate immune system, secreting various cytokines, including those responsible for control of fibrous tissue in healing and repair
Non-circulating cells				
Kupffer cells (liver sinusoids)				Fixed phagocytic cells lining sinusoids which filter large molecules/particles from blood or lymph
				Other 'fixed tissue macrophages' exist, e.g. in the lung (alveolar macrophages) or central nervous system (microglia)
Megakaryocytes in bone marrow				Produce platelets which contain serotonin, PDGF, etc. Also important in haemostasis
Hepatocytes				Produce proteins important in: • clotting and fibrinolytic system • complement system • kinin system • acute-phase proteins

Cytokines (cell-derived chemical peptide messengers released into the circulation), such as colony-stimulating factor (CSF) and interleukins, stimulate the bone marrow to generate large quantities of neutrophils, which are released into the blood – normal circulating levels of around 3–5 neutrophils $\times 10^9$/L may rise to 10–15 $\times$ 10^9/L at the height of an acute inflammatory response.

Chemicals released at the site of tissue damage or microbial invasion upregulate receptors on the endothelium lining near to blood vessels, which arrest circulating neutrophils for long enough to allow them to migrate along a chemical gradient to the site of damage.

A Russian microbiologist, Elias Metchnikoff, working at the Pasteur Institute in Paris in 1884, demonstrated that leukocytes phagocytose (engulf and destroy) bacteria and concluded that the purpose of the inflammatory response was to bring phagocytic cells to the area to kill the organisms. Neutrophils and macrophages are the main phagocytes.

THE CHEMICAL MEDIATORS

In 1927, Sir Thomas Lewis identified histamine, present in tissue mast cells, as a mediator of acute inflammation. Since then a vast array of chemical and cell-derived mediators have been identified, but not all have a proven role *in vivo*. They may be derived from the plasma, participating inflammatory cells or damaged tissue itself.

Part 2: Defence against disease

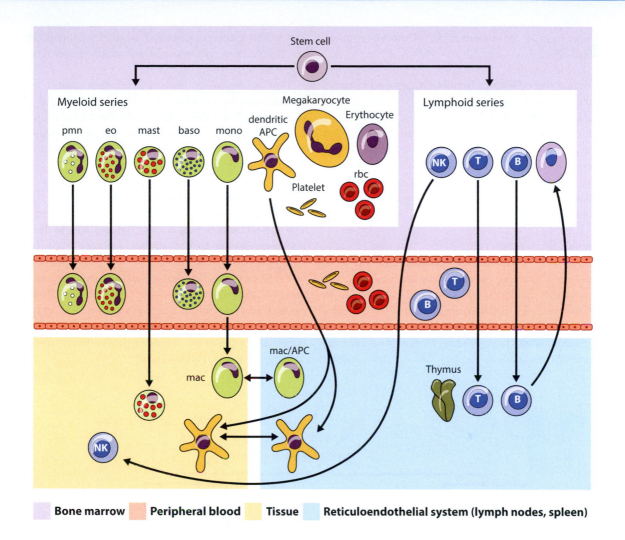

Figure 4.6 Haematopoietic and lymphoid cells originate in the bone marrow and usually circulate in the blood, lymphoid organs or tissues as indicated. APC, antigen-presenting cell; baso, basophil; eo, eosinophil; mac, macrophage; mast, mast cell; mono, monocyte, which becomes mac in the tissues; pmn, neutrophil polymorph; rbc, red blood cell.

The cell-derived products include the following:

- Vasoactive amines
- Cytokines and growth factors
- Arachidonic acid derivatives (eicosanoids)
- Platelet-activating factor (PAF)
- Lysosomal enzymes
- Oxygen-derived free radicals
- Nitric oxide.

The plasma-derived mediators include the following:

- The kinin system
- The coagulation and fibrinolytic system
- The complement system.

Some are chemical mediators important in the amplification of the inflammatory response; others play their role in the elimination of the offending agent.

Chapter 4: The acute inflammatory response

Their roles in inflammation are discussed in more detail later.

What is the difference between acute and chronic inflammation?

Acute inflammation

This is generally of short duration, lasting from a few minutes to a few days, and the cellular exudate is rich in neutrophils, with some macrophages, arriving hours after the initial insult. It is the body's normal response to injury and ideally will deal with the problem and lead to healing and repair.

Chronic inflammation

This is never a normal event and indicates that there has been a problem clearing a stimulus (e.g. chronic hepatitis B infection, or an organism resistant to destruction, such as tuberculosis), or that there is some persistent stimulus (e.g. heat damage from smoking or chemical toxicity from drinking alcohol, or continuing metabolic stress due to obesity). Chronic inflammation

may last for months or years and the chief cells involved are lymphocytes, plasma cells and macrophages (see Fig. 4.6).

The inflammatory process, whether acute or chronic, may be modified by a whole host of factors such as the cause of the damage, the nutritional status of the patient, the competence of the patient's immune system and intervention with antibiotics, anti-inflammatory drugs or surgery.

What may happen to a patient with acute inflammation?

The patient with a boil on the bottom will be less impressed than we are with the inflammatory processes taking place (Fig. 4.8). He will be well aware that the boil exhibits the cardinal features of pain, swelling, redness and heat! He may regard the fact that the injury has caused microvascular changes via mediators, leading to cellular and humoral factors accumulating at the site of injury, as of no interest. What our patient dearly wants to know is what will happen next?

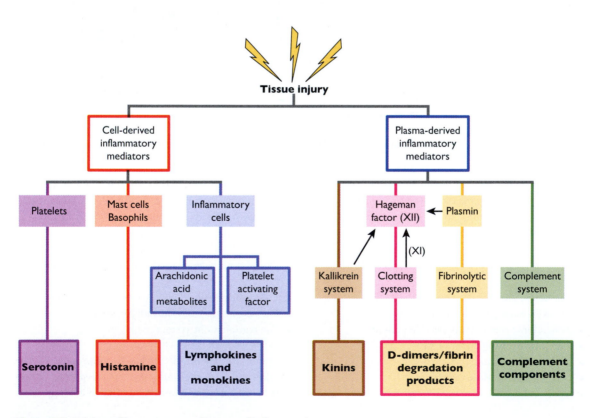

Figure 4.7 Origins of important mediators of inflammation.

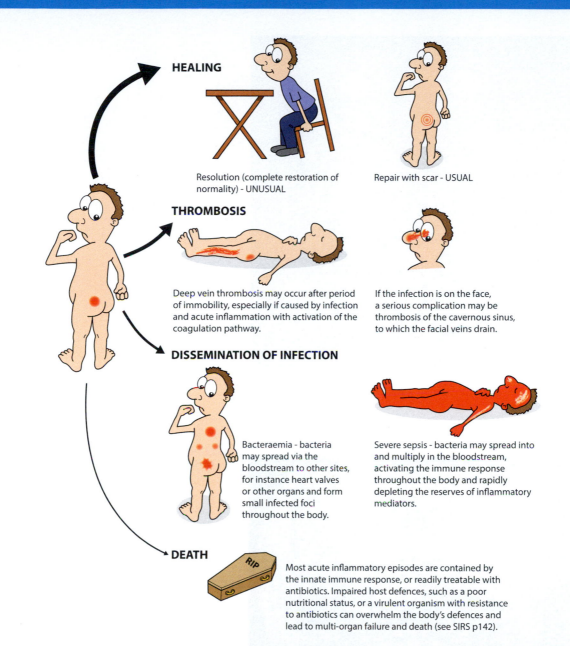

HEALING

Resolution (complete restoration of normality) - UNUSUAL

Repair with scar - USUAL

THROMBOSIS

Deep vein thrombosis may occur after period of immobility, especially if caused by infection and acute inflammation with activation of the coagulation pathway.

If the infection is on the face, a serious complication may be thrombosis of the cavernous sinus, to which the facial veins drain.

DISSEMINATION OF INFECTION

Bacteraemia - bacteria may spread via the bloodstream to other sites, for instance heart valves or other organs and form small infected foci throughout the body.

Severe sepsis - bacteria may spread into and multiply in the bloodstream, activating the immune response throughout the body and rapidly depleting the reserves of inflammatory mediators.

DEATH

Most acute inflammatory episodes are contained by the innate immune response, or readily treatable with antibiotics. Impaired host defences, such as a poor nutritional status, or a virulent organism with resistance to antibiotics can overwhelm the body's defences and lead to multi-organ failure and death (see SIRS p142).

Figure 4.8 Possible outcomes following acute inflammation. This patient has a boil – an infected focus in his skin. Likely causative organisms include Staphylococcus aureus. Most acute inflammatory foci heal with some scarring.

There are several possibilities. The inflammatory and healing process may restore the tissue to its normal state, with nothing to suggest that anything has been amiss. Healing may take place but leave a scar. The injury and the inflammation may grumble on for a long time, perhaps as an abscess, which may require surgery.

Dissemination via the bloodstream may set up additional sites of acute inflammation; damaged heart valves are particularly at risk – bacterial endocarditis carries a high mortality. Catastrophically, infection may completely overwhelm the body and lead to death from severe sepsis. This last outcome is more likely where

Chapter 4: The acute inflammatory response

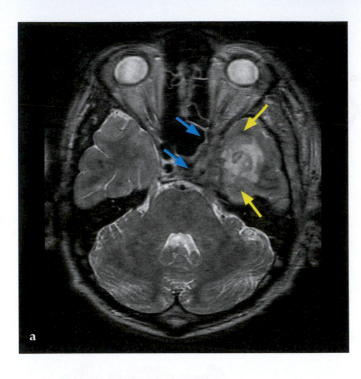

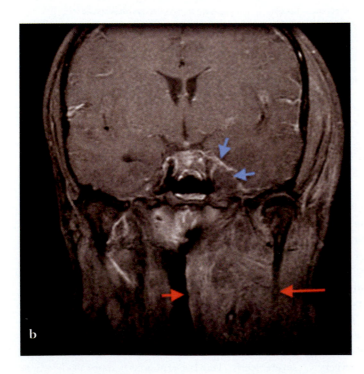

Figure 4.9 Acute cavernous sinus thrombosis is a ghastly complication of acute inflammatory disease in the head and neck, particularly sinusitis with orbital cellulitis The cavernous sinus thrombosis in this case is a result of extensive infection in the deep face (related to tonsillitis) which spread into the cavernous sinus via the foramen ovale in the skull base. (a) Proptosis of the left eyeball (it is protruding out) due to oedema of the orbit behind the eyeball, with oedema of the temporal lobe (yellow arrows) and thrombus (blue arrows) filling the left cavernous sinus. (b) A coronal image showing non-enhancing thrombus filling the left cavernous sinus (blue arrows) and diffuse inflammatory change in the infratemporal fossa or deep space (red arrows).

inflammatory defences are deficient, such as in cancer patients treated with cytotoxic drugs or patients receiving immunosuppressive drugs for autoimmune disease or after an organ transplantation.

Deep vein thrombosis is a well-known risk after surgery, partly because of immobility leading to stasis, but also because acute inflammation upregulates coagulation. A disastrous outcome of persistent acute inflammation on the face or in the nasal sinuses is cavernous sinus thrombosis. The venous drainage from the face is into the cavernous sinus, which can undergo thrombosis with severe consequences (Fig. 4.9).

The final outcome depends on the interactions between the various processes involved in inflammation. Just as the zebra is neither black with white stripes nor white with black stripes, so it is the combination of the various inflammatory components that determines the texture of the whole.

Now that we have the overall concept of acute inflammation and its clinical relevance, we must look more closely at the complex cellular and molecular events of this process. We first examine the changes in the microvasculature.

VASCULAR CHANGES

Julius Conheim (1839–84), a German pathologist (a pupil and later assistant to the 'father' of cellular pathology, Rudolf Virchow), delineated the vascular changes of inflammation using living preparations of thin membranes, such as the mesentery, and demonstrated the vasodilatation and subsequent exudation of fluid. His experiments with frog mesentery demonstrated that injury produced vasodilatation resulting in more blood entering the tissue but sometimes a slower flow of blood in capillaries. This allows white cells (leukocytes) to attach to the vessel wall (margination) and then move across the wall to the extravascular compartment (diapedesis or emigration).

Before we consider the causes of altered vascular permeability, it is worth revising the normal physiological factors that control the movement of fluid across a small vessel wall and into the extravascular tissues.

Fluid flows away from areas of high hydrostatic pressure and towards areas of high osmotic pressure (Fig. 4.10). Thus, fluid leaves the arterial end of the capillary network and is reabsorbed at the venous end (Fig. 4.10a), with any excess being removed via the lymphatics. A rise in hydrostatic pressure within the vessel, with no changes in permeability, will increase the leakage of fluid out of the vessel but it will have no protein in it (Fig. 4.10b). However, if the permeability of the vessel wall increases, then fluid can move more readily and protein molecules may also leak across. Movement of protein molecules will alter the osmotic pressure gradient, such that less fluid is reabsorbed into the blood at the venous end of the capillaries, so tissue fluid will increase (Fig. 4.10c). In areas of inflammation there is usually a rise in hydrostatic pressure and an increase in vascular permeability (Fig. 4.10d). The fluid remaining in the tissues drains into vessels termed 'lymphatics'. Lymph fluid and the lymphatic system are so important to the inflammatory defences that we spend a moment introducing them here.

Lymph and the lymphatic system

What is lymph? When we considered fluid flow across vessels (Figs 4.11 and 4.12) it became clear that more fluid moves out of tissue capillaries due to the pump pressure within the system (hydrostatic pressure), than is returned to the bloodstream due to the pull of plasma proteins (plasma oncotic pressure). The amount of interstitial fluid increases if there is nearby inflammatory activity, when plasma proteins also leak out of the excessively permeable vascular endothelium. Once within the lymphatic channels the interstitial fluid is termed 'lymph', which also contains a variety of inflammatory cells, particularly lymphocytes, DCs (APCs) and macrophages.

Lymphatic channels start as blind-ended sacs, with ultra-thin, ultra-permeable walls. They consist of little more than endothelial lining cells, supported by delicate collagen fibres initially, but as these vessels drain into larger ones they start to resemble veins, eventually developing a muscularised wall. The lymph is all eventually channelled into one large vessel, the thoracic duct, which empties into the venous system at the junction of the left subclavian and internal jugular veins. Similar to the venous system, lymphatic channels have valves and rely on the pumping action of adjacent muscles to create flow.

A distinctive feature of the lymphatic system is the presence of a chain of lymph nodes, stationed at strategic points within the body; these form checkpoints at which the contents are filtered and screened for miscreants. Doctors can examine the superficial lymph node groups, because they are close enough to the skin surface to palpate (feel) and may enlarge in response to infection or other causes of inflammation, or if infiltrated by tumour (see Chapter 6).

Chapter 4: The acute inflammatory response

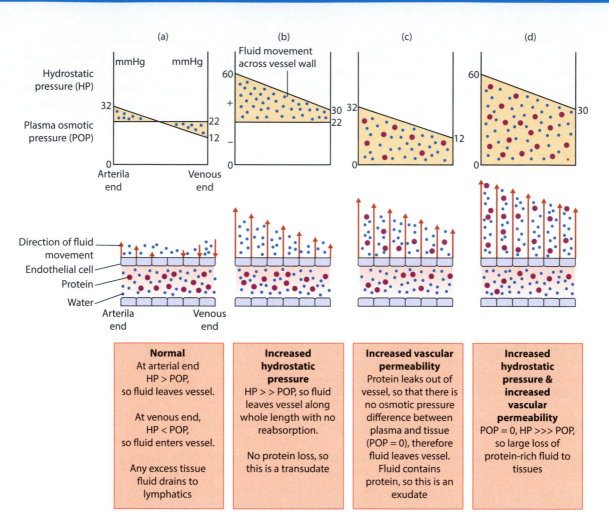

Figure 4.10 Factors affecting the movement of fluid across vessels.

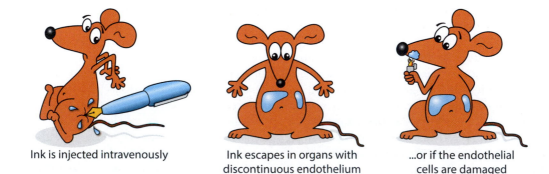

Ink is injected intravenously

Ink escapes in organs with discontinuous endothelium

...or if the endothelial cells are damaged

Figure 4.11 Vascular endothelium is continuous everywhere except the liver and spleen; here the endothelium is fenestrated. As a result of this, an intravenous injection of Indian ink will lead to the accumulation of ink particles in the liver and spleen. However, if endothelium elsewhere is damaged, e.g. by heat, gaps appear between the endothelial cells and ink particles can leak out of the circulation at these sites.

PATHOLOGICAL FORCES IN FLUID BALANCE

Now that we understand the normal physiological pressures that govern the way plasma (composed largely of water) moves between the vascular and extravascular compartments – note that we have not considered the intracellular compartment in this discussion – we must start thinking about what can go wrong – in other words the pathology of fluid balance.

This is an appropriate time to introduce a number of new words. An exudate is fluid within the extravascular spaces that is rich in protein and hence has a specific gravity >1.020. On the other hand, a transudate has a low protein content and specific gravity <1.020. Oedema simply refers to the presence of excess fluid within the extravascular space and body cavities, and may be an exudate or transudate. Pus can be thought of as a special kind of exudate – a purulent exudate. Besides protein-rich fluid exudate, it contains dead or dying bacteria and neutrophils. The consistency of the pus depends on the amount of digestion by neutrophil enzymes and the colour depends on the type of organism and the presence of neutrophil-derived myeloperoxidase, which imparts a yellow–green colour. Exudates generally contain fibrinogen, which is converted to fibrin through the action of tissue thromboplastin. The fibrin forms a mesh for cells to migrate on to and later a scaffold for healing and repair.

The fluid that forms during an inflammatory reaction is an exudate. This means that large protein molecules have leaked out of the microvasculature. What is the mechanism of the increased permeability? Most vessels are lined by endothelium, which is termed 'continuous'. In the endocrine organs, intestines, liver and renal glomeruli, the endothelium is normally more permeable because it contains 'windows', hence the name fenestrated endothelium, whereas in the spleen, liver and bone marrow, the endothelium is discontinuous. What happens after an injury has been elegantly demonstrated using simple experiments involving the intravenous injection of Indian ink (see Fig. 4.11). The ink will normally remain within the vascular compartment except in the liver and spleen, where the discontinuous endothelium allows ink to escape. If a mild injury is produced (e.g. by using heat), the damaged area will turn black. Microscopical examination would reveal that the ink particles have crossed the endothelial layer and are trapped by the underlying basement membrane. Injection of a vasoactive substance, histamine, would cause the endothelial cells of the small venules to contract, creating gaps through which the ink molecules could pass. In reality, the situation is more complicated because the vascular changes will depend on the severity of the insult.

Three types of vasopermeability reactions have been demonstrated, although they generally overlap in real situations (see Fig. 4.12):

1 Immediate–transient
2 Immediate–persistent
3 Delayed–prolonged.

Immediate–transient

This occurs immediately after injury, reaches a peak after 5–10 minutes and ceases after 15–30 minutes. This response can be produced by histamine and other chemical mediators, and is blocked by prior administration of antihistamines. The leakage occurs exclusively from small venules, which develop gaps between the endothelial cells as these cells contract. This occurs after nettle stings or insect bites.

Immediate–persistent

This results from severe injury such as burns, where there is direct damage to endothelial cells. The leak starts immediately and reaches a peak within an hour. As the endothelial cells are damaged and may even slough off, the leak will continue until the vessel has been blocked with thrombus, or the vessel is repaired. Unlike the previous example, it can affect any type of vessel.

Delayed–prolonged

This is a very interesting type of response, familiar to anyone who has over-indulged in a tropical holiday after a period under the clouds of English skies. There is an interval of up to 24 hours before the leak starts from sun-damaged capillaries and venules. Small aggregates of platelets and endothelial cells are seen in some capillaries, and it seems that the endothelial cells are damaged directly, but this takes time to manifest itself. UV light may cause this, as in this example, but it can also be seen in burns, delayed hypersensitivity reactions affecting small blood vessels and after radiation.

Chapter 4: The acute inflammatory response

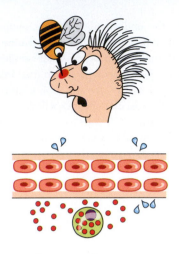

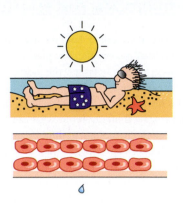

Immediate-transient response:
Immediate onset, peaking in 5–10 minutes and lasting from 15 to 30 minutes. Small blood vessels are affected. The temporary damage is caused by histamine, bradykinin, nitric oxide, complement (C5a), leukotrienes (e.g. LTB_4), platelet-activating factor (PAF). Causes include nettle sting, insect bite

Immediate-persistent response:
Caused by direct endothelial cell injury, arises immediately, peaking in about 1 hour and lasts until the vessel is plugged by thrombus or repaired. Any vessel type may be involved, the mechanism of injury being due to bradykinins, nitric oxide, complement, leukotrienes (e.g. LTB_4), platelet-activating factor (PAF), and potentiated by prostaglandins. Causes include severe direct injury, such as burns. Takes days to recover, as endothelium must regenerate

Immediate-persistent response:
Caused by more subtle endothelial cell injury; arises in 18–24 hours and progresses for up to 36 hours (sometimes longer). Capillaries and venules are affected. Causes include sunburn, DXT (radiotherapy), bacterial toxins

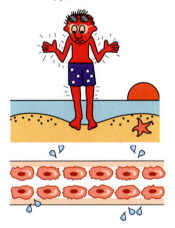

Figure 4.12 Types of vascular response.

Next we look at the cellular component of the inflammatory response.

CELLULAR EVENTS

The principal cells of the acute inflammatory response are the neutrophils and macrophages. However, neutrophils circulate in the blood and have to be recruited to the site of injury, so the first cells involved are those found within the tissues, the epithelial DCs, tissue macrophages and mast cells, and, to some extent, NK cells.

After injury, and in response to certain signals, the neutrophils migrate out of the vessels, the number recruited depending on the type of injury, e.g. infections with bacteria attract more acute inflammatory cells than purely physical injuries. After the neutrophils, a second wave of cells arrives, the macrophages. These are derived from monocytes, which circulate in the blood. The movement of leukocytes out of the blood vessels and their role in combat can be divided into discrete steps. These are as follows:

- Margination
- Adhesion
- Emigration
- Chemotaxis
- Phagocytosis and degranulation.

Margination and adhesion

When haemodynamic changes take place in the vasculature during inflammation, white cells fall out of the central axial flow and line themselves up along the wall (a little reminiscent of the school disco! Fig. 4.13). When the blood flow slows, the leukocytes start to roll along the endothelial surface. There are specific 'adhesion molecules' that stick leukocytes to endothelial cells and the number of these molecules on cell surfaces is increased (or 'upregulated') by inflammatory mediators. There are two main groups of adhesion molecules involved in the attraction and binding of neutrophils and macrophages (Fig. 4.14).

First, within minutes, P-selectin is upregulated, then, within 1–2 hours of injury, E-selectin is upregulated; both are expressed on the luminal surface of the endothelial cell. Their complementary binding molecules are present on the leukocyte surface: glycoproteins called Lewis X or A. These paired molecules bind only loosely and the leukocyte rolls along the blood vessel wall, slowing as it encounters each E-selectin molecule. The next pair of upregulated molecules, intercellular adhesion molecule 1 (ICAM-1) on the endothelium and lymphocyte function-associated antigen 1 (LFA-1) on the leukocyte, bind together similar to a 'lock and key', arresting the rolling leukocyte on the endothelial surface. These molecules have to be induced, which takes about 6 hours. Resting endothelial cells and leukocytes express very few adhesion molecules, but some inflammatory mediators increase leukocyte expression of LFA-1 (e.g. complement fragments [C5a] and leukotrienes, such as LTB$_4$), whereas other mediators enhance endothelial ICAM-1 expression (e.g. IL-1 and bacterial endotoxin). People with a genetic deficiency of these adhesion molecules have repeated bacterial infections.

Our knowledge of adhesion molecules is expanding rapidly. Broadly, there are four families, three of which are involved in inflammation: the integrins, the immunoglobulin gene superfamily and the selectins (the last family is the mucin-like glycoproteins, which are found on many cell surfaces and in the extracellular matrix, and bind CD44 on the surfaces of leukocytes). Some (but not all) of their family members are listed in Table 4.3. Their expression changes during inflammation so that different types of cells adhere at different stages. Some of these molecules have been termed 'addressins' because they act as address labels to allow cells to leave the circulation in a specific tissue. This is particularly important in the recirculation and 'homing' of lymphocytes, which is discussed on page 183.

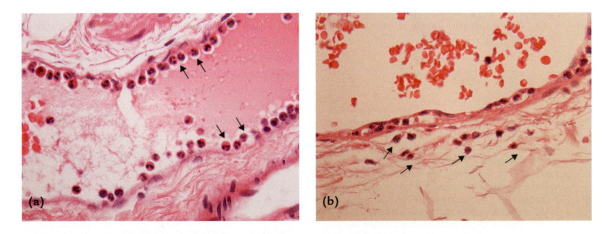

Figure 4.13 Photomicrographs illustrating (a) margination and (b) emigration.

Chapter 4: The acute inflammatory response

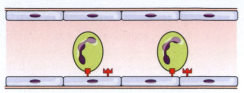

a) Laminar flow: Neutrophils normally lie in the central stream, but the flow rate slows in acute inflammation and they 'fall out' of the axial stream towards the margin. IL-1, TNF and endotoxins upregulate E-selectin on thelining endothelial cells. The neutrophils bind these loosely and roll along the endothelial surface.

(b) Next the endothelial cells express integrins (e.g. ICAM-1), which bind tightly to LFA-1 on the neutrophil, tethering it to the wall. These are also induced by acute inflammatory mediators. neutrophils and other inflammatory cells migrate into the extravascular space by extending a pseudopodium into the junction between endothelial cells. They dissolve the basement membrane with proteases.

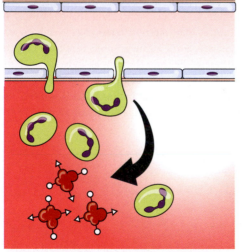

(c) The neutrophil migrates between endothelial cells and the defect in the basement membrane seals behind them. Some red cells follow, but in acute inflammation there is generally little extravasation of red cells unless there has been vascular damage. The neutrophil then moves along a chemotactic gradient set up by chemokines (which attach to matrix proteins in the soft tissues) to the inflammatory stimulus. Other inflammatory cells, particularly macrophages and eosinophils, use the same principles

Key facts

Agents that induce leukocyte adhesion molecule expression

By polymorphs

- C5a
- TNF
- LTB$_4$

By endothelial cells

- IL-1
- TNF
- Bacterial products (e.g. formyl-methionine peptides)
- Complement C3a, C5a
- Haemodynamic stress
- Lipid products
- Advanced glycation end-products (in diabetes)
- Hypoxia
- Viruses

Emigration and chemotaxis

Once the leukocytes have firmly adhered to the endothelium, they form foot-like processes termed 'pseudopodia' that push their way between the endothelial cells. This occurs in response to the binding of ICAM-1 (which triggers the leukocyte to flatten against the endothelial surface), and the presence of molecules called chemokines, which lie on the endothelium and the extracellular matrix and bind leukocytes; the process of binding stimulates the formation of pseudopodia and movement towards the point of binding. Eventually, the leukocyte lies between the endothelial cell and the basement membrane, where it releases a protease to digest the basement membrane, which allows it to leave the vessel and enter the extravascular space. Neutrophils, basophils, eosinophils, macrophages and lymphocytes

Figure 4.14 Neutrophil movement across blood vessels in acute inflammation is mediated by adhesion molecules and then follows a chemotactic gradient to the source of the inflammatory response.

Table 4.3 Adhesion molecules (Figs 4.13 and 4.14)

Family	Some family members	Principally expressed on	Main function
Integrins	β1 family, e.g. VLA-4 β2 family, e.g. LFA-1	Lymphocytes and monocytes	Mediates immune and inflammatory responses including binding immunoglobulin superfamily molecules on endothelial cells to provide firm adhesion to vessel wall before migration
Immunoglobulin superfamily	ICAM-1, -2 and -3, VCAM-1	Endothelial cells, lymphocytes and monocytes	As above by binding to integrins
Selectins	E-selectin L-selectin P-selectin	Endothelium Lymphocytes, polymorphs and monocytes Platelets and endothelial cells	Initial phase of leukocyte adhesion to vessel wall
Cadherins	B, E, M, N, P, R, T	Range of tissues	Homophilic calcium-dependent cell–cell adhesion, e.g. at sites of desmosomes and adherens junctions. Not specifically involved in inflammation

ICAM-1, intercellular adhesion molecule 1; LFA-1, leukocyte function-associated antigen 1; VCAM-1, vascular cell adhesion molecule 1; VLA-4, very late antigen 4.

all use this route. Red blood cells may also pass through the gaps, but only as passive passengers.

The cells are able to move towards a chemical signal and this specific movement is termed 'chemotaxis'. Note that this is different to chemokinesis, which is an increased and accelerated random movement. The Boyden chamber is a popular system for demonstrating chemotaxis, consisting of two chambers separated by a micropore filter. The cells go into one chamber and the putative chemical mediator is placed in the other. If cells move from the first to the second chamber, then chemotaxis is demonstrated.

Chemotactic agents can be considered as exogenous or endogenous. The former include bacterial products such as formyl-methionine (fMet) peptides, unmethylated nucleotides, lipopolysaccharide (LPS, also known as bacterial endotoxin) and bacterial proteoglycans, plus double-stranded DNA (dsDNA) from some viruses. Endogenous chemotaxins include fragments of the complement system (the most potent is C5a), products of arachidonic acid metabolism (e.g. prostaglandins and leukotrienes, particularly LTB_4) and cytokines such as IL-8.

How does it work? Well, as with so many cellular stimuli, the first stage depends on the chemotactic agents binding to specific receptors (G-protein-coupled receptors) on the leukocyte cell membrane. This leads to an increase in ionised calcium within the cytoplasm. Movement is achieved by the polymerisation, coupling and decoupling of actin, which is calcium dependent.

Leukocyte activation

There has been much interest in recent years over how white blood cells of all types are activated. Several classes of surface receptor exist and promote their effects via a variety of signalling pathways. Signalling

Chapter 4: The acute inflammatory response

by toll-like receptors can activate transcription factors such as NF-κB and signalling pathways such as mitogen-activated protein (MAP) kinase. Cell membrane receptors, such as the toll-like receptors (TLRs) or cytosolic receptors such as NODs (nucleotide-binding oligomerisation domain receptors) are two PRRs (pattern recognition receptors) for PAMPs (pathogen-associated molecular patterns) – note that PAMPs are not only found on pathogenic, but on virtually all, microbes (see Fig. 4.5). Also, depending on the type of organism, the inflammatory response may be modified, e.g. triggering the production and migration of eosinophils in preference to neutrophils in the case of helminth infection. PRRs can also be activated by PAMPs that are endogenous, non-infective agents with similar 'stressor' molecular patterns (the acronym DAMP signifying that these represent 'damage' rather than 'pathogen' has been suggested, rather than the confusing 'PAMP').

Activated leukocytes migrate along chemotactic pathways, become enlarged, produce increased lysosomal granules and their contents, in the cases of macrophages and neutrophils, and show increased activity, e.g. increased phagocytic activity in neutrophils and macrophages.

PHAGOCYTOSIS

Once activated neutrophils and macrophages arrive at the site of injury; they ingest any necrotic tissue debris, dead and live bacteria, a process termed 'phagocytosis'. This requires a number of distinct steps: the material has to be recognised as foreign or dead, it has to be engulfed and ingested and, finally, it has to be killed or degraded.

Not all the processes by which neutrophils and macrophages differentiate between normal tissue and foreign or dead tissue are known; however, it is clear that bacteria coated with certain protein substances called opsonins are ingested more readily. The process is termed 'opsonisation'. It derives from the Greek word 'opson' meaning 'relish', i.e. getting ready for eating. There are two major opsonins, for which the phagocytes (usually neutrophils and macrophages) have specific, high-affinity receptors:

1 **Immunoglobulin (IgG),** which binds to FcgRI (receptor for constant portion of IgG) (see page 188)
2 **C3b component of complement**, which binds to CR1 (complement receptor 1) (see page 139).

There are also several minor opsonins, such as **mannose-binding lectin, fibronectin, fibrinogen** and **C-reactive protein**, all of which are present in the plasma and can coat particles such as invading microorganisms (Fig. 4.15).

After the opsonised fragment attaches to the phagocyte receptor, the cell puts out a pseudopodium. This extension of cell cytoplasm encircles the particle so that it becomes wrapped in what was originally cell surface membrane. This new intracytoplasmic membrane-bound sac is termed a 'phagosome'. Another such sac, normally present in the cell and packed with destructive enzymes, is the lysosome. A lysosome fuses with the phagosome, producing a phagolysosome. This allows the enzymes to have access to the engulfed particle and it is within this vesicle that the killing takes place. If some proteolytic enzymes leak out of the phagolysosome, as may occur if the lysosome fuses with the phagosome while the phagosome is still open to the cell surface, they may damage adjacent tissue – a phenomenon described, rather poetically, as 'regurgitation during feeding'.

Fusion of a lysosome with the cell membrane with local extracellular release of toxic metabolites can be deliberate and is important for attacking large organisms, such as worms, that are too large to be ingested.

A damaging scenario may occur if the lysosomal membrane is traumatised by phagocytosed agents such as silica or urate crystals, causing the release of its contents into the tissues. The system can backfire, as happens if antibody–antigen (ab–ag) complexes lodge in capillaries (e.g. farmers' lung, a type III hypersensitivity reaction, or glomerulonephritis [in which the glomerular basement membrane is damaged because ab–ag complexes lodged within it stimulate the release of lysosomal contents]).

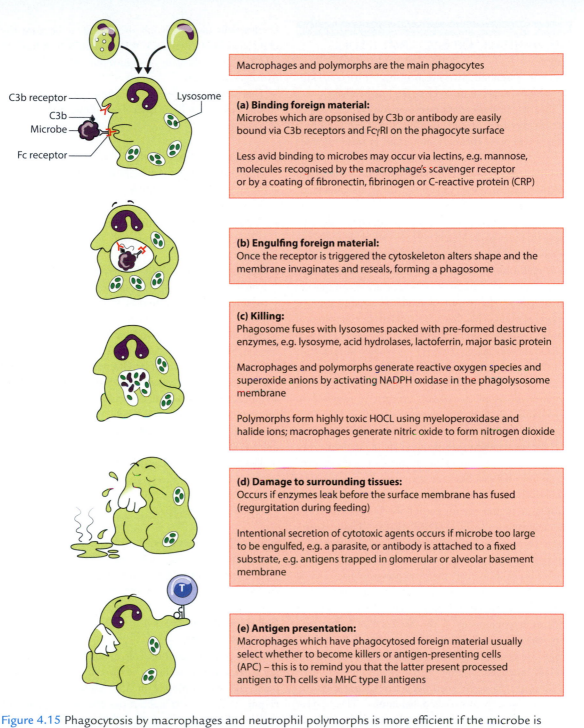

C3b receptor
C3b
Microbe
Fc receptor
Lysosome

Macrophages and polymorphs are the main phagocytes

(a) Binding foreign material:
Microbes which are opsonised by C3b or antibody are easily bound via C3b receptors and FcγRI on the phagocyte surface

Less avid binding to microbes may occur via lectins, e.g. mannose, molecules recognised by the macrophage's scavenger receptor or by a coating of fibronectin, fibrinogen or C-reactive protein (CRP)

(b) Engulfing foreign material:
Once the receptor is triggered the cytoskeleton alters shape and the membrane invaginates and reseals, forming a phagosome

(c) Killing:
Phagosome fuses with lysosomes packed with pre-formed destructive enzymes, e.g. lysozyme, acid hydrolases, lactoferrin, major basic protein

Macrophages and polymorphs generate reactive oxygen species and superoxide anions by activating NADPH oxidase in the phagolysosome membrane

Polymorphs form highly toxic HOCL using myeloperoxidase and halide ions; macrophages generate nitric oxide to form nitrogen dioxide

(d) Damage to surrounding tissues:
Occurs if enzymes leak before the surface membrane has fused (regurgitation during feeding)

Intentional secretion of cytotoxic agents occurs if microbe too large to be engulfed, e.g. a parasite, or antibody is attached to a fixed substrate, e.g. antigens trapped in glomerular or alveolar basement membrane

(e) Antigen presentation:
Macrophages which have phagocytosed foreign material usually select whether to become killers or antigen-presenting cells (APC) – this is to remind you that the latter present processed antigen to Th cells via MHC type II antigens

Figure 4.15 Phagocytosis by macrophages and neutrophil polymorphs is more efficient if the microbe is opsonised by antibody or complement. However, this is not essential and either cell can recognise foreign particles and engulf them. Factors that assist recognition include lectins, e.g. mannose, present on many organisms but not on mammalian cells. The macrophage scavenger receptors, which bind modified low-density lipoprotein (LDL), can bind to some microbes.

Chapter 4: The acute inflammatory response

MECHANISMS FOR BACTERIAL KILLING

There are essentially two mechanisms for bacterial killing: oxygen dependent and oxygen independent.

The oxygen-dependent system involves toxic oxygen radicals that have an unpaired electron (indicated by a dot). These are extremely important in many diseases and are termed 'reactive oxygen species' (ROS). ROS are important in defence mechanisms such as these and also play a normal role in cell signalling, hormone biosynthesis and fertilisation.

ROS have come to clinical prominence because some of their pathological effects, which include the metabolic syndrome (obesity, hypertension, type 2 diabetes mellitus), atherosclerosis (oxidised LDLs are one of the initiators of atheroma), heart disease and cancer. Nitrotyrosine, the result of oxidative modification of certain proteins, has been postulated as a serological marker of cardiovascular disease.

ROS include superoxide ($O^{2-\cdot}$), singlet oxygen ($O^{\cdot}$) and the hydroxyl radical (OH^{-}). These molecules are produced by the respiratory burst that occurs during the process of phagocytosis. Oxygen is reduced to superoxide ion, which is then converted to hydrogen peroxide (H_2O_2).

A brilliant mechanism protects the neutrophil from being damaged by these products: they are not produced until the cell has been activated. To achieve this, some of the chemical components are moved from the cytoplasm to the membrane of the lysosome, where they combine to form NADPH oxidase. This converts oxygen within the lysosome to H_2O_2 and the toxic oxygen radicals. Iron is required as a cofactor.

This is not, however, the most powerful bactericidal chemical. Neutrophils contain an enzyme, myeloperoxidase, that converts H_2O_2 to HOCl (hypochlorous acid, or hypochlorite) in the presence of halide ions (e.g. chloride) and nitric oxide, produced by macrophages, reacts with superoxide anion to form the strong oxidant, nitrogen dioxide.

These are powerful oxidants active against bacteria, fungi, viruses, protozoa and helminths. This system is of clinical importance because its absence produces 'chronic granulomatous disease of childhood', an inherited disease in which the neutrophils are able to ingest

bacteria but unable to kill them. This is because the child lacks the enzyme NADPH oxidase, which leads to a failure of production of superoxide anion (O^{2-}) and hydrogen peroxide.

There are a number of oxygen-independent mechanisms that are useful in microbial killing, including the following:

- Lysozyme, an enzyme that attacks the cell wall of some bacteria (especially Gram-positive cocci)
- Lactoferrin, an iron-binding protein that inhibits growth of bacteria
- Major basic protein (MBP), which is a cationic protein found in eosinophils and is active principally against parasites
- Bactericidal permeability increasing protein (BPI), which, as the name implies, causes changes in the permeability of the membranes of the microorganisms.

Also the low pH found in the phagolysosomes, besides being bactericidal itself, enhances the conversion of H_2O_2 to superoxide. Unfortunately, the leukocyte is not always successful in killing all organisms and some bacteria, such as the mycobacterium that causes tuberculosis, can survive inside phagocytes, happily protected from antibacterial drugs and host defence mechanisms.

This takes us on to consider the chemical mediators involved in inflammation.

CHEMICAL MEDIATORS

Some inflammatory mediators are derived from cells in the blood or tissues, whereas others are circulating proteins, most of which are manufactured in the liver (see Fig. 4.7).

Cell-derived mediators
Lysosomal contents
We have just discussed the role of newly generated killing chemicals in lysosomes of the phagocytes, neutrophils and macrophages in the acute inflammatory response. We now talk about the destructive power of their pre-formed lysosomal contents.

Lysosomal enzymes and accessory substances are present in neutrophils and monocytes, packaged in

Mediators of acute inflammation

Cell derived

- Arachidonic acid derivatives (thrombin, prostaglandins and leukotrienes)
- Cytokines (interleukins, colony-stimulating factors, interferons, chemokines, growth factors such as EGF, PDGF, FGF and TGF)
- PAF
- Vasoactive amines (histamine and serotonin)
- Lysosomal contents
- NO

Plasma derived

- Kinins
- Clotting and fibrinolytic systems
- Complement

membrane-bound vesicles ('granules') to prevent them from damaging their own cells. The contents vary between cell types.

In the neutrophil there are two types of granules, the smaller *specific* and the larger *azurophilic*. These contain substances that increase vascular permeability and are chemotactic. Myeloperoxidase, which gives the azurophilic granules their name, causes the greenish-yellow colour seen in pus, which is an accumulation of dead and dying neutrophils plus liquefied tissues and (often) microbes. The lysosomes contain phospholipase A_2 and plasminogen activator (see below).

Monocytes and their tissue counterparts, macrophages, also contain lysosomes with a powerful array of hydrolytic and proteolytic enzymes, phospholipase A_2 and plasminogen activator.

The lysosomal enzymes destroy many extracellular components including collagen, fibrin, elastin, cartilage and basement membrane, and can activate complement, as well as producing intracellular killing in the phagolysosome as already described.

If these processes were unopposed, there would be massive tissue destruction. So there are anti-proteases

within the serum and tissue fluids to neutralise these enzymes and therefore regulate the extent of tissue damage. Does this seem a far-fetched idea, distant from clinical practice? Well not so!

A deficiency of one such anti-protease, α_1-antitrypsin, leads to unopposed action of neutrophil elastase and hence destruction of elastic tissue, especially in the lungs (Fig. 4.16a) and liver (Fig. 4.6b), particularly if repeated episodes of acute inflammation are stimulated by smoking (lung) or alcohol consumption (liver) (Fig. 4.16c). Clinically, a patient with α_1-antitrypsin deficiency may have emphysema of the lungs and liver cirrhosis.

Arachidonic acid derivatives (eicosanoids)

These are the thromboxanes, prostaglandins and the leukotrienes. Similar to the clotting and fibrinolytic system, they play a part in thrombosis as well as inflammation. They are best thought of as local hormones. They have a short range of action, are produced rapidly, and degenerate spontaneously or are degraded by enzymes. Arachidonic acid, the parent molecule, is a 20-carbon polyunsaturated fatty acid that is derived from either the diet or essential fatty acids. It is not found in a free state but is present esterified in the cell membrane phospholipid. The two main pathways of arachidonic acid metabolism, involving the cyclooxygenase (COX) and lipoxygenase pathways, and their products, are shown in Fig. 4.17, which also depicts some of the roles of the products in inflammation.

Drugs such as corticosteroids, aspirin and ibuprofen act to reduce inflammation by inhibiting the production of prostaglandins. Unfortunately the stomach relies on the production of prostaglandins for the generation of a protective bicarbonate layer, so suppression of the COX pathway by aspirin and other non-steroidal anti-inflammatory drugs (NSAIDs) can cause gastric erosion and peptic ulceration. A therapeutic breakthrough was expected when it was discovered that a subtype of COX enzymes, COX-2, was expressed only during inflammation, whereas COX-1 is expressed as part of normal cell metabolism. Drugs were developed that selectively inhibited COX-2. This prevented the gastrointestinal tract side effects but unexpectedly led to an increase in cardiovascular diseases such as stroke and myocardial infarction in susceptible individuals.

Chapter 4: The acute inflammatory response

(a)

Lung with emphysema: the 'holes' (arrowed) are hugely distended, coalescent, alveolar spaces. If blood α_1-AT levels are less than 75% of normal, patients are at severe risk of developing emphysema. Any inflammatory stimulus exacerbates the risk – this patient was a coal miner.

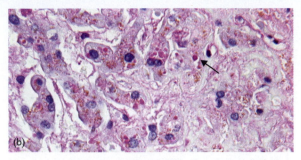

(b)

Mutations in the genes encoding α_1-AT lead to abnormally folded protein, which cannot be secreted into the blood by the liver. Instead, it accumulates within endoplasmic reticulum, shown here (arrow). These patients are at increased risk of cirrhosis. Drinking alcohol stimulates an acute inflammatory response in the liver and increases the risk of this complication.

(c)

Figure 4.16 α_1-Antitrypsin (α_1-AT) is a 394-amino-acid glycoprotein secreted by the liver that is produced during the acute inflammatory response. It inhibits serine proteases, especially neutrophil elastase, which can cleave a wide variety of extracellular matrix components, including elastin. Elastin is what gives tissues such as lung their elastic recoil. Inflammatory episodes are more severe and prolonged than normal, with more tissue damage leading to scarring. The severity of α_1-AT deficiency varies because the disease can be caused by several different mutations.

Cytokines and chemokines

Most cytokines are polypeptides mainly produced by macrophages and activated lymphocytes; some are also made by endothelial and epithelial cells and even cells in the connective tissue. At least 150 of these polypeptides have already been identified. They act principally to regulate immune and haematopoietic cell proliferation and activity (Table 4.4). In addition they have effects in the inflammatory response. The two most important are IL-1 and TNF. They have a variety of important effects as shown in Fig. 4.18, among which are the release of chemokines.

Cytokines can be grouped broadly as follows:

- Interleukins
- Colony-stimulating factors (CSFs)
- Interferons
- Growth factors.

Interleukins are secreted by leukocytes, and predominantly act on other leukocytes. There are many interleukins, of which IL-1, IL-6 and IL-8 are most important in acute inflammation and IL-10 in damping down acute inflammation. IL-1β and IL-18 are

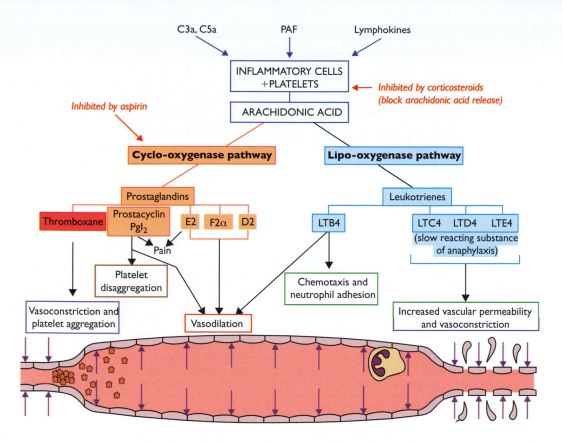

Figure 4.17 Arachidonic acid derivatives (eicosanoids) and their main effects.

essential for coordinating fibrous tissue deposition in healing and repair.

CSFs act on the bone marrow to stimulate the production of various haematopoietic cell lines, e.g. neutrophil production is greatly increased during the acute inflammatory response.

Interferons are secreted by many cell types, particularly activated macrophages. Various forms exist, of which IFN-α, -β and -γ are best characterised.

Growth factors have a role in chemotaxis as well as inducing healing and repair of tissues and playing a part in the development of malignant tumours.

The principal growth factors are as follows:

Epidermal growth factor (EGF)
Platelet-derived growth factor (PDGF)
Fibroblast growth factor (FGF)
Vascular endothelial growth factor (VEGF)
Transforming growth factor (TGF).

Most of these can be produced by macrophages, which are numerous in areas of inflammation.

Chemokines (short for 'chemotactic cytokines') are responsible for attracting inflammatory cells from the circulation to the site where they are needed. There are numerous chemokines, which are mainly produced by activated macrophages or endothelial cells. They attract specific types of inflammatory cells, e.g. IL-8 attracts neutrophils rather than macrophages or eosinophils, IL-5 eosinophils, whereas MCP-1 (monocyte chemoat-tractant protein-1), macrophage inflammatory protein-1 (MIP-1) and RANTES (regulated and normal T-cell expressed and secreted) attract monocytes, eosinophils, basophils and lymphocytes rather than neutrophils.

Platelet-activating factor

PAF is secreted by activated, antigen-stimulated, IgE-sensitised basophils, macrophages, neutrophils, platelets and endothelial cells, and is usually derived

Chapter 4: The acute inflammatory response

Table 4.4 The main actions of some of the most important cytokines

Cytokine	Origin	Action
IL-1	Secreted by macrophages, antigen-presenting cells and B cells	Activates T cells and NK cells, causes liver to produce acute-phase proteins, elevates core temperature (fever)
IL-1β, IL-18	Secreted by activated dendritic cells, macrophages	Important in control of collagen deposition in healing and repair. Imbalance may lead to fibrosis
IL-2	Secreted by activated Th cells	Stimulates proliferation of activated T and B cells and NK cells
IL-3	Secreted by Th cells and NK cells	Stimulates haematopoiesis by activation of stem cells Stimulates mast cells to proliferate and release histamine
IL-4, -5, -6 cells	Secreted by Th cells	Stimulates B cell proliferation and differentiation and antibody production by plasma cells
IL-5	Secreted by Th cells	Stimulates bone marrow production of eosinophils, e.g. parasitic infection, induces IgE class switch in B cells
IL-6	Secreted by Th cells, macrophages, stromal cells	Elevates core temperature (fever), release of acute-phase proteins, induces stem cell differentiation
IL-7	Produced by stroma of marrow and thymus	Induces lymphoid stem cells to form B- and T-cell precursors
IL-8	Made by macrophages and endothelial cells	Important for neutrophil chemotaxis
IL-10	Made by Th cells	Stimulates B-cell activation and macrophages to secrete cytokines
IL-12	Made by macrophages and B cells	Activates NK cells and acts synergistically with IL-2 to induce Ts/c cells to differentiate into cytotoxic T lymphocytes
Interferons (IFN-α, -β, -γ)	IFN-α is produced by most white cells, IFN-β is produced by fibroblasts, and IFN-γ is produced by Th, Tc and NK cells	All act to induce MHC expression and increase viral replication in a variety of cells. IFN-γ has more extensive effects on T and B cells and macrophages and induces class switch to IgG
Tumour necrosis factor (TNF)	Can induce the death of tumour cells. Receptors for the TNF family include Fas	When this is bound by Fas ligand on Tc cells, apoptosis is induced. CD40 on activated B cells belongs to this group
TNF-α	Secreted by macrophages, mast cells and NK cells	Acts on macrophages to increase cytokine production and cell surface receptor molecule expression, induces class switch to IgA
TNF-β	Secreted by Th and Tc cells	Stimulates phagocytosis by macrophages and polymorphs and NO production by endothelial cells

IL, interleukin; NK, natural killer; NO, nitric oxide; Tc, T-cytotoxic; Th T-helper; Ts, T-suppressor.

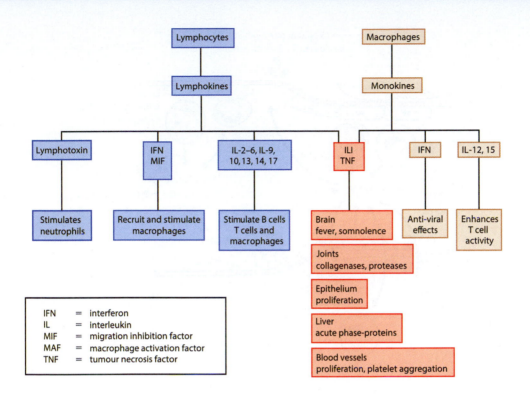

Figure 4.18 The sources and effects of the major cytokines involved in inflammation.

from phosphatidylcholine. In addition to activating platelets and causing them to aggregate, it can cause bronchoconstriction and vasoconstriction changes (although at low levels it causes vasodilatation and a marked increase in venular endothelial permeability!), leukocyte adhesion and chemotaxis. It can also stimulate the production of other mediators, in particular the arachidonic acid metabolites, and stimulate the 'respiratory burst' to produce oxygen-dependent inflammatory metabolites in the lysosomes of neutrophils and macrophages. PAF activity is inhibited by specific PAF acetylhydrolase enzymes (Fig. 4.19).

Vasoactive amines (histamine and serotonin)

Mast cells and platelets (cytoplasmic fragments of megakaryocytes, rather than true cells) are often the first to be involved in the acute inflammatory response (along with the DCs, tissue macrophages and NK cells). They are certainly the fastest to act.

Histamine is stored in and released from mast cells, basophils and platelets. Mast cells are found in the tissues lying close to blood vessels, where the effects of histamine and other vasoactive mediators are greatest. Mast cells degranulate (release their granule contents by fusing the lysosomal membranes of the granule with the plasma membrane) in response to many types of stimulus. These stimuli include physical trauma such as injury by force, cold or heat, antibody binding, complement fragments C3a and C5a, releasing factors produced by neutrophils, monocytes and platelets, and IL-1, neuropeptides such as substance P and some cytokines, such as IL-1 or -8.

Serotonin (5-hydroxytryptamine or 5-HT) is present in platelets and enterochromaffin cells in humans (and mast cells in some other animals).

Platelets may release histamine and 5-HT after activation by contact with collagen, exposed by endothelial cell damage. Platelet degranulation can also be stimulated by the activation of mast cells via the cross-linking of bound IgE: mast cells release PAF, causing platelet aggregation and release of platelet granule contents.

Chapter 4: The acute inflammatory response

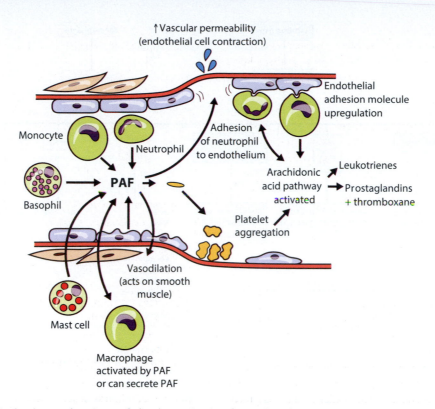

Figure 4.19 Activation and actions of platelet-activating factor (PAF).

Those of us who have asthma or allergy to inhaled allergens, such a dander from cats or dogs, will appreciate the symptomatic aspects of the release of histamine: wheezing due to bronchoconstriction, swelling from tissue oedema, with laryngeal stridor, caused by vasodilatation of arterioles (but constriction of arteries) and increased permeability of venules. The action of histamine on vessels is mediated via H_1-receptors on endothelial cells, although some of its other actions (e.g. bronchoconstriction) are effected via H_2-receptors (see type I hypersensitivity in Chapter 1, page 80 and Chapter 6, page 202).

The role of these amines is thought to be in the early phase of inflammation because it has been shown that H_1-receptor blockade by antihistamines has no effect on the permeability after 60 minutes.

Read more about asthma in Pathology in Clinical Practice Case 2

Nitric oxide
NO is produced by endothelial cells, macrophages and specific neurons in the brain, and has roles in smooth muscle relaxation, reduction of platelet aggregation and adhesion. It acts as a toxic radical to certain microbes and tumour cells. NO can also decrease leukocyte adhesion to endothelium and thus can diminish the inflammatory response (under-production of NO by damaged endothelial cells in people with diabetes or vessels with atherosclerosis may contribute to the vascular damage common in these diseases). NO has a half-life of seconds, so it acts only on immediately adjacent cells.

Macrophage NO production occurs only when induced by cytokines such as interferon-γ, whereas endothelial and neural NO is produced constitutively. Uncontrolled production of NO by macrophages can lead to massive peripheral vasodilatation and shock.

Plasma-derived mediators
Kinin system
Bradykinin is the major active product of this system. It is a polypeptide that is one of the most powerful vasodilators known to humans, increases vascular permeability and also induces pain when injected into the skin.

The kinin cascade is activated by the Hageman factor (clotting factor XII), which activates prekallikrein to form kallikrein. Kallikrein in turn activates high-molecular-weight kininogen (HMWK) to form bradykinin. HMWK is made in the liver.

The relationship to the coagulation system is shown in Fig. 4.20. As with other cascades, this contains an amplification step because kallikrein itself acts to stimulate production of the Hageman factor. Kallikrein is a highly active substance that is a chemotaxin in its own right, but can also cleave complement component 5 to make C5a (a highly potent chemotaxin); it also converts plasminogen to plasmin, which dissolves fibrin (the end-product of the coagulation cascade).

The clotting and fibrinolytic systems

This system is not only important in inflammation but is also central to blood clotting (see also Chapter 7, Fig. 7.3, page 218). The end-products, fibrinopeptides, act as a chemical mediator in inflammation. The fibrinopeptides increase vascular permeability and are chemotactic for neutrophils. As in the kinin system, the cascade is activated by the Hageman factor – this is activated by contact with collagen and basement membrane, exposed when vascular endothelial damage occurs (the 'tissue factor pathway' or 'extrinsic pathway'), or by HMWK and kallikrein from the kinin cascade – and includes an amplification loop so that plasmin stimulates the Hageman factor.

The liver is the source of most of the clotting factors (I (fibrinogen), II (prothrombin), V, VII, VIII, IX, X and XI), and also protein C, protein S and antithrombin. It also makes HMWK, from which bradykinin is produced, and most of the complement factors (although some of these can also be made by other cells, such as macrophages).

The body likes to keep everything balanced, so, as the coagulation cascade is stimulated to coagulate, generating fibrin, the enzyme that can break down fibrin is manufactured. Plasmin is a multifunctional protease that also lyses fibrin clots to produce fibrin degradation products, which themselves induce permeability changes and also trigger the complement system by cleaving C3.

The complement system

The system comprises 9 liver-derived plasma proteins (C1–9) which are split into about 20 cleavage products; these are involved in increasing vascular permeability, chemotaxis, opsonisation and direct lysis of organisms

(Fig. 4.21). The most important components concerned with the inflammatory reaction are as follows:

C3a and C5a, which increase vascular permeability and are chemotactic for neutrophils and macrophages (particularly C5a)
C3b and C3bi, which opsonise microbes for phagocytosis
C5b–9, which form the membrane attack complex (MAC); this causes cell lysis by assembling a porthole in the microbial membrane, which causes unregulated movement of ions and fluids and cell death.

There are three main pathways by which complement can be activated. The classic pathway is rapidly initiated by antigen–antibody complexes; IgM is the best at activating C1, because its pentameric structure clusters many antigens together in one place.

Activation through the alternative pathway is slower because its initiation requires a little bit of luck. C3 is abundant in the blood, and there is some in tissue fluid as well. C3 spontaneously breaks into its active subparts, C3b and C3a, all the time. C3b is inactivated by binding to water in less than a second, but, if a microbe lies nearby, it attaches to amino or hydroxyl groups in its wall and this stabilises C3b. The bound C3b binds another complement protein, B; this becomes cleaved by protein D to make complex C3bBb. Together this complex can cleave more C3 (preventing the need for spontaneous events) and can also cleave protein C5 into C5a and -b. The actions are summarised in Fig. 4.21.

The third complement-activating pathway is the mannose- or lectin-binding pathway (see also the earlier description of soluble PRRs, see Fig. 4.5). Mannose is a carbohydrate constituent of many bacterial cell walls, especially Gram-positive bacteria, and also some important fungi such as *Candida albicans* (the organism that causes 'thrush'), and some viruses and parasites. Lectins are proteins that bind carbohydrates. Mannose-binding lectin (also called mannose-binding protein or MBP) is present in the blood in a protected state, bound to another protein. When its lectin binds mannose on the surface of a microbe, the other protein functions with it to become a C3 convertase. Thus C3b is formed and the pathway continues.

Protection of human cells from the actions of complement C3b includes the secretion by all our cells of defensive proteins, such as decay-accelerating factor, which speeds the breakdown of the C3bBb complex by

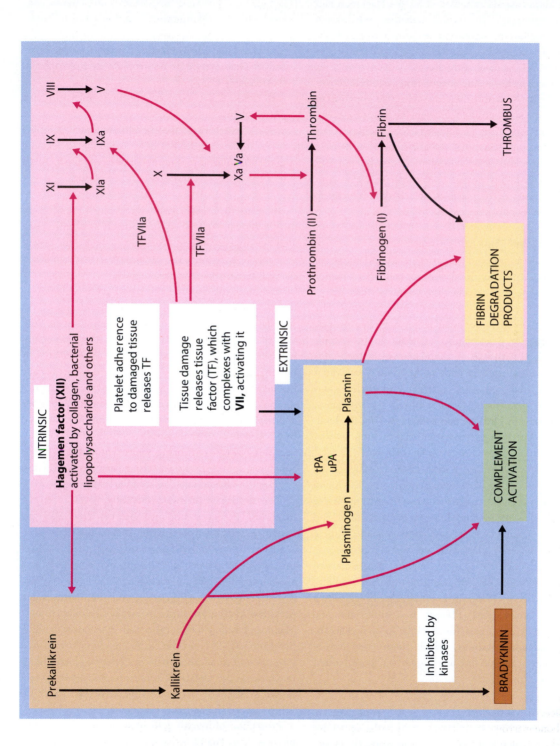

Figure 4.20 The coagulation cascade and fibrinolytic pathways and their interactions with the kinin and complement systems. Red arrows denote activation of other systems. Inhibition pathways are not shown.

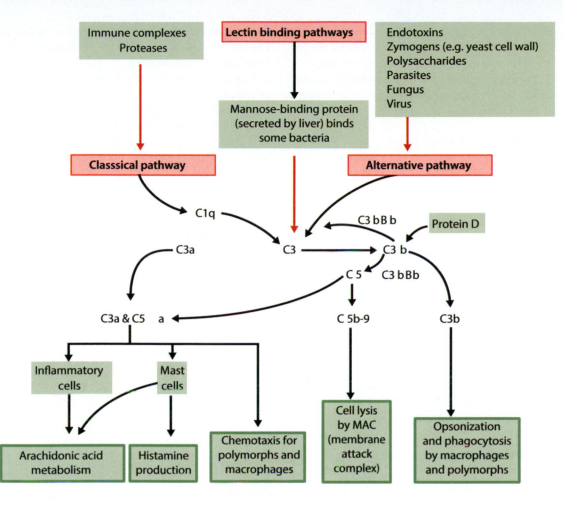

Figure 4.21 The complement system.

other blood constituents. CD59 on the cell surface of human cells prevents the MAC complex from forming and penetrating the cell wall.

THE SYSTEMIC EFFECTS OF INFLAMMATION

Having considered acute inflammation, we have an idea of the *local* effects involved, and alluded to the fact that, at the same time, there are many *systemic* effects that may take place, such as the stimulation of leukocyte production by the bone marrow, fever, rigors, tachycardia, hypotension (a drop in blood pressure), loss of appetite, vomiting, skeletal weakness and aching. These are known collectively as acute phase reactions. If these are severe and

uncontrolled, these may result in the systemic inflammatory response syndrome (SIRS), see page 142.

Fever is a regular accompaniment to inflammatory responses and occurs due to the 'resetting' of the thermoregulatory centre in the anterior hypothalamus (Fig. 4.22). The hypothalamus reacts to the new setting by causing a rise in the body's core temperature via constriction of vessels in the skin, so reducing blood flow and limiting heat loss, and promoting heat production in muscles by shivering. Biological substances that induce fever are called pyrogens. Many bacteria and viruses produce molecules that act as pyrogens and these are called exogenous pyrogens. Endogenous pyrogens are produced by the body. It is thought that exogenous pyrogens stimulate leukocytes to release the endogenous pyrogen IL-1 or IL-6, which acts on the hypothalamus by raising local prostaglandin E_2 (PGE$_2$) levels. Aspirin is useful for lowering the temperature because it interferes with PGE_2 production. Paracetamol reduces temperature by directly acting on the temperature centre.

It has been suggested that fever evolved as a mechanism to disadvantage organisms unused to operating at high temperatures.

THE ACUTE-PHASE PROTEINS

Localised inflammatory responses lead to changes in plasma proteins due to alterations in liver metabolism. These proteins are called acute-phase proteins and this change is thought to be mediated by IL-1, -6 and -8, and TNF-α. There is increased production of clotting factors and complement, which are important because they are consumed during the inflammatory process. Plasminogen breaks down blood clot. Mannan-binding lectin activates complement.

Several acute-phase proteins act as opsonins, such as complement factors, C-reactive protein (CRP) and serum amyloid P component. Serum amyloid A both recruits inflammatory cells to the site of damage and can induce enzymes that degrade the extracellular matrix, facilitating their movement. Ferritin binds iron so that microbes cannot utilise it. The enzymes α_1-antitrypsin and α_1-antichymotrypsin downregulate inflammation, by breaking down elastases.

Clinically, CRP and the erythrocyte sedimentation rate (ESR) are the two most useful tests of current inflammatory activity. The ESR increases with fibrinogen

---

Normal temperature regulation

Skin and spinal thermoreceptors alert the temperature control centre in the pre-optic neurons of the hypothalamus of changes in core temperature from normal 37-37.5 °C, via stimulation of cold- or warm-sensitive neurons.

If too hot: cutaneous veins dilate, radiating heat from skin; sweating assists heat loss by evaporation; behaviour changes cause shedding of clothing and seeking of shade or immersion in cool water.

If too cold: shivering generates heat and peripheral vasoconstriction reduces heat loss; brain initiates use of warm clothing and the subject seeks a warmer environment.

Disorders due to impaired ability to exercise normal temperature regulation include:
• **Heat exhaustion**, which can lead to heat stroke (eg due to sunstroke)
• **'Malignant hyperthermia'** (eg an idiosyncratic reaction to certain anaesthetics)
• **Hypothermia** (eg unprotected exposure to cold environment)

Actions of anti-pyretic drugs
• Steroids: block PLA2
• Aspirin and NSAIDs: block COX
• Paracetamol: appears to have several different mechanisms of action; unlike the other drugs mentioned, it does not have significant anti-inflammatory effects.

Is fever of value?
• It may inhibit the proliferation and survival of some organisms.
• It can diminish some endotoxin effects.
• It encourages speedier phagocyte activity.
• Fever may enhance T cell proliferation.

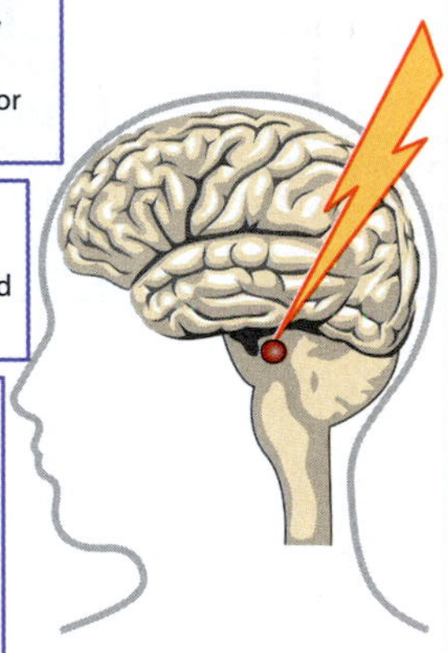

Fever – temperature regulation centre is re-set by endogenous or exogenous pyrogens

Causes:
Infection by bacteria, TB, fungi, viruses
Connective tissue disorders (CTD) eg rheumatoid arthritis
Tumours, eg lymphoma, renal carcinoma

Mechanisms:
• 'Exogenous' pyrogens, from outside the body, include lipopolysaccharide (LPS) and mannose.
• Infection also stimulates the release of 'endogenous' pyrogens, which act on the temperature regulation centre.
• Endogenous pyrogens are **cytokines.** The most important are IL-1 and IL-6, also TNF, IFNα, β, γ.
• Infections, CTD and tumours can cause fever via cytokine release.
• 'Neoplastic fevers' caused by tumours are often related to IL-6 production; IL-6 serum levels have independent prognostic value in some conditions, eg diffuse large B cell lymphoma.
• Infective agents can directly stimulate **toll-like receptors** (TLR) and cause endogenous pyrogen release.
• The final common pathway for both cytokines and TLR is the release of PGE2 in the temperature regulation centre through the stimulation of arachidonic acid breakdown by phospholipase A2 (PLA2) and cyclo-oxygenase (COX).
• PGE2 is recognised by **receptor EP3**
• This causes a **'re-set'** from the previous normal to a higher core temperature. The usual 'warming' mechanisms come into play to raise the core temperature to the revised setting – largely by shivering and peripheral vasoconstriction.
• Malaise, loss of appetite and tiredness are side-effects of cytokine release.

Figure 4.22 Normal control of body temperature and the mechanism of fever.

Radiology

Imaging and the 'PUO'

The 'PUO' is a common clinical detail on many request slips in the radiology department. Pyrexia of unknown origin applies to a patient who is sick with sepsis and a fever, and the source of sepsis (if sepsis is a cause of the fever – remember fever is a consequence of an inflammatory response, of which there are many causes apart from infection) is not obvious from the history and clinical examination.

The patient will already have had a battery of tests on body fluids and solids – blood, urine, sputum, cerebrospinal fluid (CSF) and faeces.

In the meantime all patients should have a chest radiograph – a simple investigation that may reveal an occult pneumonia.

If this is not helpful one needs to delve back into the patient's history and clinical findings to glean clues that may decide the next form of imaging.

1 Pain/change in bowel habit/neurological signs may be a great localiser:

   a If the patient has abdominal pain or had recent surgery the investigation of choice is a CT of the abdomen to look for an intra-abdominal collection or conditions such as appendicitis (ultrasound may be as helpful and radiation free in children), diverticulitis, pyelonephritis or colitis.

   b If the patient has specific pain in the right upper quadrant an ultrasound scan may reveal cholecystitis and gallstones, bile duct dilatation in cholangitis or a liver abscess.

   c Back pain and fever should initiate an MRI of the spine to look for discitis ± an epidural abscess.

   d Headaches or delirium may lead to MRI of the brain to look for encephalitis or a cerebral abscess.

   e In children who develop a limp or a painful hip ultrasound is used to identify a hip effusion in a transient viral synovitis. An isotope bone scan may help in identifying osteomyelitis.

2 A history of cardiac surgery, valve surgery, clinical signs of endocarditis or an abnormal ECG will lead to echocardiography and in some centres cardiac MRI to look for valve vegetations, pericarditis or myocarditis.

If the septic focus can still not be localised there are a few options still left up the radiologist's sleeve:

3 Gallium scan: a whole-body nuclear medicine scan in which gallium-labelled radioisotope is injected intravenously, localising to areas of inflammation that can be imaged with a gamma camera.

4 An FDG–PET (positron emission tomography)–CT scan: a whole body scan in which the ^{18}F-labelled fluorodeoxyglucose (FDG) radioisotope localises to areas of inflammation.

5 Whole-body CT or MRI – in times of desperation and when one does not have access to the techniques in (1) and (2).

It should be said, however, that it is quite rare for a patient to finally have a gallium scan, whole-body MRI or a FDG–PET scan to localise infection or one of the many other causes of fever, because most patients will have had their septic focus identified by this stage.

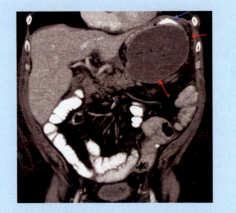

Figure 4.23 A 37 year old man presents with fevers and abdominal pain one week after gastric surgery. On CT there is a large collection (or abscess) in the lesser sac (red arrows), displacing and effacing the fundus of the stomach which contains oral contrast (blue arrow).

production, but there is a lag phase of around a week before changes, particularly a decrease, can be discerned. CRP has a half-life of 6–8 hours and is extremely useful in tracking known inflammatory disease. In those patients with chronic diseases, e.g. Crohn's disease, a chronic inflammatory bowel disease with acute exacerbations, a rise in the level may be an indication of an acute exacerbation (Fig. 4.24).

Some clinicians use CRP as a screening tool for infective inflammatory disease but this is not always helpful, e.g. a patient with volvulus or a torted ovarian cyst may report severe abdominal pain, but

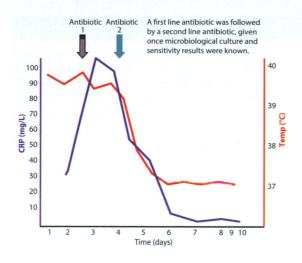

Antibiotic 1   Antibiotic 2   A first line antibiotic was followed by a second line antibiotic, given once microbiological culture and sensitivity results were known.

Figure 4.24 C-reactive protein (CRP) fluctuations are an accurate reflection of inflammatory disease activity. CRP levels rise within 4–6 hours and have a half-life of 12 hours. This graph, from a patient with bronchopneumonia, shows the parallel between the patient's temperature, CRP levels and treatment. (Courtesy of Dr Jo Sheldon, St George's Hospital.)

the CRP (which would rise in screening for infective enterocolitis, inflammatory bowel disease or appendicitis) may be negative.

A normal CRP is often encountered in systemic lupus erythematosus (SLE), in which the ESR is usually raised.

Transport proteins, such as haptoglobins, may be important in regulating the amount of amines and oxygen free radicals. Iron is an important constituent of lactoferrin and a cofactor in the respiratory burst, which probably explains the increase in plasma transferrin in inflammation. Ferrous iron is also an essential factor in the cross-linking of collagen, which occurs during wound healing.

The acute-phase reaction varies depending on the cause of inflammation. Viral infection is a poor inducer of acute-phase proteins, whereas bacterial infections produce a major response, probably by bacterial endotoxins acting indirectly through raised TNF-α levels.

The number of leukocytes in the peripheral blood increases in many forms of inflammation so that they are available to fight infection. Cytokines act through

---

🔑 Key facts

Systemic Inflammatory Response Syndrome (SIRS)

Definition:
Body-wide activation of pro- and anti-inflammatory cytokines, which synergise to create a destructive spiral of cytokine activation (SIRS is one component of the 'Cytokine storm' phenomenon).

Causes:

- Infection (in which case the term 'severe sepsis' should be used, rather than SIRS)
- Non-infectious: haemorrhage, ischaemia, pulmonary embolism, trauma, acute pancreatitis, burns, anaphylaxis, drug reaction and several other causes.

Clinical signs:

These are features of acute inflammation; two or more are present in patients diagnosed with SIRS. There are extra factors to consider in the childhood diagnosis of SIRS.

- High or low temperature (above 38.5°C or below 36 °C)
- Tachycardia of more than 90 beats per minute

- Tachypnoea (respiratory rate above 20 breaths per minute or arterial pCO2 low)
- Very raised or very low white cell count
- Hyperglycaemia which is not due to diabetes mellitus.

Effects:

- SIRS patients may develop disseminated intravascular coagulation or multi-organ failure.

Related terms:

- 'Cytokine storm': Massive hyperactivation of immune cells, which can be localised to one organ, eg lung in Bird Flu, or systemic, as in SIRS or severe sepsis, and leads to tissue damage and organ failure. Both pro- (TNFα, IL-1, IL-6) and anti- (IL-10) inflammatory cytokines are elevated in the serum.
- 'Critical illness immune dysfunction' (p146) is predominantly a complication of over-suppression of the immune system and, like SIRS, is probably a subset of cytokine storm.

colony-stimulating factors to increase the production and release of cells from the marrow. Again there are differences depending on the type of infection, and this may be helpful in making a diagnosis. Bacterial infection provokes an increase in neutrophils, viral infections cause a rise in lymphocyte numbers, and allergic reactions or parasitic infections result in more eosinophils.

Trauma or stress of any kind also affects the hypothalamus–pituitary–adrenal axis, resulting in the production of growth hormone, prolactin, antidiuretic hormone (ADH), adrenocorticotrophic hormone (ACTH) and adrenaline. These hormones are responsible for the breakdown of glycogen, changes in fatty acid metabolism and sodium–potassium transport. It is these metabolic changes that are responsible for the malaise, weakness, loss of appetite and other varied systemic effects observed during injury.

LOBAR PNEUMONIA

Although it is now rarely seen clinically because antibiotic treatment halts the usual course of the disease, lobar pneumonia is often used to illustrate the process of acute inflammation (Figs 4.25–4.28). This is because it exemplifies the process of resolution – complete restoration of normality after a period of intense inflammation.

It is so named because the lung parenchyma is involved in continuity so that the process affects a whole lobe or contiguous lobes. *Streptococcus pneumoniae*, a Gram-positive diplococcus bacterium, is the most common cause of lobar pneumonia.

The first the body may know of this is when the organism invades the lung, or is encountered by a macrophage lining the alveolar space, leading to changes in the microvasculature. These result in a massive outpouring of fluid into the alveolar spaces resulting in congestion (Fig. 4.25a). This fluid is rich in fibrin and red blood cells.

Soon afterwards, neutrophils follow and the fibrin-rich fluid and cells spread from alveolus to alveolus via the pores of Kohn. The neutrophils attack the organisms and phagocytose them, leading to the death of many organisms and neutrophils. Not surprisingly, the alveoli are now airless and the lung is firm and red, with the texture of liver – the stage is termed 'red hepatisation' (Fig. 4.25b).

As this process progresses, the macrophage is recruited not only to phagocytose dead neutrophils and bacteria but also to digest the fibrin mesh. The lung is still firm but the large inflammatory cell infiltrate and reduction in vasodilatation give it a grey colour and, hence, the term 'grey hepatisation' (Fig. 4.25c).

The final outcome will depend on the competence of this system and whether the basic framework of the lung tissue is intact. Ideally, the degenerate cells and dead organisms in the alveolar spaces will be cleared and re-aerated and resolution (Fig. 4.25d) will take place.

If the alveolar framework has been destroyed or the exudate has not been cleared, organisation will occur, leading to scar formation. The infection may persist, destroying lung tissue, but become localised so that an abscess is formed. This is a collection of pus walled off by fibrous tissue. Alternatively it may spread to the rest of the lung, involve the pleura, cause an empyema or disseminate via the bloodstream to other areas of the body, and can lead to death due to severe sepsis or respiratory failure (see also fig 4.8).

DOWNREGULATION OF ACUTE INFLAMMATION

Resolution of the acute inflammatory response begins when the stimulus is removed and there is no further recruitment of inflammatory cells, or any need for the production of proinflammatory cytokines. Effector cells, such as neutrophils or macrophages, have a short lifespan, particularly after they have been activated.

However, resolution is a more active process than this. At around day 3, regulatory T cells (CD4+ and CD25+) enter the war zone and exert three major immunomodulatory effects: first they inhibit proinflammatory cytokine secretion by cytotoxic cells (especially IL-2 and TNF-α), second they promote the secretion of anti-inflammmatory cytokines, particularly TGFβ and IL-10, and third they directly kill cytotoxic cells.

Another event is the emergence of r-macs (resolution macrophages), which have anti-inflammatory activity and promote the repopulation of tissues by their normal lymphocyte subtypes.

Resolution of acute inflammation is primarily initiated by the phagocytic cells themselves – the

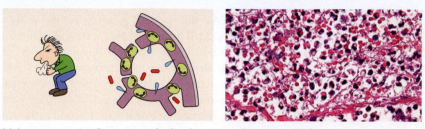

(a) Acute congestion: Bacteria invade alveolar spaces of lung. Acute inflammatory response characterized first by increased vascular permeability, with the formation of a fibrin-rich exudate

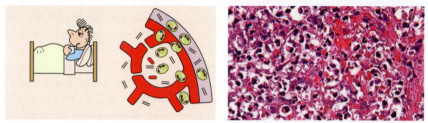

(b) Red hepatization: Neutrophils are quickly attracted to the site, accompanied by red blood cells. The fluid spreads between alveolar spaces via pores of Kohn and soon the entire lung lobe is consolidated (solidified) due to a mixture of fibrin, red and white blood cells. Neutrophils phagocytose the bacteria. The texture and colour of the lobe resembles fresh liver

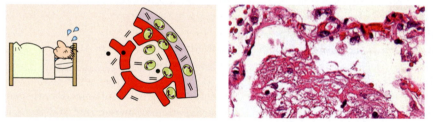

(c) Grey hepatization: Macrophages are attracted to the site, also lymphocytes. Further phagocytosis occurs. Bacteria and dead red and white cells are removed and the fibrin mesh starts to be digested. The grey/white colour of the lobe is due to the high fibrin and white cell content; the texture resembles cooked liver (ugh!)

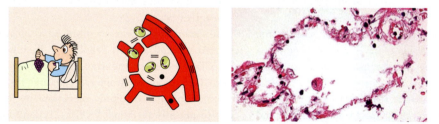

(d) Resolution: The last few fibrin strands and white cells are removed, together with any remaining bacterial corpses and the lung returns to normal. This is possible because the basic skeleton of the lung (formed by reticulin, a type of collagen) is not damaged, unlike bronchopneumonia, in which the inflammatory process is centred on infected bronchioles and is characterised by destruction of the adjacent lung framework (this is usually followed by scarring of lung tissue)

Figure 4.25 The stages of lobar pneumonia.

neutrophils and macrophages undergo apoptosis, which facilitates resolution of the inflammatory process and permits the processes of healing and repair to be carried out. Apoptosis induces immunoregulatory changes that cut short the inflammatory process, promote resolution and prevent autoimmunity developing. The uptake of

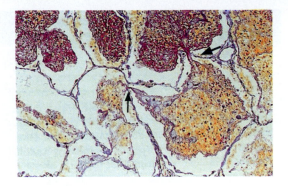

Figure 4.28 Photomicrograph of lung alveoli filled with inflammatory exudate and fibrin passing through the pores of Kohn (arrows).

Figure 4.26 Left lung with consolidation (grey hepatisation) of the lower lobe.

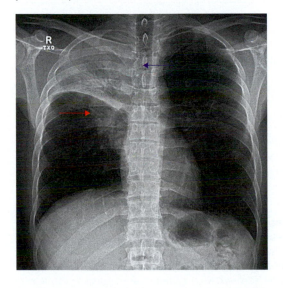

Figure 4.27 Chest radiograph showing consolidation in the right upper lobe (blue arrow) and an enlarged right hilum due to lymphadenopathy (red arrow).

apoptotic cells causes macrophages and dendritic cells to release anti-inflammatory cytokines such as IL-10, TGF-β, NO and the proapoptotic eicosanoid PGE$_2$ (PGD is anti-apoptotic) (Table 4.5).

Apoptotic neutrophils and macrophages, replete with phagocytosed organisms and antigenic fragments, must be cleared efficiently by phagocytosis, a process often termed 'efferocytosis'. This must be achieved before

the fragments break up and release their entrapped organisms, toxic chemicals and enzymes into the tissues to trigger a further wave of acute inflammation.

Efferocytosis may be impaired in severely ill patients, such as those with overwhelming septicaemia, severe burns, acute lung injury or severe trauma with sepsis; in this case for some reason the macrophage appears unable to recognise the apoptosis 'eat me' phosphatidylserine surface marker.

The process of clearing acute inflammatory cells, some of which have been induced to undergo apoptosis by local cytokines, can leave the body critically deficient in defence against new infections in the immediate aftermath of the initial event. This is thought to underlie the life-threatening complications that may occur after a patient has survived severe illness (see Chapter 9), and is thought to reflect anergy of the acute inflammatory response due to an excessive amount of, or response to, anti-inflammatory cytokines, such as IL-10. This has been termed 'critical illness immune dysfunction' (CIID) and explains the high incidence of nosocomial (hospital-acquired) infections in patients who have the double problem of reduced immunity and exposure to often multi-drug-resistant aggressive bacterial organisms. Post-influenza pneumonia may be an example of CIID (Fig. 4.29; see also Fig. 3.3), although in this case the patient is usually fit and well before developing influenza A, after which a bacterial

Read more about critical illness in Pathology in Clinical Practice Case 43

Chapter 4: The acute inflammatory response

Table 4.5 Pro- and anti-inflammatory cytokines

Cytokine type	TNF	CSF	IL	IFN	TGF	PG[a]
Constitution	Protein	Protein	Protein	Protein	Protein	Lipid, derived from arachidonic acid in cell membranes
Proinflammatory	TNF-α	GM-CSF	IL-1β IL-2 IL-6 IL-12 IL-18	IFN-γ		
Anti-inflammatory			IL-4 IL-10 IL-13	IFN-α	TGF-β	PGE$_2$

CSF, colony-stimulating factor; GM, granulocyte–macrophage; IFN, interferon; IL, interleukin; PG, prostaglandin; TGF, transforming growth factor; TNF, tumour necrosis factor.
[a]Also known as 'eicosanoids'.

Resolution of acute inflammation requires:
• removal of the initial stimulus
• cessation of production of pro-inflammatory mediators
• removal of apoptotic phagocytes, their enclosed toxic chemicals and dead or dying organisms
• repopulation of the healed tissue with its normal cells
If resolution does not occur, chronic inflammation develops

As inflammmation is downregulated there may be a period of 'immunoparesis' in which the patient is vulnerable to new infection or repeat infection with the original agent
Examples of illness which may follow excesssive down regulation of acute inflammation include:

Susceptibility to normally innocuous and ubiquitous organisms (eg Candida albicans, often an oral commensal, causes oesophgeal candidiasis, with ulceration and dysphagia)

Recrudescence of previous infection, such as Shingles (VZV) or TB reactivation

Hospital aquired (nosocomial) infection, eg Klebsiella or Pseudomonas pneumonia, a wound infection with a multi-drug resistant pathogen, eg MRSA, or an agent such as C.difficile which may cause life-threatening diarrhoea

Critical illness immune dysfunction syndromes may supervene after an initial insult, particularly if this is severe and systemic, as in burns or septicaemia (severe sepsis). Previous (viral) influenza can predispose to swiftly fatal bacterial pneumonia with overwhelming septic shock (See also SIRS page 142).

Figure 4.29 Critical illness immune dysfunction (CIID): causes and effects; VZV = Varicella zoster; TB = Tuberculosis.

infection with, typically, *Streptococcus pneumoniae* or *Haemophilus influenzae,* or occasionally a staphylococcus, may supervene with overwhelming and often fatal consequences.

The clearance of inflammatory cells, downregulation of pro-inflammatory signals and upregulation of anti-inflammatory signals paves the way for healing and repair, discussed in Chapter 5

CHAPTER 5

HEALING AND REPAIR, CHRONIC AND GRANULOMATOUS INFLAMMATION

WOUND HEALING AND REPAIR

Wound healing is an orderly, controlled process with transient stages, the first of which is the acute inflammatory response, with its procession of pro- and anti-inflammatory cytokines and inflammatory cells to the site of injury or infection.

These events are followed by regression of the inflammatory cells, the resolution phase, which is an active, rather than a passive, event, involving the expenditure of energy in the induction of apoptosis of neutrophils and the phagocytosis by macrophages of the apoptotic cellular material and necroinflammatory debris from the battleground.

Next there is a phase in which **proliferation** of reparative cells occurs, as epithelial cells try to bridge the wound gap, fibroblasts, myofibroblasts and new blood vessels proliferate, and extracellular matrix and collagen are secreted. The collagen requires **remodelling** to return the tissue as closely as possible to normality. If bone repair is involved, healing requires the development of cartilage and bone in addition to fibrous tissue.

Reconstitution of the original inflammatory cell content to the tissue, including lymphocytes, is important to normal function and is mediated by modified macrophages. The desired result is a repair that includes the minimal possible amount of collagen scar tissue.

These processes are tightly regulated by cytokine signalling, which is in turn regulated by the intercellular signalling pathways, often involving the inflammasome, a temporary structure that comes into being when needed (a bit like a genie from a lamp, and almost as mysteriously), and then disassembles. The inflammasome is assembled in different cell types, particularly dendritic cells and macrophages, in response to endogenous or exogenous 'stressors' (also known as 'alarmins') – these are listed in Table 4.1 – and lead to the activation of caspases and the production of IL-1β and IL-18, which are held in a delicate balance to allow control over the type and extent of collagen deposition. Defects in this mechanism may explain some of the unwanted complications of excessive fibrosis described later (see page 159).

CLINICAL SCENARIO: ACUTE TRAUMA

Edward, a 19-year-old student, is rushed to the accident and emergency department (A&E). He has been rescued from the wreck of a small car, entangled with a juggernaut, the result of a miscalculated overtaking manoeuvre on a rainy night. He is unconscious on admission, but he almost immediately regains consciousness and appears lucid and in pain. He has obviously broken his left thigh, which is swollen and shows early bruising. A grating sensation is palpable when the thigh is gently pressed. He has a gaping wound on his right thigh, about 6 cm diameter and 1 cm deep, through which the subcutaneous fatty tissue can be seen, along with much oozing and crusted blood.

His blood pressure is low, at 95/40 mmHg, and his pulse rate is 120 beats/min: these features are signs of shock and indicate that he is probably bleeding internally (see page 258). He may have lost a litre of blood or so into the tissues surrounding his broken femur. The presence of tenderness and guarding over his upper left abdomen raises the possibility of a ruptured spleen.

While his blood is being cross-matched for transfusion, fluids and plasma expanders are infused to maintain his circulation. An emergency scan indicates that he does indeed have a ruptured spleen.

Edward undergoes an emergency splenectomy. His fractured femur is set at the same time, and the gaping wound in his right thigh is debrided (the injured tissue is scraped away), and the site packed with gauze impregnated with iodine (a disinfectant). By the time his parents have been tracked down and retrieved from a party by the police, Edward is settled in intensive care, in traction, and is beginning to regain consciousness after the anaesthetic.

Once his parents have recovered from the shock of the events they begin to fire questions at the consultant orthopaedic surgeon. Edward was unconscious on arrival – will he suffer any permanent brain damage? He is a keen rugby player – when, if ever, will he be able to play again? And what about the loss of his spleen – isn't it important? How long will his operation scar take to heal? He supplements his student grant by stacking supermarket shelves late at night, a job that involves lifting heavy boxes and crates. When will he be able to return to work?

The surgeon tackles the questions when he can get a word in. It is too early to tell whether Edward has any lasting brain injury, but the speed of his return to consciousness bodes well.

The spleen *is* an important organ, with a particular role in removing certain bacteria from the blood (see page 174). Without going into too much detail, he tells them that Edward must be inoculated against encapsulated bacteria and influenza, and will have to take lifelong antibiotic prophylaxis against other bacterial infections, but should otherwise manage very well.

Edward will not be fit for shelf stacking for at least 3 months, not just because of the fractured femur but also because his abdominal wound will take that long to regain sufficient strength. As the situation was so urgent he underwent a laparotomy, with a large abdominal incision, rather than laparoscopic surgery. His fracture is uncomplicated and should heal well. As long as all goes according to plan, he should be walking on crutches in 6 weeks and be ready for rugby after several months of rehabilitation.

Edward's parents are content with this, but let us examine the surgeon's statements about wound and bone fracture healing and splenectomy in a little more detail.

We know already that any tissue injury will initiate acute inflammation (see Chapter 4) and neutrophil polymorphs will soon arrive at the scene of the damage, along with macrophages, to start phagocytosing the debris. If any microorganisms have penetrated the skin wounds, they will be immobilised and phagocytosed, opsonised by a combination of circulating antibody and complement. But what happens next?

CELL CAPACITY FOR REGENERATION

What happens when injury causes loss of normal tissue and leaves a defect, e.g. a cut in the skin? In this situation the end-result depends on the size of the defect and the capacity of the tissue to regenerate. Not all tissues of the body have the same capacity to regenerate and cells can be divided into three major types: labile, stable and permanent (Fig. 5.1).

Labile cells include epithelial and blood cells; these divide and proliferate throughout life and the cells have a set lifespan. Stable cells, such as liver hepatocytes, normally divide only when stimulated to do so, e.g. in response to cell death due to viral hepatitis. If you remove half the liver, its resident stem cells will generate new tissue and return the liver to its original size! Other examples of stable cells are fibroblasts, vascular endothelial cells, smooth muscle cells, osteoblasts and renal tubular epithelial cells. Permanent cells cannot divide but may be capable of some individual cell repair if the nucleus and synthetic apparatus are intact. Examples include neurons and cardiac muscle cells. This statement may be challenged in the near future,

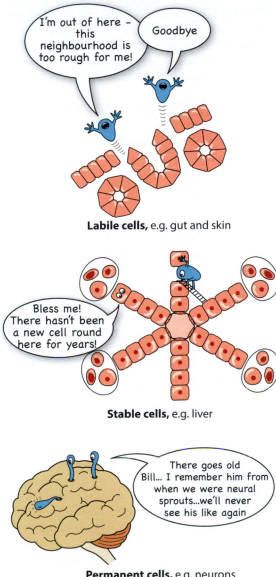

Labile cells, e.g. gut and skin

Stable cells, e.g. liver

Permanent cells, e.g. neurons

Figure 5.1 Labile cells at wear-and-tear sites constantly regenerate from a population of stem cells, e.g. an enterocyte may live only 1–2 days. Stable cells live for many years, only dividing when necessary to replace cells that have been damaged. Hepatocytes can divide, but, if there is major liver damage, the liver stem cell will regenerate new cells. New knowledge from research into stem cells and cell plasticity suggest that some cells previously thought to be 'permanent' and unable to divide or regenerate are capable of limited regeneration.

because there is some evidence to suggest that there is more 'plasticity' in the central nervous system than previously thought. Whether this is from redeployment of existing cells or genuinely new growth is not yet clear. The role of stem cells is under review — it is already clear that marrow-derived stem cells can migrate to damaged tissue and differentiate into appropriate tissues. (Examples include finding the Y chromosome in hepatocytes from a female bone marrow transplant recipient of a male donor marrow.) If a permanent cell is damaged but not destroyed, as in injury to a nerve axon, there may be regrowth of the damaged portion. However, if the whole cell is destroyed, it will be replaced by a small scar, because its neighbouring cells are incapable of proliferating to replace it.

When injury takes place and the processes of inflammation are set in motion, the elements of repair and healing are also activated. Briefly, the processes that take place during and after the injury are as follows:

- Removal of dead and foreign material
- Regeneration of injured tissue from either cells of the same type or stem cells
- Replacement of damaged tissue by new connective tissue.

Wound healing requires the following:

- Haemostasis
- Inflammation
- Cell proliferation, angiogenesis and repair
- Adequate nutrition.

Ideally, adequate tissue repair will occur within 3 weeks. This process of restoring the tissue to pristine condition is called resolution. Resolution requires the inflammatory process to deal quickly with the insult, the tissue to have not lost its basic scaffolding and any damaged specialised cells to be capable of regeneration. The size of the defect is very important, because any destruction of the tissue scaffold will result in scarring.

Although the basic mechanisms involved in wound healing are the same, by convention the healing of cleanly incised wounds, where the edges are in close apposition, is considered separately from those in which there is extensive loss of epithelium, a large subepithelial tissue defect that has to be filled in by scar tissue and where the edges cannot be brought together with sutures. These two

Chapter 5: Healing and repair, chronic and granulomatous inflammation

circumstances are described as 'healing by primary intention' or 'healing by secondary intention'. These terms first appeared in a surgical treatise published in 1543, although Thomson (1813) in *Lectures on Inflammation* gives Galen the credit for introducing these concepts.

HEALING BY PRIMARY AND SECONDARY INTENTION

HEALING BY PRIMARY INTENTION

Let us return to Edward and consider his abdominal surgical incision, made at the time of the splenectomy. This is about the cleanest type of wound that you can get, in both senses. Not only is surgery performed using aseptic technique, but also a sharp knife makes the wound, with the minimum of tissue trauma. The healing in this type of instance, in which the two sides need merely to be pushed back together and held still (by sutures) in order to heal, is known as 'healing by primary intention' (Fig. 5.2).

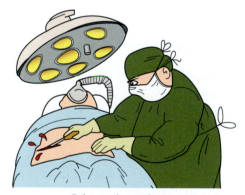

Primary intention
'Clean' wounds e.g. incisions
which heal with little scarring

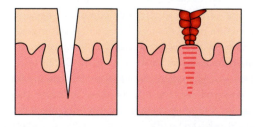

Figure 5.2 Wound healing by primary intention.

The skill of the surgeon has a role to play in the outcome, of course, not least in the selection of an incision site that will not come under undue tension, which will tend to pull the wound sides apart as soon as the sutures have been removed.

HEALING BY SECONDARY INTENTION

What about the large gash on Ed's thigh? You will recall that this was not sutured, but left open, packed with gauze. How will this wound heal? And why not suture the sides together, so that the skin can form a barrier to infection by marauding bacteria?

This was a very large wound — almost the size of the palm of Edward's hand; it would have been impossible to appose the sides without generating unbearable tension on the tissues. Also, there was ragged and crushed tissue at the wound site, not the clean edges of a surgical excision. Dead tissue may form a nidus (from the Latin for 'nest') for infection, and it is quite possible that microorganisms had already entered the wound site by the time Ed arrived at A&E. The surgeon cleared away the obviously dead tissue. By allowing this wound to granulate up, i.e. heal by what is called 'secondary intention', the patient will stand a good chance of an uncomplicated repair, although at the expense of an unsightly scar (Fig. 5.3).

The processes involved in healing by secondary intention are the same as those for primary intention, but the amount of scar tissue generated is greater. One feature that helps to speed up the healing process, and is not seen in relation to healing of incised wounds, is wound contraction. It is interesting that contraction, which diminishes the size of the wound bed, begins at about 1–2 days, which is before collagen deposition has been fully established. Myofibroblasts within the dermis at the wound edges, possibly controlled by sympathetic nerves, contract and draw the wound edges closer. Wound contraction continues at just under 1 mm/day. It peaks at 5 days but continues for up to 15 days.

As scar contraction is so effective, even a 6-cm gash such as Edward's will reduce to about half the diameter or less once it has scarred. Although not a problem on his thigh, in a wound at a more strategic site — e.g. over a joint — scarring of this degree could interfere with function and an alternative approach, such as skin grafting, may have to be employed.

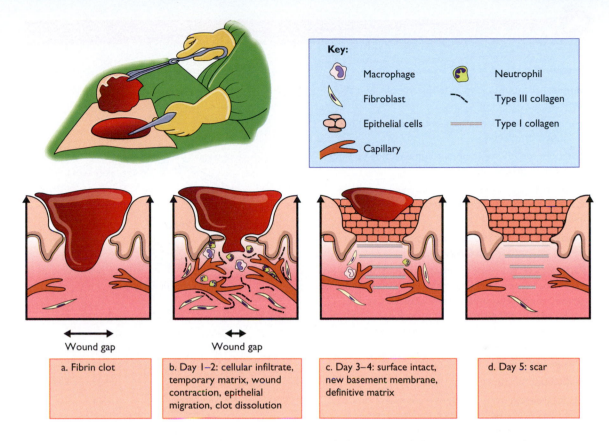

Key:

Macrophage   Neutrophil

Fibroblast   Type III collagen

Epithelial cells   Type I collagen

Capillary

Wound gap   Wound gap

| a. Fibrin clot | b. Day 1–2: cellular infiltrate, temporary matrix, wound contraction, epithelial migration, clot dissolution | c. Day 3–4: surface intact, new basement membrane, definitive matrix | d. Day 5: scar |

Figure 5.3 Wound healing by secondary intention: repair of a large tissue defect generates scar tissue.
(a) 0–6 hours: fibrin clot plugs wound. Growth factors released. Acute inflammatory mediators produced. Bone marrow stimulated to make inflammatory cells.
(b) Days 1–2: cellular infiltration, new blood vessels bud into wound site, temporary matrix forms, epithelial cells migrate and clot dissolves.
(c) Days 3–4: surface intact; new basement membrane, definitive matrix and acute inflammation regresses.
(d) Days 5–14: scar formation, further wound contraction and tissue remodelling.

Wounds of this size can alternatively be closed in a two-stage operation. First, the wound is cleaned by washing and debridement, and left to granulate for 5–7 days. When it is clear that it is healing well, the surgeon will then scrape the wound base and sides until there is pinpoint bleeding, indicating good vascularity, and the edges, now under less tension due to diminished tissue oedema, can be apposed and sutured together. (This will still result in healing by secondary intention, with the production of a moderate amount of scar tissue, because there has been appreciable tissue loss.)

Most wounds, whether of skin or internal organs, will heal in this way but an interesting exception is bone. This breaks the rules and does not heal with a fibrous scar. Even if the 'scaffold' is completely distorted, as with a traumatic fracture, it will remodel to resemble the original, both in appearance and in the effectiveness of its structure and function. If it didn't the bone would remain flexible at the breakpoint. A sequence of radiographs showing phases of fracture repair is shown on page 157. Some problems of bone healing are outlined on page 162.

Minor wounds are generally closed in 10–14 days, but major wounds may take months to a year. Full strength is usually never again achieved in larger wounds, which, even a year later, may be only 80% of the original.

Key facts

Fracture types and general treatment (Fig. 5.4)

Simple: no communication between bone and skin surface

Compound: bone fragments breach skin, thus has potential for infection

Comminuted: multiple bone fragments due to more than one breakage site

Greenstick (*typical in young children*): incomplete fracture, with part of a fractured long bone remaining intact

Stress: the result of abnormal stresses applied to bone (similar to metal fatigue)

Pathological: the bone is abnormal and fractures easily: four types: congenital disease, such as osteogenesis imperfecta; inflammatory disease, such as osteomyelitis; neoplastic disease, such as primary or metastatic tumour in bone; metabolic disease, such as osteoporosis

Fracture management

Reduction: to restore bony alignment

Immobilisation: to maintain the alignment until union has occurred; if this cannot be achieved by external fixation in a plaster cast, internal fixation with pins, screws or plates may be required

Rehabilitation: to restore normal function if possible, or cope with residual disability

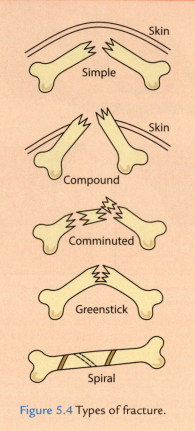

Figure 5.4 Types of fracture.

WHAT DOES WOUND HEALING LOOK LIKE AT THE TISSUE LEVEL?

As healing by primary and secondary intention involves the same fundamental processes, they are discussed together.

Removal phase and acute inflammation

A skin wound causes injury to epidermal and connective tissue cells, sparking off the first steps in the acute inflammatory response (see page 110).

First there is bleeding (haemorrhage), due to damage to the small blood vessels of the skin. This exposes clotting factors and platelets to collagen in the basement membrane and the extravascular tissues, which stimulates the formation of a blood clot. This is achieved by the activation of the Hageman factor (factor XII) in the intrinsic pathway and tissue factor (factor III) in the extrinsic pathway, found in extravascular cells, which becomes activated in the presence of factors III and Va. Fibronectin and von Willebrand's factor (vWF) bind extracellular tissues and are molecules to which platelets can bind (see Figs 7.2 and 7.3).

Fibrin and fibrinogen formed by the coagulation cascade help to 'glue' the wound edges together and provide the protective dry surface scab. We have already discussed how the combination of tissue damage and the activation of the coagulation cascade can kick off acute inflammation and the way that activation of Hageman factor also activates the kinin system and more indirectly leads to complement activation (see Fig. 4.20).

Reflex vasoconstriction for the first 5–10 minutes after injury (mediated by serotonin, prostaglandins, adrenaline, noradrenaline and thromboxane released by tissue cells such as mast cells and macrophages, and also by platelets) slows the blood flow and thus makes it easier for platelets to aggregate and form a primary haemostatic plug. Activated platelets release many agents, including thrombospondin, serotonin, platelet-derived growth factor (PDGF), proteases and other growth factors. (Haemostasis is discussed further in Chapter 7.)

After 5–10 minutes of vasoconstriction there is vasodilatation, due to the actions of histamine (mast cells), other prostaglandins, leukotrienes (inflammatory and endothelial cells and platelets) and kinins (plasma). This leads to an increase in vascular permeability, allowing the exudation of protein-containing fluid from the plasma into the tissues, carrying supplies of plasma proteins such as fibrinogen and complement. Swelling occurs.

The upregulation of endothelial cell selectins and intercellular adhesion molecule 1 (ICAM-1) attracts neutrophils to the vessel walls. The neutrophils marginate and then migrate from the vessels to the damaged area, following a chemotactic pathway of cytokines and complement fragments (see Figs 4.13 and 4.14).

Neutrophils are very important if pathogenic microorganisms have been introduced at the time of the injury, but otherwise have a relatively minor role in wound healing, confined to their phagocytic activity (removal of microbes and debris) and the production of inflammatory mediators. They are short-lived and disappear by apoptosis after about 2 days, unless an inflammatory stimulus persists. Macrophages often phagocytose neutrophils that have phagocytosed organisms, removing both from the tissues.

There are fewer monocytes in the blood than there are neutrophils, but these are recruited in a similar fashion and begin to move to the damaged tissues. Don't forget that the macrophages already present in the tissues will migrate to the site of damage and immediately begin work. Unlike neutrophil polymorphs, which are 'made to order' by the bone marrow influenced by G-CSF, macrophages can be derived from monocytes, which leave the circulation and enter the tissues, but can also proliferate locally. Their roles are the phagocytosis of debris, the destruction of microbial pathogens that have breached the tissue defences, and the secretion of inflammatory cytokines such as PDGF, vascular endothelial growth factor (VEGF) and proteases.

Later on, at about 3 days, T lymphocytes are attracted to the wound site. They may participate in wound healing by interacting with macrophages, other antigen-presenting cells, B cells and other T cells and, if required, processes involved in the generation of a specific immune response to particular microbial pathogens will take place.

The regenerative phase

Meanwhile epidermal cells begin to re-cover the surface. Within a few hours of the injury, a single layer of epidermal cells starts to migrate from the wound edges to form a delicate covering over the raw area exposed by the loss of epidermis. Normal keratinocytes are non-motile and tethered to their neighbouring epithelial cells by junctions such as desmosomes. Keratinocytes and other epithelia can alter their phenotype to produce contractile actin microfilaments in response to growth factors, such as epithelial growth factor (EGF), to re-epithelialise a surface defect.

The distance that the cells need to cover is reduced by an early phase of wound contraction, initiated mainly by myofibroblasts. Epidermal cell movement can provide an initial covering for very small wounds (such as sutured incisional wounds), but in most instances new cells are derived from the stem-cell compartment in the basal layer of the epidermis, stimulated by growth factors secreted by platelets and damaged endothelial cells (and the absence of inhibitory growth factors from neighbouring epithelial cells).

From about 12 hours after wounding, new epidermal cells begin to proliferate and they grow under the protective fibrin/fibronectin clot. Epithelial cells secure their grip by attaching to fibronectin in the matrix that has formed at the wound site. Matrix metalloproteins are also important for epithelial cell migration. They secrete collagenases and plasminogen activator, dissolving the clot and the matrix. They also grow for a little distance down the gap between the cut edges to form a small 'spur' of epithelium, which afterwards regresses. If the wound has been sutured, a similar downgrowth of new epidermis occurs in relation to the suture tracks and, on occasion, these may form the basis of keratin-forming cysts within the dermis, the so-called 'implantation dermoid cysts'. (This ability of epidermal cells to grow along tracks created by sutures or other foreign material is of course the basis for piercing ears for earrings.)

Chapter 5: Healing and repair, chronic and granulomatous inflammation

Once re-epithelialisation is complete, basement membrane proteins form and the epithelial cells revert to their normal non-migratory phenotype, and attach to both each other and the basement membrane. Covering a wound not only prevents infection but also keeps it from drying out, and this enhances epithelial cell migration.

Thus, a switch to a migratory epithelial cell phenotype, with a capacity to dissolve surrounding tissues and interact with connective tissue matrix proteins, is a normal response to wound healing. (One can begin to understand how cancer cells, which can express inappropriate or mutated genes, can revert to this phenotype and invade and migrate through the tissues with such ease.)

The replacement phase: granulation tissue formation and repair

Meanwhile, there is an influx of macrophages, a proliferation of fibroblasts and production of collagen within a connective tissue matrix gel, and ingrowth of many fine capillaries (granulation tissue). This combination of a richly vascularised gel in which both inflammatory cells and collagen-producing fibroblasts are present is known as granulation tissue. The term is derived from the observation that the raw surface of a wound shows a granular appearance, rather like that seen on the surface of a strawberry. Each of these 'granules' contains a loop of capillaries and hence bleeds easily if traumatised. The production of granulation tissue is a fascinating process common to all forms of repair.

Macrophages and fibroblasts are key cells in wound healing and are responsible for the demolition and removal of tissue debris and inflammatory exudate, and for restoring the tensile strength of the subepithelial connective tissue.

Macrophages secrete chemoattractants, which recruit fibroblasts to the wound site. Macrophages also expand the existing small fibroblast population by stimulating them to proliferate. Several agents do this, including fibroblast growth factor (FGF), TGF-α, PDGF and complement C5a. Some of these may come from sources other than the macrophage, e.g. platelets and endothelial cells. As with the proliferating epithelial cells, migrating fibroblasts attach to fibronectin in the connective tissue matrix to facilitate movement.

Fibroblasts are vitally important cells in tissue repair. They may metamorphose into myofibroblasts.

They proliferate locally in response to FGF. Fibroblasts synthesise and secrete the collagen and elastins required for tissue repair. They also secrete fibronectin and matrix components. Fibroblasts secrete procollagen, which is split to form tropocollagen; this aggregates to form fibrils, which align to form collagen fibres. Collagen deposition requires the presence of a framework of fibronectin. Surplus collagen is removed by collagenases secreted by several cell types, including fibroblasts and macrophages.

Ingrowth of new small blood vessels (angiogenesis) into the area undergoing repair is initiated by macrophages, which secrete angiogenic factors such as VEGF in response to a low oxygen tension and the accumulation of lactic acid in the tissues. Angiogenesis involves the budding of new endothelial cells from small intact blood vessels at the edges of the wound, and chemoattraction of these new endothelial cells into the fibrin/fibronectin gel within the wounded area.

Growth factors such as FGF and VEGF can stimulate the endothelial cells of capillaries and post-capillary venules to secrete proteases that digest the surrounding basement membrane. The endothelial cells then proliferate to produce a bud of cells protruding through the gap in the wall towards the source of the stimulus and into the matrix gel in the wound area. At first the bud of endothelial cells is solid, but eventually it canalises behind the 'tip' cells to allow the flow of blood, although quite how the circulatory loop is completed is not known.

There are antiangiogenic factors that control and limit the extent of new vessel formation. As the inflammatory process subsides, those new vessels that are no longer useful regress by undergoing apoptosis. In diabetic retinopathy and cataract formation these mechanisms fail to prevent the ingrowth of new vessels into the retina and lens, respectively.

Angiogenesis is an important aspect of tumour growth and some successful new treatments for cancer involve the use of anti-angiogenic drugs, such as thalidomide (yes – there is some good in this much vilified medication) and anti-VEGF.

The late repair phase: scar tissue formation and remodelling

We talk about repair because the healed site is rarely as good as new (achieving this is called resolution). Usually, the repair site is marked by a collagen scar.

At around day 3 following a skin injury, there is a temporary matrix of type III collagen in the wound site. Epithelial migration over the wound surface halts by contact inhibition and a definitive matrix with type I collagen is laid down. The vessels and inflammatory cells reduce in number. Usually by day 5, bundles of collagen have been laid down across the damaged tissue to form a scar and the epidermis has returned to normal thickness. Continuing collagen production may occur for ≥2 weeks.

Remodelling of the collagen starts at about 3 weeks, when fibronectin is reduced, the matrix proteins have altered in favour of proteoglycans, the proportions of type I and other collagens are roughly normal and much of the excess fluid has been reabsorbed so that the collagen fibres lie closer together. Collagenases are secreted in tandem with new collagen fibres. Replacement and remodelling of the collagen formed early in wound healing are an important part of the healing process.

Initially the scar is red, because of the increase in small vessels, but it will pale over the next few weeks as the vessels regress and the collagen thickens.

If the cut is fine and there is good wound apposition, the scar tissue is limited and the cosmetic result good, but how strong is the repair? Immediately after surgery, the sutured wound has around 70% of the strength of normal skin, but this is principally conferred by the sutures. When these are removed, after 7–10 days, the wound strength drops to 10% of normal – a point to emphasise to patients. Strength then increases rapidly over the next month to reach a maximum at around 2–3 months, when a well-healed scar will have 70–80% of the tensile strength of uninjured skin.

The strength of a scar does not correlate with the amount of collagen but may be related to the type, with type I being stronger than the type III deposited early in the repair process. The ultimate development of tensile strength in a wound depends on the production of adequate amounts of cross-linked collagen and the final orientation of that collagen. Type I collagen must be cross-linked by hydroxylation of proline or lysine residues; this requires adequate supplies of oxygen, vitamin C and ferrous iron. Tensile strength is also referred to, somewhat alarmingly, as 'wound-bursting strength' (Fig. 5.5)!

Ed, our trauma patient, will have scars at the sites of his thigh wound and laparotomy incision, but his

Figure 5.5 "Wound bursting strength" is the tensile strength of a repaired wound. Even after 3 months this is only 80% of normal skin.

fractured femur should show resolution, a restoration of normality (see below).

Unlike TB or bronchopneumonia, in lobar pneumonia, in which there is little or no tissue destruction because the inflammation takes place in the alveolar spaces, a return to normality is the usual outcome.

Another exception is that of damage to a fetus (e.g. by intrauterine surgery), when complete resolution is usual. Much interesting research is being directed towards trying to find a way of achieving this in adults, possibly by switching on dormant embryological genes.

BONE HEALING

We have not forgotten Ed's fractured femur. The femur is a tubular bone, which means that it is formed of a hollow sheath of lamellar bone, filled with fat, bone marrow and a delicate meshwork of bony trabeculae. The tubular shaft imparts the strength of the bone, and the lamellae (or layers) indicate the direction in which the bone has been laid down, in line with the direction of stress. The direct trauma imparted on the femur by the car crash caused it to fracture transversely across its shaft. About a litre of blood will have oozed from vessels damaged at the site, causing a localised haematoma. Early bruising and swelling of the thigh were evident by the time he reached hospital, because extracting him from the car crash took well over an hour. The two free fracture ends grated together when the leg was moved, which caused Edward great pain (periosteal stretching is a potent stimulator of pain) and produced the sensation of crepitus at the fracture site, as noted in A&E.

The fracture was reduced, i.e. pulled back into alignment, and maintained in this position by traction. The site was immobilised in a cast to allow optimal conditions for healing. Inside the leg, the body is beavering away at much the same thing. At the fracture site a kind of internal splint is formed by the haematoma, formed from clotted blood, and also the localised tissue swelling, secondary to the release of inflammatory mediators (see Fig. 5.6). Neutrophils and macrophages from the blood and surrounding tissues appear within hours and increase in number over the next 2 days. They phagocytose the debris: bony fragments, blood clot and damaged connective tissue. Under the influence of inflammatory mediators, fibroblasts invade the wound site and the ingrowth of new capillaries is stimulated, carrying nutrients to the site. If pressed, the site is still slightly mobile at this stage, but 'sticky' and still tender. As the healing process progresses, fibrous tissue and cartilage are laid down, and these ossify to form woven bone. In Ed's case, this takes 6 weeks.

At this point the fracture is considered to be united, and is no longer tender or mobile under pressure, although it remains swollen. Radiologically, there is a clear difference between the original tubular lamellar bone, still obviously sundered but linked by a cuff of woven bone around the wound site, which appears as a loose bony meshwork on a radiograph (Fig. 5.7).

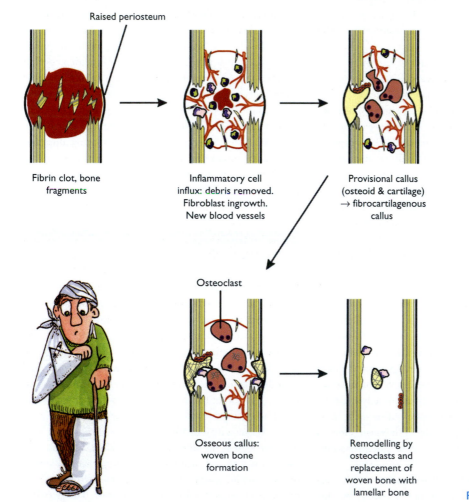

Raised periosteum

Fibrin clot, bone fragments

Inflammatory cell influx: debris removed. Fibroblast ingrowth. New blood vessels

Provisional callus (osteoid & cartilage) → fibrocartilagenous callus

Osteoclast

Osseous callus: woven bone formation

Remodelling by osteoclasts and replacement of woven bone with lamellar bone

Figure 5.6 Bone healing.

Part 2: Defence against disease

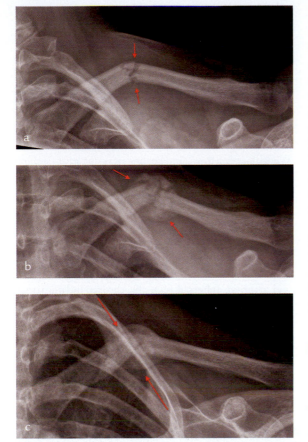

Figure 5.7 The natural evolution of a fracture (red arrows): (a) a 25-year-old rugby player who sustained an injury during a tackle which left him with a slightly angulated midclavicular fracture. (b) The follow-up radiograph at 6 weeks shows the callus forming in the hematoma around the fracture ends. (c) At 12 weeks the fracture has almost healed but still requires remodelling to achieve normal alignment.

Several factors influence the rate at which fractures heal, particularly the type and site of the fracture (upper limb fractures heal faster than lower limb ones, oblique or spiral fractures more quickly than transverse ones) and the age and nutritional state of the patient, e.g. a child with a simple spiral fracture of the humerus may heal in 3 weeks, whereas an elderly patient, an adult with a comminuted fracture or any patient with a fracture at a poor-healing site, such as the lower leg, may take 24 weeks to achieve union.

It will take several weeks/months more for the woven bone to be remodelled by osteoclasts within the bone, and for osteoblasts to lay down lamellar bone along the lines of stress; the patient must be gently mobilised as soon as possible to help this process. Once union has occurred and the patient can weight bear, the lumpy new cortical bone gradually becomes resorbed and smoothed out, and the excess medullary new bone is removed, restoring a normal medullary cavity. Woven bone, which is quite rapidly formed and much less efficient at weight bearing, is resorbed completely and is replaced by lamellar bone. This restoration to normality may take up to a year.

Fortunately, Ed is young and well nourished, and heals quickly. He is back on the rugby pitch 9 months later. He returns to his supermarket job rather more quickly than is sensible, and develops a small incisional hernia in his abdominal scar. It is causing him no problem at present and he has decided to put up with it.

 Read more about fracture clinics in Pathology in Clinical Practice Case 28

HEALING IN OTHER SPECIALISED TISSUES

Central nervous system
Cerebral infarction is a typical example of neuronal loss. Neurons are a 'permanent' tissue and it has always been thought that, once lost, they are gone forever. Encouragingly, there is now some evidence to suggest that a limited degree of regeneration can take place in the hypothalamic–neurohypophyseal system. Scarring is not a feature after necrosis within the CNS – dead brain tissue liquefies and is gradually cleared, leaving empty spaces often known as 'lacunae'. The process stimulates 'gliosis' – the brain's equivalent of fibrosis. This consists of proliferation of glial cells which, together with the ingrowths of capillaries, may form a physical barrier to the regeneration of new neuronal fibres. There is encouraging new work in this field, particularly involving the use of stem cells.

Peripheral nervous system
Severed axons can regenerate. New fibrils sprout from the proximal end of the severed axon, each invaginating the surrounding Schwann cells as they grow, at a rate of about 1 mm/day. If a fibril grows down an existing endoneurial sheath, its function may be recovered.

If the sprouts grow away from the correct pathway, substances secreted by stromal cells in the connective tissue bind to the tips and prevent them growing further. Often, this does not happen and instead the regenerative efforts result in a tangle of fibres embedded in fibrous scar, a 'traumatic neuroma'.

Liver

The liver is an amazing organ, capable of regenerating from half to two-thirds of its volume after an acute insult. In fact it is so good at regeneration that it is possible for a live donor to supply a liver transplant by division of his own liver, and survive the event. Possible, but not particularly recommended!

The liver can regenerate in three ways. In day-to-day insults, in which a single hepatocyte might die, the adjacent cell will divide and take its place. In a more extensive injury, stem cells that lie in the liver tissue next to the portal tracts ('oval cells') proliferate to form new liver; sometimes they are confused and make new bile ducts as well – they can form either. A really major insult, such as a paracetamol overdose, may overwhelm the system and stem cells from the bone marrow may migrate to the site to assist.

The new liver produced is microscopically indistinguishable from the original, except that it contains less age-related pigment (lipofuscin).

WHAT CAN GO WRONG WITH THE HEALING PROCESSES?

When it comes to wound healing, it is not any one thing but rather a complex and dynamic interplay between many factors within an intricate network that determines the final outcome. Failure to heal satisfactorily can be the result of either systemic or local factors.

Chronic inflammation and/or fibrosis may ensue if the initiating stimulus persists or directly activates the adaptive immune system. In some cases (e.g. rheumatoid arthritis, tuberculosis), the chronic inflammatory process may cause severe local tissue destruction. This is discussed further in the next section. The reason is not always apparent, but in some cases it occurs because of persistent damage (e.g. irradiation, or the presence of indestructible irritants such as silica or asbestos) or continual stimulation may be due to the harmful chemicals

History

Friedrich Schwann (1810–82)

Friedrich Schwann (Fig. 5.8) was a German anatomical professor who made two major contributions: he showed that fermentation was associated with living organisms and he developed the cell theory with Schleiden and Müller. He submitted his manuscript describing how cells with their nucleus and protoplasm are the building blocks of all animal and plant tissues to a Catholic bishop for approval before publication.

He discovered the axon sheath cells that bear his name, the striped muscle in the upper part of the oesophagus, the importance of bile for digestion and the enzyme pepsin. He also studied muscle contraction and demonstrated that the tension of a contracting muscle varies with its length.

Figure 5.8 Frederick Schwann (1810–82). (Reproduced with permission from the Wellcome Trust.)

in cigarette smoke. To understand why it is that chronic inflammatory cells are generated as part of the adaptive immune response, read on to Chapter 6.

Failure of wound healing may reflect problems with nutrition or oxygen supply (e.g. anaemia, peripheral vascular disease, post-irradiation ischaemia, pressure sores), or mechanical forces may interfere with immobilisation of the wound site. There may be interference with normal healing processes by drugs or other iatrogenic (medically generated) factors. Massive tissue swelling in some trauma patients can mean that, after emergency surgery, the abdomen cannot be closed for several weeks ('abdominal compartment syndrome'), and must be kept intact and sterile with special dressings.

Read about wound healing problems due to immobility in motor neurone disease in Pathology in Clinical Practice Case 42

At the opposite end of the spectrum, abnormally exuberant and extensive fibrosis may be simply unsightly and annoying, or life threatening.

Fibrosis is the result of excessive, often disordered, collagen deposition following persistent or dysregulated stimulation of the healing response. We now have more understanding of the roles played by inflammasomes, which assemble in response to activation of pattern recognition receptors (PRRs) which have been stimulated by stressor molecules – 'pathogen-associated molecular patterns' (PAMPs; see Table 4.1, page 113). The result is caspase activation and the generation of IL-1β and IL-18, important in wound healing and collagen deposition. Instead of arriving in waves as a wound is repaired, IL-1β and IL-18 are produced continuously.

Sometimes the problem lies with excessive collagen production (e.g. hypertrophic scar formation within the bounds of the original wound site); in other cases, defective remodelling appears to be most important (keloid scars both contain excessive collagen and extend in a crab-like fashion into adjacent tissue, beyond the edge of the original wound site).

The clinical manifestation of disordered fibrosis or fibrous scarring depends on whether this occurs at a localised site (as with keloid scars or Dupuytren's contracture in the hand), affects an entire organ (e.g. cirrhosis of the liver, chronic renal damage or interstitial lung disease) or is generalised throughout the body (e.g. systemic sclerosis).

The effect is usually permanent, leading to impaired function of the tissue involved. Curiously, in some cases, such as idiopathic pulmonary fibrosis (Fig. 5.9),

Read more about idiopathic pulmonary fibrosis in Pathology in Clinical Practice Case 15

disabling scar tissue develops without demonstrable previous inflammation.

It has been estimated that unregulated excessive fibrosis is thought to contribute to almost half of all deaths globally. The administration of agents such as steroids may slow the rate and reduce the extent of a fibrotic process. A range of cytokine antagonists and anti-VEGF and other agents has been used in the prevention and treatment of keloid scars. The person who discovers an effective way to remove established fibrosis must surely win a Nobel Prize!

SYSTEMIC FACTORS THAT INTERFERE WITH EFFECTIVE WOUND HEALING

Nutrition

Deficient protein intake may inhibit collagen formation and so inhibit the regaining of tensile strength. Sulphur-containing **amino acids** such as methionine seem to be particularly important. Lack of **vitamin C** has been found to inhibit the secretion of collagen fibres by fibroblasts and adversely affect the deposition of chondroitin sulphate in the extracellular matrix of granulation tissue. **Vitamin A** has functions in relation to epithelial proliferation and epithelial differentiation, both important in wound healing. A role for **zinc** in wound healing was discovered more or less by accident. In the course of a study on the effects of certain amino acids on wound healing, a phenylalanine analogue that had been expected to impair healing instead accelerated it. Careful study of this analogue revealed that the sample used had been contaminated by zinc. Zinc deficiency, such as is found in patients

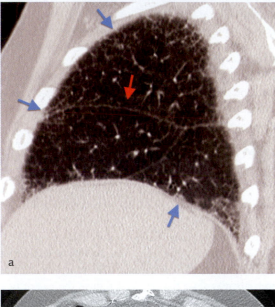

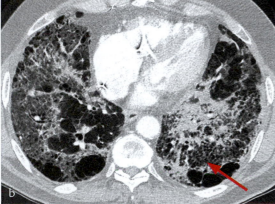

Figure 5.9 Idiopathic pulmonary fibrosis (IPF) – early and late stages, an example of an autoimmune disease with the body's inflammatory mediators attacking pulmonary tissue to an unknown insult. High-resolution CT exquisitely depicts the pathological processes involved at an almost microscopic level. (a) A sagittal reconstruction with fine peripheral lines extending into the lung from the subpleural regions (blue arrows) – early interstitial fibrosis. (b) Unchecked the process results in inflammation producing changes that reflect a chronic process, with architectural distortion of the lung parenchyma produced by marked fibrosis and destruction of lung parenchyma; this produces small dark air spaces – 'honeycombing' (red arrow).

who have been on parenteral nutrition for long periods and in patients with severe burns, is associated with poor healing.

Steroid treatment, chemotherapy and radiotherapy

It is well known that steroids damp down the inflammatory response and are of great use in diseases where this response is causing more harm than good. The effect on healing may be a secondary phenomenon related to this effect on inflammation, and is probably due to a reduction in macrophages entering the wound and, hence, a reduction in macrophage-derived factors important in healing. There may also be a direct effect on fibroblasts to reduce collagen production. Steroids are therefore administered in situations where inappropriate scarring is taking place, such as in interstitial fibrosis in the lung.

Chemotherapy and radiotherapy also reduce the number of circulating monocytes and so probably cause a reduction in wound macrophages. Radiotherapy can lead to the fibrotic stenosis or occlusion of small- to medium-sized blood vessels.

LOCAL FACTORS

Foreign material and/or infection

The presence of infection or of a foreign body will increase the intensity and prolong the duration of the inflammatory response to injury. It is worth remembering that fragments of dead tissue, such as bone, and other elements of the patient's own tissues that have become misplaced, such as hair or keratin, act as foreign bodies which can form a nidus – a focus and breeding point for bacterial or fungal infection.

Poor immobilisation of wound edges

Excess mobility in any tissue will impair healing and prolong the time to full recovery. Anything that leads to undue tension or excess movement at the wound site, such as poor siting of an incision, may interfere with healing.

In bones, movement at the fracture site may cause non-union or the development of pseudarthrosis.

Part 2: Defence against disease

Vascular supply

If the arterial perfusion of a wound site is compromised by stenosis or occlusion, as, for example, in atherosclerosis or radiotherapy, healing may be delayed or completely inhibited. Adequate venous drainage is also important, and impairment of this may play a part in the genesis of chronic ulcers, which often occur on the anterior surface of the legs in elderly patients (Fig. 5.10). Poor oxygenation of adequately perfused tissue, e.g. in severe anaemia, will also impair healing.

Abnormal scar formation

The scar may remain well vascularised and contain an exuberant excess of collagen. A hypertrophic scar bulges from the original wound site, develops immediately and often regresses over time, whereas a keloid scar gradually protrudes out of and then beyond the original site, starting to develop up to a year after the wound has healed. Keloids do not regress.

Both hypertrophic scars and keloids tend to occur in dark-skinned people and particular sites are involved, typically the ear lobes, shoulders and over the sternum, or at a site that is under tension, such as the natural facial creases. The exact mechanisms are not known but there is an immature, hypervascularised collagen matrix present and it may be that there is abnormal expression of regulatory genes controlling interactions between epithelial and mesenchymal cells. Genetic factors may be important.

Scar contractures

This is usually a complication of severe burns or injuries in which a large amount of tissue has been lost. Marked contraction may cause deformities of joints because of involvement of the underlying tendons.

Cirrhosis of the liver

The liver is amazingly good at regenerating after its cells have been killed by one-off insults, e.g. hepatitis A infection or paracetamol overdose. However, if the insult persists over many months and years, as with alcoholic or non-alcoholic fatty liver disease, eventually haphazard bands of fibrous scar tissue are laid down (Figs 5.11 and 5.12). The delicate irrigation system required for normal blood flow and normal liver function is interrupted. Back-pressure in the portal vein

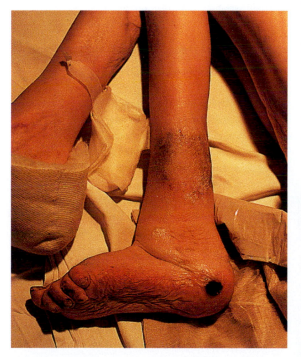

Figure 5.10 Poor blood supply, oedema of the lower limbs, poor nutrition and infection may all contribute to the failure to heal of a pressure sore in a bed-ridden patient.

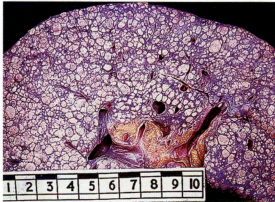

Figure 5.11 Cirrhosis of the liver: the irregular pattern is visible to the naked eye.

Chapter 5: Healing and repair, chronic and granulomatous inflammation

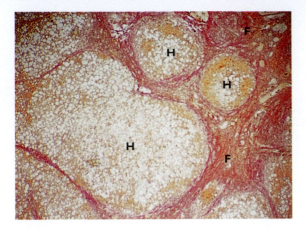

Figure 5.12 Photomicrograph of liver with cirrhosis showing nodules of hepatocytes (H) separated by fibrous bands (F). The hepatocytes are distended with fat, which together with inflammation was the injurious stimulus.

leads to the re-opening of fetal shunts between the portal blood supply to the liver and the systemic veins that drain the rest of the body, sometimes with catastrophic consequences, such as when engorged veins in the oesophagus (oesophageal varices) rupture and can cause death from massive haematemesis.

PROBLEMS SPECIFIC TO BONE HEALING

Non-union is often a result of excess movement in the fracture ends; the gap is bridged by fibrous scar tissue rather than the usual bone-forming process. This can result in a useless bone, which cannot resist the pull of attached muscles. If the fibrous segment is extremely short, a good union may be achieved, at the cost of bone strength.

Pseudarthrosis is analogous to non-union, but here a joint develops, sometimes with a synovial space and cartilaginous lining. Obviously this creates a useless bone.

In mal-union the bony ends are poorly apposed, and may overlap. Mal-union may result in a shortened, deformed bone with a degree of normal function. In other forms of mal-union, adhesions may form between adjacent muscle and the bone, as a result of scarring after inflammation.

In osteomyelitis a focus of infection may persist after a compound fracture. However, osteomyelitis more commonly results from blood-borne spread of bacteria.

IMPROVING WOUND HEALING

The discussion of factors modifying wound healing emphasises that a clean, uninfected, immobile wound with the sides closely apposed in a healthy patient is most likely to heal quickly and neatly. There are several new approaches to wound healing that are under investigation.

Wound healing may be improved by the local use of ultrasound or laser therapy, which are thought to increase vascular permeability.

Synthetic growth factors, applied topically, may stimulate the healing of chronically ulcerated sites and there has been particular interest in the topical application of keratinocyte growth factor and TGF-β.

Stubborn bone fractures may be persuaded to unite by the passage of electrical currents through the bone, and skin wounds may also benefit.

CHRONIC INFLAMMATION

Chronic inflammatory disorders are some of the most common, fascinating, devastating and mysterious diseases to affect humans. They include tuberculosis, sarcoidosis, syphilis, leprosy, Crohn's disease, rheumatoid arthritis, systemic lupus erythematosus (SLE) and the pneumoconioses. Despite the availability of treatment and some information on prevention, tuberculosis remains a significant worldwide problem. For this reason (and because it turns up in exams with frightening regularity!), we discuss tuberculosis in a moment.

First we had better consider what chronic inflammation is; the definition is not as clear-cut as one might like. You might expect that acute inflammation and chronic inflammation are the ends of a spectrum, and that, after a certain length of time has elapsed, an inflammatory process is considered to be chronic. To some extent this is true.

Acute inflammation peaks at about 3 days and resolves within 5 days; if the stimulus persists, chronic inflammatory cells (lymphocytes, plasma cells and macrophages) begin to accumulate at the site, and this is now a chronic inflammatory process. Thus,

Read more about *Heliobacter pyloris* gastritis in Pathology in Clinical Practice Case 20

we are defining chronic inflammation by the type of cell present as well as by the length of time involved. Sometimes the acute phase is so transitory that the disease is not usually detected during this phase and is thought of as a chronic inflammatory process; examples are helicobacter gastritis and tuberculosis.

Viral disease is a special case, because even the acute event is characterised by an infiltrate of lymphocytes. When, then, is the disease classified as chronic? Definitions vary with disease types, but, as an example, viral hepatitis is considered chronic after it has persisted for more than 6 months, or there is fibrous tissue deposition.

Some categories of autoimmune disease are also examples in which chronic inflammation is the starting point – consider T-lymphocyte-mediated diseases such as atrophic gastritis (also known as pernicious anaemia) or coeliac disease.

Curiously, there are instances in which acute inflammatory cells persist, e.g. when neutrophils become walled off by fibrous tissue to form abscesses. Some of you will have experienced a dental abscess, in which bacteria enter the jaw through a cracked or carious tooth and thrive in the oxygen-poor environment of the dental root and surrounding bone. This is an example of chronic suppurative inflammation. Microscopic examination of the abscess would show pus and acute inflammation in the centre of the abscess, with surrounding chronic inflammatory cells and fibrous tissue.

Osteomyelitis is a fascinating example of what the body does when it cannot rid itself of a focus of infection that causes persistent acute inflammation. In this case, infectious organisms such as *Staphylococcus aureus* from skin or wound infections, or *Salmonella typhi* (which causes typhoid and may persist in bile in the gallbladder for months or years) can spread to bone via the blood. You would think that this would be the worst place for a bug to survive, given that bone marrow is where the inflammatory cells are made! A suppurative focus in the bone becomes walled off by new bone formation (the 'involucrum') and multiple sinuses may form between the purulent focus and the skin. As numerous anastomosing canals full of pus form within the bone and these are poorly vascularised, it can be very difficult to cure with antibiotics, and in extreme circumstances the area may need to be excised. A focus of infection within the body always carries the potential to break into the bloodstream and cause septicaemia or the formation of multiple abscesses in far-flung parts of the body, such as the heart valves or other organs.

However, much of the time we find that chronic inflammatory diseases appear to have either a negligible or no acute component and that the disease is typified by a chronic inflammatory cell infiltrate from the start, or the acute inflammatory phase is transient. Some diseases are characterised by recurrent bouts of acute and chronic inflammation, such as the idiopathic chronic inflammatory bowel diseases, ulcerative colitis and Crohn's disease. Gout is an example of a destructive arthritis caused by recurrent episodes of acute and chronic inflammation in response to the precipitation of uric acid crystals within joints.

Rheumatoid arthritis is an inflammatory condition in which there is a transient acute phase, but the key feature is that of abundant plasma cells and lymphocytes, which expand the synovium of joints. This becomes greatly thickened and covered in a thick membrane, or 'pannus', of fibrin and chronic inflammatory cells. The pathogenesis of rheumatoid arthritis is not entirely understood, but about 70% of patients have circulating immunoglobulin IgM antibodies directed against the Fc component of IgG. Why these patients form antibodies against their own body constituents ('autoimmune disease') and why joints and connective tissues are the major targets are some of the puzzles that are still to be wholly resolved (see page 203). The inflammatory process destroys bone, leading to severe joint deformities, typically in the knee, wrists and the small bones of the hands.

We discuss other autoimmune diseases, such as SLE, a little later when antibodies have been explained in more detail. Figure 5.13 details the main types of chronic inflammation and Fig. 5.14 illustrates the radiological effects of chronic inflammation in rheumatoid arthritis.

Read more about Rheumatoid arthritis in Pathology in Clinical Practice Case 38

Chapter 5: Healing and repair, chronic and granulomatous inflammation

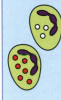

Non-specific: e.g. supervenes after acute infection, as in *Helicobacter*-associated gastritis, characterized by lymphocytes and plasma cells

Persistent viral infection, e.g. hepatitis B, continues to stimulate local immune response in liver

Chronic suppurative: e.g. osteomyelitis, pilonidal abscess—continuing stimulus to neutrophil production and recruitment; often walled off by fibrous tissue and poorly vascularised, so difficult to treat with antibiotics. May require surgical clearance

Eosinophil-rich: often indicates underlying parasitic infection

Granulomatous: usually a response to agents which are difficult to destroy using lysosomal enzymes or lymphocyte-mediated immune responses. Epithelioid or multinucleate giant cells form by fusion, e.g. tuberculosis, reaction to silica or talc, parasitic infection. Sometimes the causative agent is unknown, e.g. Crohn's disease or sarcoidosis

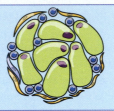

Autoimmune: e.g. rheumatoid arthritis (IgM anti-IgG antibodies in 70 percent), primary biliary cirrhosis (anti-mitochondrial antibodies). Plasma cells may be a prominent feature in the inflammatory infiltrate

Figure 5.13 Main types of chronic inflammation.

Figure 5.14 Chronic inflammation: end-stage rheumatoid arthritis in the left hand and wrist with erosion of the ulnar styloid (green arrow), collapse of eroded carpal bone (red arrow) and disorganisation of eroded, chronically inflamed, metacarpophalangeal and interphalangeal joints (yellow arrows).

Part 2: Defence against disease

GRANULOMATOUS INFLAMMATION

Granulomatous inflammation is a special type of chronic inflammation, characterised by the presence of granulomas. A granuloma is a collection of activated macrophages and is frequently surrounded by a rim of lymphocytes (Fig. 5.15). The macrophages have been modified and are larger than normal macrophages, with more abundant pink cytoplasm. They resemble epithelial cells and are referred to as 'epithelioid' macrophages.

Typical situations in which granulomatous chronic inflammation occurs are summarised on page 174. Two main scenarios are encountered: a response to the phagocytosis of inert foreign material and an immune-driven response. In the former, an inert substance such as a splinter of glass, which cannot be destroyed by the lysosomal contents, causes multiple macrophages to fuse during the attempt to phagocytose and destroy the substance.

Immune-mediated granuloma formation occurs when an antigenic substance, usually microbial, is ingested. Selected components from the microbe are processed by the macrophage and presented via its type II MHC molecules to helper T cells. These become activated and secrete IFN-γ, which activates the macrophage and stimulates it to transform into an epithelioid giant cell, or several macrophages may fuse to form a multinucleate giant cell.

The factors influencing granuloma formation are largely unknown but a variety of cytokines appear to be involved. Animal experiments suggest that IL-1 is important in the initiation of granuloma formation and that TNF is responsible for their maintenance. IL-2 has been shown to increase their size and IL-5 can attract eosinophils to the granulomas, as seen in parasitic disease. IL-6 is believed to have an important role in tuberculous granulomas. We discuss tuberculosis (TB) as an example of a chronic granulomatous inflammatory disease. Depending on the cytokines in the local environment, macrophage, dendritic or B-cell antigen-presenting cells (APCs, more on this in Chapter 6) may facilitate Th1 or Th2 helper T-cell responses.

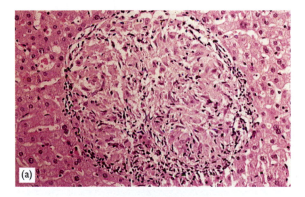

(a)

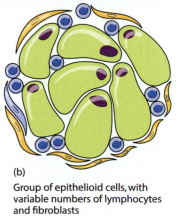

(b)

Group of epithelioid cells, with variable numbers of lymphocytes and fibroblasts

Figure 5.15 (a) Non-caseating granuloma from a patient with sarcoidosis. (b) Components of a non-caseating granuloma.

History

Robert Koch (1843–1910)

The great German bacteriologist, Robert Koch, was the first to show that tuberculosis is an infective disease, a fact that we now take for granted. Koch's original investigations (1876) were performed with anthrax, which he demonstrated to be the cause of what was then known as 'splenic fever'.

On 24 March 1882, Koch announced the discovery of the tubercle bacillus to the Berlin Physiological Society. His work proved that 'pulmonary consumption' was not a disorder of nutrition but an infective disease that ran a chronic course. Although Koch is best remembered for his contribution to tuberculosis, he did not confine his interests to that disease. He discovered *Cholera vibrio* which had created havoc from India to Egypt, investigated bubonic plague in India, researched diseases caused by the tsetse fly in East Africa and studied malaria in Java.

'Koch's postulates' are still referred to today (Page 166).

Chapter 5: Healing and repair, chronic and granulomatous inflammation

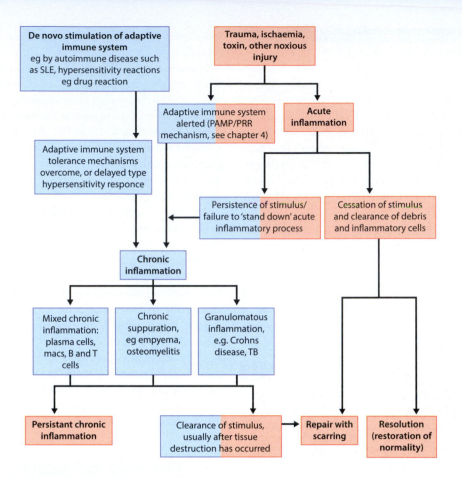

Figure 5.16 Chronic inflammation becomes clinically apparent if the initial inflammatory stimulus persists. This diagram compares 'de novo' chronic inflammation with that following acute inflammation which has failed to remove the offending agent.

Key facts

Koch's postulates

It was clear to Koch that in order to implicate a particular organism as the cause of a disease, he must:

- demonstrate the organism in the lesions in all cases of that disease
- be able to isolate the organism and cultivate it in pure culture outside the host
- produce the same disease by injecting the pure culture into a healthy individual.

These three criteria are known as Koch's postulates.

TUBERCULOSIS

Tuberculosis (TB) is a worldwide problem and it is caused by *Mycobacterium tuberculosis*. Two strains infect humans: *M. tuberculosis hominis* and *M. tuberculosis bovis*. Bovine tuberculosis is passed from cattle to humans in milk so that it enters via the gastrointestinal tract to produce abdominal TB. It is uncommon in developed countries now that dairy herds are generally free of mycobacteria and milk is pasteurised. Infection with *M. tuberculosis hominis* is the common form and it usually produces pulmonary TB.

Patients with TB may present with a cough producing blood-stained sputum, or more subtly, with night sweats, weight loss and vague symptoms of ill-health. Figure 5.17 illustrates some manifestations of TB.

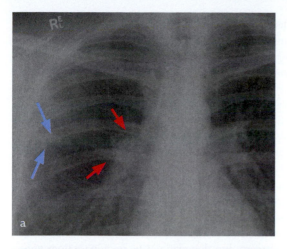

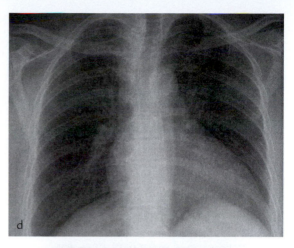

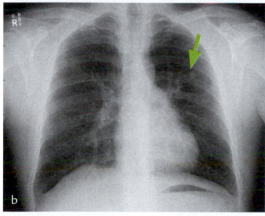

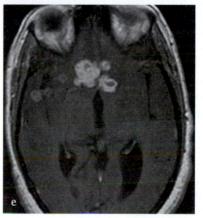

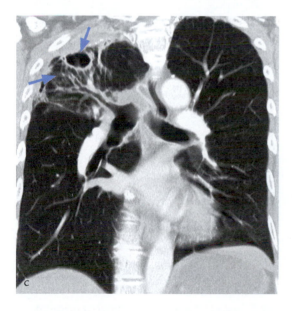

Figure 5.17 Manifestations of TB: (a, b) the Ghon complex: an area of tuberculous consolidation in the left midzone (blue arrows) associated with an enlarged hilar node (red arrows). This may heal spontaneously without treatment, with calcification of the TB nodule (green arrow). (c) Secondary or reactivated TB: TB after the primary infection often lies dormant and reactivates, most often in the lung apices. Coronal CT shows a cavity within an area of inflammatory change at the right apex associated with fibrosis and pleural thickening (blue arrows). This represents progression of secondary TB to cavitary fibrocaseous disease. (d) Miliary TB: TB may disseminate throughout the body via the blood and present in the lungs with 'millet seed'-type spots scattered throughout the lung. This type of TB infection is associated with a high mortality. (e) TB meningitis with MRI showing multiple meningeal masses with TB granulomas scattered through the meninges and CSF spaces. Patients often present with fevers and headache but may also have cranial nerve palsies and visual loss.

Chapter 5: Healing and repair, chronic and granulomatous inflammation

The WHO declared TB to be a public health emergency in 1993 and in 1995 set targets for reducing its global impact by 2015. In 2012, deaths had fallen by 45% and the prevalence of TB had fallen by 37%, against a 2015 target of 50%.

What is the TB disease burden? Globally, 8.6 million people developed TB in 2012 and 1.3 million people died of TB that year; approximately 13% of those with TB were HIV positive. This can be compared with 14 million people with TB in 2005, of whom 12% were HIV positive. In some countries the incidence of co-infection of TB and HIV is very high, e.g. 43% of TB patients in the WHO African region tested positive for HIV. This has profound implications for treatment success: 73% cure rate in TB/HIV versus 87% in those with just TB.

Both TB and HIV are treatable diseases and it is an outrage that so many people still die because of a lack of access to the appropriate medical services for diagnosis, be it due to a lack of resources or to war or other conflict, or to the resources needed to fund medication and oversee patient compliance (i.e. ensuring that the drugs are taken appropriately and that the course is completed).

TB is seen in only 5/100 000 of the population in the UK white ethnic group and is generally seen as reactivation of TB in migrants to the UK, born in countries with a high prevalence of TB. The disease also flourishes in socially deprived areas, and poverty and malnutrition appear to be important predisposing factors. There is a higher incidence in men and in people with alcohol problems, chronic lung diseases and conditions causing immunosuppression, e.g. cancer and AIDS.

About 12% of deaths from TB are in patients who are also infected with HIV. TB is the major cause of death worldwide in patients with HIV/AIDS; however, in those countries in which AIDS patients have access to highly active antiretroviral therapy (HAART), the lifespan can be normal and the cause of death is usually pneumonia, often streptococcal. The risk of contracting TB is much higher in HIV/AIDS patients because the organisms act synergistically: both infect and proliferate in macrophages and HIV also infects helper T cells (CD4+ Th cells). TB assists HIV by stimulating a T-cell-mediated immune response; when the Th cells proliferate, so does HIV virus and this leads to Th-cell death. HIV infection depresses immune system function by causing the death of Th cells, which makes it difficult for the body to generate a T-cell-mediated immune response against TB. CD4+ Th cells are at the heart of all immune responses (see Fig. 6.13).

M. tuberculosis is a slender, rod-shaped organism, approximately 4 μm in length. It is not visible on haematoxylin and eosin-stained sections but is stained using the Ziehl–Neelsen method (Fig. 5.18). This reaction relies on the fact that, once stained, the organisms are resistant to decolourisation with acid and alcohol (hence 'acid- and alcohol-fast-bacilli', AAFB). Mycobacteria grow very slowly in culture and may not be apparent for 6–12 weeks. Observing the bacilli in excised tissues will allow faster diagnosis and treatment, but they will not be seen in sections unless there are approximately a million bacteria per millilitre of tissue. Use of a polymerase chain reaction to detect mycobacterial nucleic acid is both reasonably fast and more sensitive, but technically more difficult and not generally available.

M. tuberculosis does not possess any toxins with which to harm its host, although a number of cell membrane glycolipids and proteins, including bacterial stress proteins, act to provoke a hypersensitivity reaction. It is the hypersensitivity reaction that causes the tissue destruction so characteristic of this disease.

PRIMARY TUBERCULOSIS

This occurs in individuals who have never been infected with *M. tuberculosis*.

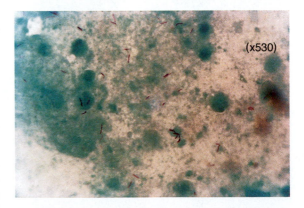

(x530)

Figure 5.18 Ziehl–Neelsen stain showing acid-fast (red) bacilli.

CLINICAL SCENARIO: TUBERCULOSIS

Let us consider the case of Amanda O'Brain, an Irish junior doctor. For some weeks or months her colleagues have subliminally absorbed the sound of her chronic dry cough into their collective consciousness, joking that they can recognise that she is coming by the sound of her footsteps and persistent slight cough. She has been too busy to take too much notice – the cough is barely productive of sputum and she has not felt particularly breathless. However, recently she has felt more tired than usual and one night she feels feverish, with a drenching sweat that soaked her bedsheets. She takes her temperature and finds it has spiked to 38°C, only to return to normal in the morning. Alarmed, because she knows that this phenomenon can be a 'B' symptom of lymphoma, she reports to the occupational health consultant in her hospital. A chest radiograph shows mottled shadowing in both lungs, radiologically suggestive of TB. She manages to cough up some sputum, which has a faint tinge of blood, and this is sent for microscopy and culture. A fluorescent antibody test for TB shines clearly positive in the direct microscopy sample and a diagnosis of TB is made. A further sputum sample is sent for subtyping and culture is undertaken to determine the organism's sensitivity to antibiotics. Fortunately, the organism is sensitive to the standard triple therapy of isoniazid, rifampicin and streptomycin, and she starts treatment.

TB, in common with several other infections, is a notifiable disease. The nightmare of tracing Amanda's contacts begins. She has seen hundreds of patients in the months that have elapsed since she first developed her persistent cough. Amanda is a gregarious individual with many friends and an extensive family. All must be tracked down and, if appropriate, offered treatment along with their immediate contacts.

TB is almost always contracted by inhaling aerosolised *M. tuberculosis*, coughed up by a person with 'open' TB.

Inhalation of the organism first produces a small lesion (approximately 1 cm in diameter), usually in the subpleural region in the lower part of the upper lobe or the upper part of the lower lobe of the lung. This is referred to as the Ghon focus. Lesions occur in these sites because the bacterium is a strict aerobe and prefers these well-oxygenated regions. When the tissue is first invaded by the mycobacterium, there is no hypersensitivity reaction, but an initial, transient, acute, non-specific, inflammatory response with neutrophils predominating. This is followed rapidly by an influx of macrophages, which ingest the bacilli and present their antigens to T lymphocytes, leading to the proliferation of a clone of T cells and the emergence of specific hypersensitivity. The lymphocytes release lymphokines, which attract more macrophages. These accumulate to form the characteristic granuloma, containing a mixture of macrophages, including epithelioid cells and Langhans-type giant cells. Tissue destruction leads to necrosis in the centre of the granuloma called caseous necrosis (Figs. 5.19 and 5.22) because, macroscopically, the necrotic area resembles cheesy material! Tubercle bacilli, either free or contained in macrophages, may drain to the regional lymph nodes and set up granulomatous inflammation, causing massive lymph node enlargement. The combination of the Ghon focus and the regional nodes is called the primary complex (Fig. 5.20).

Thus, the development of hypersensitivity results in tissue destruction, but also improves the body's resistance to the mycobacterium by promoting phagocytosis

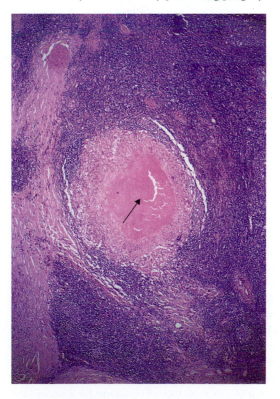

Figure 5.19 Photomicrograph showing lymph node with caseous necrosis (arrow).

Chapter 5: Healing and repair, chronic and granulomatous inflammation

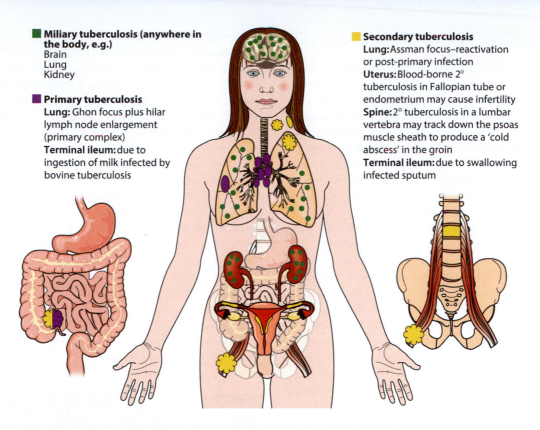

■ **Miliary tuberculosis (anywhere in the body, e.g.)**
Brain
Lung
Kidney

■ **Primary tuberculosis**
Lung: Ghon focus plus hilar lymph node enlargement (primary complex)
Terminal ileum: due to ingestion of milk infected by bovine tuberculosis

■ **Secondary tuberculosis**
Lung: Assman focus–reactivation or post-primary infection
Uterus: Blood-borne 2° tuberculosis in Fallopian tube or endometrium may cause infertility
Spine: 2° tuberculosis in a lumbar vertebra may track down the psoas muscle sheath to produce a 'cold abscess' in the groin
Terminal ileum: due to swallowing infected sputum

Figure 5.20 Typical sites involved by primary, secondary and miliary tuberculosis.

and reducing intracellular replication of bacilli. It is not known why the granulomatous response to mycobacteria produces caseation, whereas most other granulomatous reactions do not. Another puzzle is why the attraction of macrophages in most inflammatory responses produces a dispersed infiltrate, whereas in granulomatous reactions they form well-demarcated collections – granulomas.

To return to our patient with TB: as stated above, in about 90% of otherwise healthy people the primary Ghon complex (Ghon focus plus enlarged hilar lymph nodes) will heal. There will be replacement of the caseous necrosis by a small fibrous scar, and the lesion will be walled off. Calcification may also occur in these lesions. Despite this, the mycobacterial organisms may survive and lead to reactivation infection at a later date, especially if the host defences become lowered, as can occur with cancer or steroid treatment for diseases such as rheumatoid arthritis.

There are, however, alternative outcomes. If the hypersensitivity reaction is severe, it will lead to a florid inflammatory response and the patient may present with a systemic illness. If the caseous necrosis is extensive, the tissue destruction may erode major bronchi and allow airborne spread of organisms to produce satellite lesions in either lung. Alternatively, tubercle bacilli may enter the bloodstream. If they enter a small pulmonary arteriole, then the bacilli will lodge in lung tissue. However, if they enter a pulmonary vein, the bacilli may disseminate throughout the systemic circulation. If this occurs, numerous small granulomas may be encountered in almost any organ, including the meninges, kidneys and adrenals. This type of disease is called miliary TB (Fig. 5.17d), so called because the lesions look like scattered millet seeds. This is a disastrous complication, associated with a high risk of death. Fortunately, systemic spread is not a common event in primary disease.

SECONDARY TUBERCULOSIS

Secondary, or post-primary, TB refers to infection occurring in a patient previously sensitised to the mycobacterium. Most of these cases are due to reactivation of latent mycobacteria following an asymptomatic primary infection (Fig. 5.17c). The latency period can vary tremendously and reactivation may not occur for many decades. Reactivation of latent TB tends to occur if the immune system becomes compromised, as happens in patients treated with steroids or immunosuppressives (e.g. transplant recipients, patients with autoimmune disease, sarcoidosis or chronic inflammatory bowel disease). A particular risk group is patients with HIV/AIDS. Some immune modulation drugs that depress T-cell function have been linked with the reactivation of TB. (Occasionally there is reinfection – usually with a new strain of TB from an outside [exogenous] source.)

The favoured site for reactivation is the upper lobe of a lung, which is thought to be due to the higher oxygen concentration in this part. (If you remember, tubercle bacilli are obligate aerobes.) Macrophages and other inflammatory cells are programmed to work in inflamed, often poorly oxygenated, tissues and perform less well if the oxygen concentration is higher than normal. This focus of reactivation in the upper lobe is called the Assman focus. Although the reasons for this are far from clear, it is believed that, at the time of primary infection, some tubercle bacilli spread to other parts of the lung and body via the bloodstream. Most die off as a result of the immune response but a few may survive. Recent studies suggest that the bacillus can survive within macrophages in several ways. First, it may interfere with the fusion of the phagosome with the lysosome. This appears to happen if the tubercle bacillus can form protein kinase G (PKG).

Second, it may become dormant in response to low oxygen states. Low levels of secretion of nitric oxide (NO) by macrophage cell-signalling pathways appear to induce tubercle bacillus dormancy genes (high levels cause the organism to be stressed).

There are three main outcomes of secondary tuberculous infection: healing, cavitation or spread. As there has been previous infection, the host will have some degree of immunity. If this is sufficient, the reactivated infection will heal with scarring and subsequent calcification. Intervention with anti-tuberculous drugs will also enhance and modify the healing process. If, on the other hand, the host's degree of hypersensitivity is high and/or the organisms are particularly virulent, there may be considerable lung tissue destruction and caseous necrosis (Fig. 5.21), which can lead to the formation of a cavity. The infected caseous material may spread through destroyed tissue or via bronchi to adjacent parts of the lung, and extend the local disease. Patients with cavitation and bronchial erosion are usually highly infectious, because they can cough up large numbers of organisms. (In survivors, the cavity may be colonised by fungi such as *Aspergillus* spp., which may later invade the lung and disseminate via the blood.) The pleura may become involved, with the production of an effusion, which can contain caseous and necrotic material, a tuberculous empyema. Infected sputum may be swallowed, spreading the disease to the gastrointestinal tract. Systemic spread to produce miliary TB is more common in post-primary TB because tissue destruction is greater and there is more likelihood of venous disruption and dissemination of organisms.

The natural history of the disease will depend on the host's immunity, hypersensitivity and factors such as nutritional status, associated disease, previous BCG (Bacille Calmette–Guérin) inoculation and the treatment of active disease with anti-tuberculous drugs. TB is a treatable disease, yet, despite this, it remains a major problem in less developed countries and there is a worrying trend towards the development of antibiotic-resistant strains in developed countries. Triple therapy, using three antibiotics to

Figure 5.21 Cut surface of the lung showing white areas of caseation due to reactivation of tuberculosis in the upper and lower lobes.

Chapter 5: Healing and repair, chronic and granulomatous inflammation

which the organism is sensitive, should be used in order to prevent the emergence of resistant strains.

Inoculation with the BCG vaccine greatly reduced the incidence of TB meningitis in children with primary infection, and also the risk of developing post-primary TB, although it has less effect on reactivation TB. BCG does not prevent the organism from causing an initial infection. Its function is to simulate a primary infection, thus priming the immune system for a rapid response within days of encountering the organism, unlike the 3–4 weeks required to generate new cell-mediated immunity. Interestingly, governments such as those in the UK and Australia no longer offer BCG vaccination, except to those who are highly likely to be exposed to TB in childhood.

It is important that those people with active TB are isolated until they have been treated, because it has been estimated that one untreated individual spreads the disease to an average of twelve other people.

Traditional anti-tuberculous therapy requires the use of multiple drugs for 6 months to 1 year. Many patients fail to complete the course, partly because it is so long and also because some drugs have side effects such as loss of colour vision and tinnitus (buzzing and ringing in the ears). A worrying feature has been the emergence of drug-resistant and multi-drug-resistant strains of TB; the latter is defined as TB that is resistant to the two most important anti-tuberculous drugs, rifampicin and isoniazid. Multi-drug resistance is a particular problem in Russia.

The WHO launched a drive (the 'Stop TB strategy') to eradicate TB in 2006 and recommended treatment using 'directly observed therapy, short-course' (DOTS). This has been found to be highly effective with an average success rate of around 75–84%.

SARCOIDOSIS

Sarcoidosis is a baffling systemic disease of unknown aetiology and is characterised by the presence of granulomas which, unlike TB, do not exhibit caseous necrosis and so are termed 'non-caseating granulomas' (Fig. 5.22).

It is a systemic disorder of variable severity, which is why it can present in numerous ways. Many patients

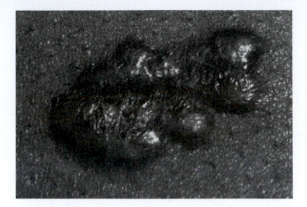

Figure 5.22 Raised skin lesion in sarcoidosis. Microscopy would show epithelioid granulomas.

are asymptomatic and in some the diagnosis is only made *post mortem*. Almost every organ in the body may be affected, the most common being lung, liver, spleen, skin and salivary glands, with the heart, kidneys and CNS slightly less commonly affected. Most patients present with respiratory symptoms (shortness of breath, haemoptysis and chest pain), but some have a more rapid course with fever, erythema nodosum and polyarthritis (Fig. 5.23). Patients may also present with signs and symptoms of hypercalcaemia, and lytic bone lesions, especially in the phalanges, are strong supportive evidence for the disease.

OTHER TYPES OF CHRONIC GRANULOMATOUS DISEASE

Some other causes of granulomatous inflammation are listed below. Many are due to infections (if eosinophils are also present, parasitic disease is most likely), some are of autoimmune aetiology and sometimes we do not know the cause, as in the cases of sarcoidosis and Crohn's disease. A tissue biopsy can help when trying to establish the cause of granulomatous disease. The only way to be certain of the cause is either to see it under the microscope, as, for example, with fungal hyphae or acid- and alcohol-fast bacilli, or to culture the organism. Some macrophage variants are more common in specific conditions and the type seen may offer a clue to the underlying cause. We discuss the origin and activity

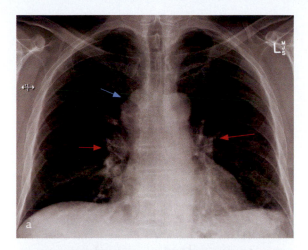

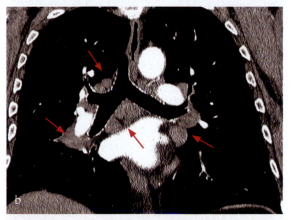

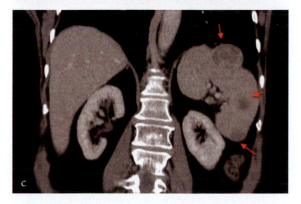

Figure 5.23 Manifestations of sarcoidosis. (a) Chest X-ray showing bilageral enlarged hilar nodes (red arrows) and an enlarged right paratracheal mediastinal node (blue arrow). (b) Coronal CT showing extensive mediatinal lymphadenopathy (red arrows). (c) Systemic sarcoidosis with sarcoid granulomas in the spleen (red arrows).

of this remarkable cell in the next section, but for now consider briefly the ways in which macrophages may adapt to fight particular diseases, becoming morphologically distinctive in the process.

Having read about acute inflammation, chronic inflammation and a little about the immune system, you will appreciate the complex interplay that exists between cells and mediators. It is a veritable orchestra! The helper T cell is probably the conductor, but a key player (the first violin, to continue the analogy) is undoubtedly the macrophage and it is useful to summarise inflammation and repair by reviewing macrophage function.

Read more about granulomatous disease in Pathology in Clinical Practice Case 16

MACROPHAGES

Macrophages are involved in all inflammatory processes due to their ability to phagocytose particles, process and present antigens, and secrete an array of mediators. Macrophages are derived from the same stem cells in the bone marrow that give rise to polymorphonuclear leukocyte precursors. These marrow cells produce monocytes that circulate in the blood for a day or two before migrating into the tissues, where they are called macrophages or (a more old-fashioned term) histiocytes. Macrophages can proliferate, slowly, within the tissues. Their role is a mixture of simple scavenger, litter collecting and generally keeping the extracellular tissues tidy, and they scout for trouble in the form of invading pathogenic organisms.

Apart from the 'free-range' tissue macrophages there are also populations of relatively fixed macrophages in the epithelium (e.g. Langerhans' cells in skin) and lining the endothelial aspects of vessels in the liver (Kupffer cells), spleen, bone marrow and lymph nodes – the so-called reticuloendothelial system (RES). Other macrophages of the RES line part of the CNS (microglial cells).

Particularly in the spleen, sinusoidal macrophages remove from circulation those red blood cells that have become altered or damaged. The life span of a normal red cell is 120 days. This can be markedly

Chapter 5: Healing and repair, chronic and granulomatous inflammation

Key facts

Some causes of granulomatous inflammation

Bacterial	Tuberculosis
	Leprosy
	Syphilis
Fungal	Cryptococci
	Coccidioides sp.
Protozoal	Toxoplasmosis
	Pneumocystis sp.
Parasitic	Schistosomiasis
Inorganic material	Silicosis
	Berylliosis
	Foreign material
Autoimmune	Rheumatoid arthritis
	Primary biliary cirrhosis
Unknown	Sarcoidosis
	Crohn's disease

possible to vaccinate the patient against *H. influenzae* type b (Hib), *Neisseria meningitidis* and *Streptococcus pneumoniae* 2 weeks before the removal of the spleen, it is advised that this be carried out 2 weeks after surgery. Annual vaccination against the prevalent strain of influenza is also recommended.

A crucially important function is a macrophage's ability to scavenge and phagocytose; it does this in its resting state and also in areas of tissue damage. Macrophages possess scavenger receptor molecules, which are important in self-/non-self-discrimination and appear able to bind to a wide range of modified molecules. These include modified low-density lipoprotein (LDL) and modified albumin, and other protein molecules can also be bound and phagocytosed. Unlike other tissue cells that bear LDL receptors, in macrophages these surface receptors are not down-regulated as the LDL content of the cytoplasm increases. This means that macrophages keep taking up lipid, etc. and become foamy macrophages, seen in areas of tissue damage and atheromatous plaques (see page 184).

Just like the neutrophils in acute inflammation, macrophages emigrate and become activated under the

reduced in inherited diseases of the red cell membrane (e.g. hereditary spherocytosis, in which the red cells are round and incompressible, or sickle cell anaemia in which a point mutation alters the shape of the red cell under conditions of poor oxygenation, and causes sludging of the sickled cells within blood vessels) or autoimmune diseases, in which antibodies are produced against the body's own red cells. The sinusoidal macrophages also phagocytose bacteria that have been opsonised with antibody. A common misconception is that the spleen is redundant and can be removed with impunity: splenectomy patients are particularly susceptible to severe infection by encapsulated bacteria, such as pneumococcal or meningococcal infection or *Haemophilus influenzae*. Splenectomised patients must receive lifelong antibiotic therapy. In children, daily antibiotics are given. Adults must keep a course of antibiotics available, because even relatively mild infection may cause fatal sepsis or meningitis without the spleen to screen for, and kill, organisms in the blood. If it has not been

Key facts

The role of the macrophage in chronic inflammation

Antigen presentation to T cells and B cells leading to clonal expansion

Scavenging abnormal cells, particles, large molecules, immune complexes, etc.

Chemotactic factors for other leukocytes and fibroblasts

Stimulate endothelium for adhesion molecule activation and granulation tissue formation

Fusion with other macrophages to form giant cells

Phagocytosis and killing of some organisms

Harbouring of some organisms

Exocytosis for attacking parasites or damaging tissues

Macrophage products involved in tissue injury

Toxic oxygen metabolites

Nitric oxide

Proteases

Eicosanoids

Coagulation factors

Neutrophil chemotactic factors

influence of chemotactic factors, adhesion molecules, cytokines and microbial components. An activated macrophage increases its size and its lysosomal enzyme content and speeds up its metabolism and ability to phagocytose and kill microbes. When macrophages become hyperactivated they start to secrete TNF, which is a multifunctional cytokine that can activate other immune cells or cause virally infected or tumour cells to die by apoptosis.

Hyperactivation is a red-alert status that is usually achieved when microbial molecules such as lipopolysaccharide (LPS), found in Gram-negative bacteria, or mannose, found in many Gram-positive bacteria and other organisms, are encountered.

Macrophages are also important because they synthesise and secrete factors that promote granulation tissue formation and enhance healing and repair, e.g. macrophage-derived growth factors such as PDGF, FGF and TGF-β stimulate cell growth, angiogenic factors stimulate new vessel formation, and fibrogenic stimuli and collagenases permit the remodelling and laying down of scar tissue. This, together with factors released from other cells, results in new vessel formation, and fibroblasts for collagen and extracellular matrix production.

After acute, transient insults macrophages will assist in healing and repair and then depart, continuing their nomadic, scavenging lifestyle. However, in **chronic inflammation** (see pages 132, 176), macrophages produce a variety of cytokines that maintain the inflammatory process and attract and stimulate other inflammatory cells, contributing to local tissue injury

and fibrosis. This is potentially harmful to the body. Their continued presence is due principally to continued emigration from the blood, a reduction in movement out of the tissues and some local proliferation, all of which is in response to cytokines, inflammatory mediators or microbial products. They are naturally longer-lived cells than the neutrophils; they can live for weeks or months in the tissues, whereas neutrophils die within hours of activation and phagocytosis.

MACROPHAGES IN SPECIFIC DISEASES

Epithelioid cells occur in all types of granulomatous disease. They have some resemblance to epithelial cells because they possess abundant cytoplasm, packed with endoplasmic reticulum, Golgi apparatus and vesicles. Thus, the cells are well adapted to synthesise macrophage products, such as arachidonic acid metabolites, complement, coagulation factors and cytokines. However, they are less mobile and less proficient at phagocytosis than ordinary macrophages. The multi-nucleate giant macrophages principally form by fusion of epithelioid cells. Each may have 50 nuclei or more and it is the arrangement of these nuclei that distinguishes the types. The Langhans' giant cell is typical of (although not exclusive to) TB or sarcoidosis, and has its nuclei arranged as a horseshoe at the periphery. The foreign-body giant cell, predictably, often contains identifiable foreign material (e.g. suture material or shards of glass) in its cytoplasm. Its nuclei are randomly arranged throughout the cell. Foreign-body giant cells are also often seen in parasitic infections, such as schistosomiasis ('bilharzia').

These are the most important macrophage variants but, for completeness, we will mention two others (Fig. 5.24). These are the Warthin–Finkeldey cell, pathognomonic of measles and characterised by the presence of eosinophilic nuclear and cytoplasmic inclusions, and Touton giant cells, which have a central cluster of nuclei surrounded by foamy lipid-laden cytoplasm. Touton cells occur in xanthomas, which are benign tumorous collections of lipid-laden macrophages in the skin.

Chapter 5: Healing and repair, chronic and granulomatous inflammation

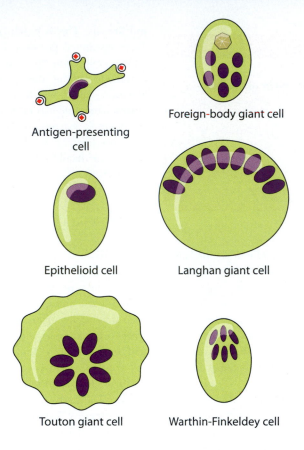

Antigen-presenting cell

Foreign-body giant cell

Epithelioid cell

Langhan giant cell

Touton giant cell

Warthin-Finkeldey cell

Figure 5.24 Some types of macrophage: tissue macrophages (macs) tend to work mainly as scavengers, clearing detritus from dead cells or foreign material from the tissues. Examples are Touton giant cells or foreign-body giant cells, which contain cholesterol/lipid or foreign material, respectively. Some tissue macs have more specific functions, e.g. splenic sinusoidal macs, which remove old or damaged red blood cells from the circulation. If pathogens are detected by the pattern recognition receptors (PRRs) of macs, they become activated, enlarge and secrete cytokines, which attract and activate other cells and present antigen to T cells. Activated macs may become 'epithelioid' cells, which can aggregate to form granulomas.

Part 2: Defence against disease

Case study: alcoholic liver disease

Clinical

A 60-year-old man with long-standing history of alcohol abuse presented with a 3-week history of general malaise, weight loss, loss of appetite and productive cough.

Examination

He was noted to be short of breath with an increased respiratory rate and was jaundiced. He had a fever and tachycardia of 110 beats/min. He also had supraclavicular lymphadenopathy, and a mildly enlarged and tender liver. Auscultation of his chest revealed coarse crackles over both his lung fields.

Pathological

He had a long history of alcohol abuse and this predisposes to many illnesses. People with alcohol problems tend to be malnourished because they derive most of their calories from alcohol and are therefore deficient in many vitamins, especially the B group. They damage the liver by episodes of hepatitis, which heal with scarring, producing fibrosis and eventually cirrhosis. They are also predisposed to infections because of depression of the immune system by the alcohol, and, in this man, clinical examination revealed signs of a chest infection. TB is a particular problem in people with alcohol problems.

Investigations

The liver function tests (LFTs) were abnormal with a raised bilirubin of 45 μmol/L and a raised γ-glutamyl transpeptidase (γ-GT) of 90 IU/L (IU = international units). His chest radiograph showed bilateral consolidation with a small right pleural effusion. A lymph node and liver biopsy were carried out.

The lymph node showed numerous caseating granulomas with calcification. The Ziehl–Neelsen stain showed abundant acid-fast mycobacteria.

The liver biopsy showed an acute alcoholic hepatitis with marked fatty change and liver fibrosis, but without cirrhosis.

Management and progress

He was started on anti-tuberculous therapy and was counselled for his alcohol abuse.

While in hospital, he had a bout of abdominal pain and diarrhoea. Endoscopy showed gastritis but sigmoidoscopy was unremarkable. The rectal biopsy revealed the presence of amyloid in the mucosa. He was discharged on anti-tuberculous therapy and sent for rehabilitation for his alcohol abuse, but he defaulted from his appointments and was lost to follow-up.

The LFTs were in keeping with alcoholic damage, with raised bilirubin and liver cell enzymes. The chest radiograph confirmed a pneumonic process involving both lungs.

Caseous necrosis with granuloma formation is a classic picture of TB and special stains revealed mycobacteria. Immune suppression due to alcohol abuse is responsible for the reactivation of secondary TB. In areas of necrosis, calcification is common and this type is called dystrophic calcification. The serum calcium levels are normal, as opposed to those seen in metastatic calcification, which are raised.

His liver showed the classic fatty change common in alcohol abuse with some hepatitis, i.e. inflammation of the hepatocytes with liver cell necrosis. The result is healing by scarring with resultant fibrosis. He did not have cirrhosis, which is irreversible, unlike fatty change which is.

People with alcohol problems are predisposed to gastritis, i.e. inflammation of the gastric mucosa. They may also have gastric ulcers. Other complications include oesophageal varices (if portal hypertension has developed due to cirrhosis). Patients with long-standing chronic inflammatory diseases such as rheumatoid arthritis and TB are prone to reactive amyloidosis (AA). Rectal biopsy is a good way of diagnosing amyloid. The amyloid may be responsible for the diarrhoea but infective causes should be excluded.

Lack of compliance is a common problem with people with alcohol problems.

🔗 Read more in Pathology in Clinical Practice Cases 19 and 47

Whereas the acute inflammatory response, properly choreographed, is a normal function of injury and happens before healing takes place, chronic inflammation is abnormal. At the end of Chapter 5 we discussed several types of chronic inflammation and referred to the adaptive immune system. We know which cells are important for this refined, highly targeted system – the T and B cells (lymphocytes), plasma cells and, of course, linking the innate and adaptive immune systems, the macrophages and their vigilant cousins, dendritic cells.

In Chapter 4 we discussed how the acute inflammatory response is generated by the innate immune system, as an immediate response to a variety of infective or non-infective inflammatory stimuli. Usually, triggering the innate immune response is all that is required to remove the stimulus and healing and repair, discussed in Chapter 5, take place.

If needed, further assistance is offered by the more targeted adaptive immune response, which is already hovering in the wings like an understudy in a Shakespearian play. This is because the detection of exogenous or endogenous 'stressors' (see Table 4.1) leads to the immediate activation of innate inflammation and simultaneously triggers the adaptive immune response (Fig. 6.1).

Dendritic cells engulf and process antigen and migrate from the epithelium to the regional lymph nodes (e.g. Langerhans' cells from the epidermis of the skin),

processing the antigen and then presenting small peptide chains via their major histocompatibility complex (MCH) II receptors to T-helper (Th) cells. Macrophages from the lamina propria or dermis do likewise. Both types of cells release cytokines such as interferon γ(IFN-γ),which will recruit other macrophages and lymphocytes to the site of damage or microbial invasion. B cells can recognise and present to T cells a much wider range of substances than the phagocytes: material such as lipopolysaccharide (LPS), lipids, chemicals and polysaccharide. T cells can recognise only protein molecules, presented to them on activated B cells or other antigen-presenting cells (APCs) via their MHC class II receptors.

There are two main drawbacks to the innate immune system: first, there is often considerable associated 'collateral' tissue damage as a result of the release of toxic chemicals and enzymes by neutrophils and macrophages, as they phagocytose or die. Second, there is no generation of 'memory' and the same organism, e.g. the skin pathogen *Staphylococcus aureus*, will provoke the same sequence of events in a subsequent infective episode.

Here we discuss the highly orchestrated events that result in specific, highly targeted immunity against an antigen by the adaptive immune response. This is a combination of humoral immunity (B cells that differentiate into plasma cells and produce antibody) and cell-mediated immunity (T cells), designed for

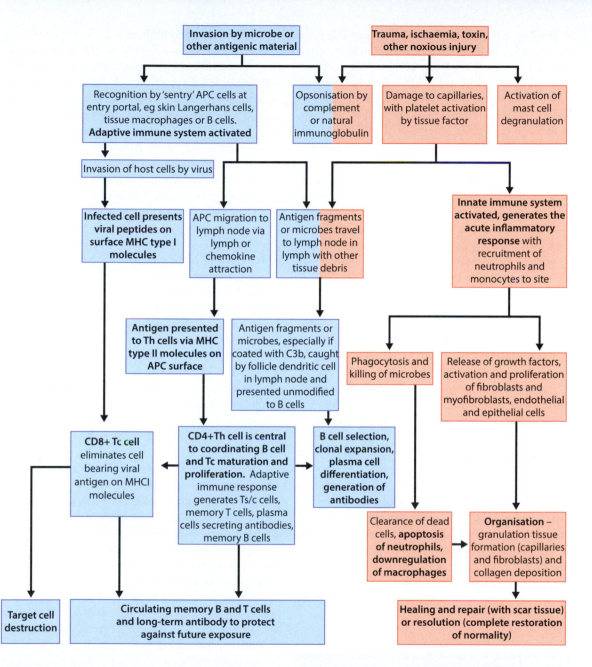

Figure 6.1 Algorithm of events in inflammation. Innate immune system events in pink boxes, acquired immune system in blue boxes.

intracellular organisms. Adaptive immunity has two main components – first the identification of a pathogenic stimulus and second the generation of cells primed to deal with the stimulus. These take place in 'inductive' sites, such as the epithelial surfaces or the lamina propria of the mucosa, and the 'effector' sites, such as lymph node, spleen or mucosa-associated lymphoid tissue (MALT).

Chapter 6: Chronic inflammation and the adaptive immune response

The key cells involved in adaptive immunity and chronic inflammation are the T cells (T lymphocytes), particularly the T-helper (CD4+ Th cells – predominantly Th1 and Th2 cells) and effector T-suppressor/cytotoxic (CD8+ Ts/c) lymphocytes and the three main antigen presenting cell (APC) types: dendritic cells, macrophages and B cells (B lymphocytes). As in the innate immune response, a host of cytokines and chemokines is also essential.

Adaptive immunity takes days to weeks to be fully established on first encountering an antigen (so it is no use as a first line of defence), but it results in continued antibody secretion and the generation of long-lived memory B and T cells, which are capable of mounting a quick and effective response on re-exposure to an antigen. This is, of course, the principle of vaccination.

We must discuss how the body trains its cells not to turn on itself – 'tolerance' – and the failure of tolerance, autoimmune disease.

The recruitment of the cells of the adaptive immune response can lead to clinically manifest chronic inflammation. Other pathological complications of the adaptive immune system are the hypersensitivity disorders (see Chapter 1, page 80), which cover diseases as diverse as asthma, eczema, allergy (type 1), multiple sclerosis, blood transfusion reactions and rhesus disease of the newborn (type 2), some forms of chronic renal failure (type 3), contact dermatitis and the destructive effects of the body's reaction to tuberculosis (type 4)! Tissue transplantation is now a standard form of treatment, raising the issue of transplant rejection (a form of type 2 hypersensitivity) and graft-versus-host disease (GVHD), in which the recipient of a bone marrow transplant is recognised as 'foreign' by the transplanted immune cells, usually a type IV cell-mediated immune response.

THE BODY'S ORGANISED LYMPHOID TISSUE

Before we discuss the adaptive immune response, we must consider the role of the lymphatic system in inflammation. The lymphatic system comprises several collections of organised lymphoid tissue – including lymph nodes, spleen and MALT – and their interconnecting network of vessels, the lymphatics.

The lymphatic vessels comprise a one-way circulatory system from the tissues back to the blood circulation. Ultimately, the lymphatics drain into the thoracic duct, which returns lymph fluid to the venous system. We alluded to the lymphatics in Chapter 4, page 121, because they are essential for clearing excess fluid from the tissues. However, lymphatics are also vital conduits that carry APCs, invading microbes, and debris from trauma or inflammation to the lymph nodes, where an immune response can take place. The spleen is well placed to trap and remove from the circulation marauding bacterial organisms that have broken into the bloodstream (bacteraemia or septicaemia).

See Fig. 6.2 for the outline of the lymphatics, lymph nodes, spleen and MALT.

Doctors can palpate the superficial lymph node groups, which can enlarge due to inflammation or neoplastic change (this may be primary, as in malignant lymphoma, or secondary to metastatic disease from a primary site in the region drained by the lymph node group) (see Chapter 14, page 356). The superficial lymph nodes are accessible to fine-needle aspiration or needle core biopsy. The deep lymph node groups are not usually palpable, but visible on CT scan and MRI, and can often be biopsied under radiological guidance (Fig. 6.3).

The lymph node is organised into zones to optimise the immune response: B lymphocytes, which can develop into plasma cells that produce antibodies, are gathered into aggregates called follicles. These are situated in the cortex of the lymph node. T lymphocytes reside outside the follicles in the interfollicular zone (Fig. 6.4). Bacterial infections are characterised by the enlargement of the B-cell areas (follicular hyperplasia). If infection is due to a virus then the T-cell response would be more important than antibody production and there would be paracortical hyperplasia.

With any luck, the defence systems activated in the lymph nodes will stop the infection from spreading to the rest of the body. If the systems fail, infection may reach the blood. Infection of the blood (septicaemia) can also occur if tissue organisms directly enter blood vessels instead of the lymphatics.

The main protection against blood-borne infection is the spleen. A septicaemic patient is gravely ill with fevers, shivering attacks (rigors) and a dangerous lowering of the blood pressure ('shock', see page 260). The spleen is all-important here; its structure and functions are

Lymphatics begin as blind sacs lined by an 'overlapping' endothelium, porous to fluid, particles, microbes and cells such as dendritic APC, macs and cancer cells.

Thymus

Thoracic duct drains into the subclavian vein

Lymphatic vessels

Thoracic duct

Spleen

MALT (mucosa associated lymphoid tissue)

Lymphatics in the gut (lacteals) specifically take up fat and fat–soluble vitamins.

Larger lymphatics are lined by continuous endothelium and have valves to prevent backflow.

Lymph nodes filter the lymph, removing microbes and antigens and cancer cells. The filtered lymph passes on to the next node in the chain and through several other lymph node stations before it re-enters the bloodstream via the thoracic duct.

Bone marrow

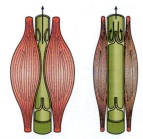

Muscle action massages lymph through the vessels.

Figure 6.2 The reticuloendothelial system.

analogous to those of the lymph node but, rather than filtering lymph, it has sinusoids lined by tissue-specific macrophages. As in the lymph node it is divided into specific B- and T-cell areas. It exists to trap organisms, debris and particulate matter, and to promote immune cell proliferation. The most serious risk of infection is from encapsulated bacteria, such as the pneumococcus (*Streptococcus pneumoniae*) and *Haemophilus influenzae*, which are better at evading the body's innate defences than other bacteria. Other important encapsulated bacteria include *Neisseria meningitides*, *Escherichia coli* and

Klebsiella pneumoniae. The capsules are made from polysaccharide and macrophages can bind only protein, so the bacteria need to be opsonised by IgG or complement C3b in order that the splenic macrophages can remove them.

Patients who undergo splenectomy are often immunised against encapsulated bacteria, but even so require lifelong antibiotic treatment, usually with co-trimoxazole for protection. Without the splenic macrophages, microbes that are coated with Ig and C3b are difficult to remove from the bloodstream and the purpose of immunisation is to prevent them reaching it. Splenectomised patients

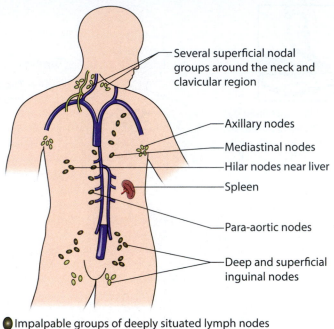

Several superficial nodal groups around the neck and clavicular region

Axillary nodes

Mediastinal nodes

Hilar nodes near liver

Spleen

Para-aortic nodes

Deep and superficial inguinal nodes

● Impalpable groups of deeply situated lymph nodes

● Superficial lymph nodes – often palpable and easy to biopsy

Figure 6.3 Distribution of lymph node groups.

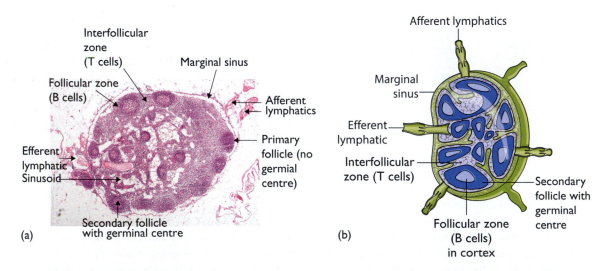

Interfollicular zone (T cells)

Marginal sinus

Follicular zone (B cells)

Afferent lymphatics

Efferent lymphatic Sinusoid

Primary follicle (no germial centre)

Secondary follicle with germinal centre

(a)

Afferent lymphatics

Marginal sinus

Efferent lymphatic

Interfollicular zone (T cells)

Secondary follicle with germinal centre

Follicular zone (B cells) in cortex

(b)

Figure 6.4 Diagram of a secondary follicle and adjacent paracortex to illustrate the anatomical relationship between B cells and follicular dendritic cells (FDCs), and T cells and antigen-resenting cells (APCs). Blood vessels are not shown.

are at risk of overwhelming post-splenectomy infection (OPSI), usually with either meningitis or sepsis, and often fatal. *Streptococcus pneumoniae* is often the culprit. Fortunately the risk of this complication is extremely low – less than 5% of splenectomised patients will suffer this.

The thymus plays a huge role in T-cell development in fetal and neonatal life, and at that time is a relatively large organ compared with other intrathoracic structures such as the heart. By young adult life the thymus has atrophied to a streak of tissue supported by fat.

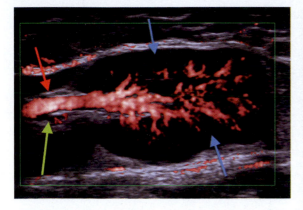

Figure 6.5 A reactive node on ultrasound: nodes are normally 'kidney' shaped with a dark outer layer – the cortex (blue arrow) and a lighter central hilum (yellow arrow), which contains the supplying vessels. A reactive or inflamed node retains its kidney shape but becomes larger and demonstrates increased flow (red arrows) through the hilum.

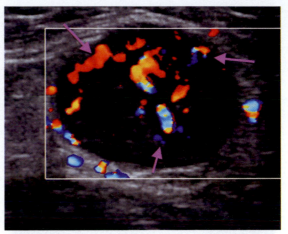

Figure 6.6 A metastatic node on ultrasound: the node has lost its kidney shape and is rounded. The bright hilum has been infiltrated by tumour. The tumour stimulates angiogenesis with the growth of blood vessels peripherally.

CLINICAL SCENARIO: STREPTOCOCCAL SORE THROAT

Let us consider the case of Bertie, a 10-year-old boy, complaining of a severe sore throat that is making it painful to swallow. Examination of his pharynx shows red, swollen tonsils with a purulent exudate on the surface. He also complains of painful lumps in his neck; these are the lymph nodes on either side of the sternomastoid muscle, which have enlarged in response to the throat infection (Fig. 6.7). He has a fever of 38.1°C.

A swab taken from the tonsil grows *Streptococcus pyogenes*. On further typing this is a Lancefield group A streptococcus. As with other Gram-positive bacteria, mannose, a carbohydrate moiety in the outer wall of the bacterium, stimulates the innate inflammatory response and is the cause of the boy's red, swollen, painful pharyngeal lymphoid tissue and of his fever. But why are the lymph nodes swollen too?

If, for the sake of illustration, one of the nodes were to be excised for microscopical examination, it would show a number of changes.

The changes arise because lymph fluid, containing cellular and particulate matter such as bacteria, secreted bacterial antigens, dendritic cells, macrophages and complement fragment C3b, drains into the lymph nodes of the neck. Figure 6.4 shows how the lymph enters a node through multiple afferent lymphatics, which empty into the subscapular sinus, then trickles through the cortex and paracortex in which the B and T cells reside. Together with antigen presentation by dendritic cells, this initiates a specific adaptive immune response that causes proliferation and differentiation of T and B cells.

T- AND B-LYMPHOCYTE ORIGIN AND RECIRCULATION

Lymphocytes are produced and mature in the bone marrow and thymus, where they are trained to tolerate 'self'-antigens (Fig. 6.8). They migrate in the blood to populate the lymph nodes, spleen, and lining of the gut and respiratory tract, the MALT, but constantly circulate through the tissues into the lymph, which trickles through the various lymph nodes in the chain, then into the bloodstream via the thoracic duct.

Lymphocytes home to the tissues from which they have just come and the cycle is repeated. The recirculation

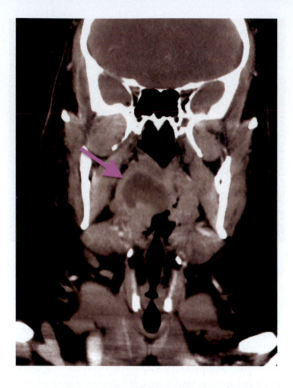

Figure 6.7 Tonsillar abscess on a coronal CT. The tonsil is enlarged and has a necrotic centre with pus in it.

of lymphoid cells is important, because it allows information about invading organisms in one part of the body to be shared with other areas of lymphoid cell production, and the entire repertoire of lymphoid cell surface receptors is available to most of the body (each B and T cell carries a unique receptor that recognises just one antigenic epitope). B cells tend to home to B-cell-rich lymphoid tissues such as Peyer's patches (the MALT of the terminal ileum), whereas T cells tend to home to the lymph nodes. A lymphocyte that has come from a specific area, such as the gut, bears 'homing receptors' that are adhesion molecules – these recognise particular surface molecules on the endothelial cells of that area (addressins) which allow it to 'home' back to its original tissue.

The repopulation of the normal tissue lymphoid population is a feature of normal healing and repair. This is facilitated by 'r-macrophages.

The actual number of lymphocytes in the tissue at any one time is very variable – this varies with the tissue type, e.g. the mucosa of the colon normally contains a considerable population of lymphocytes, whereas the mucosa of the stomach contains no organised lymphoid tissue. The number of lymphocytes in the tissues increases if the tissue is acutely inflamed, because the

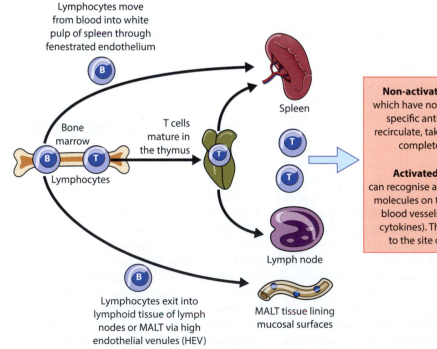

Figure 6.8 Lymphoid stem cells differentiating to form B and T lymphocytes, which circulate to the lymph nodes and spleen. MALT = mucosa associated lymphoid tissue.

Lymphocytes move from blood into white pulp of spleen through fenestrated endothelium

Spleen

Bone marrow

T cells mature in the thymus

Lymphocytes

Non-activated lymphocytes which have not encountered their specific antigen continue to recirculate, taking 12–24 hours to complete a circulation

Activated lymphocytes can recognise and bind to adhesion molecules on the endothelium of blood vessels (upregulated by cytokines). They exit and move to the site of inflammation

Lymph node

Lymphocytes exit into lymphoid tissue of lymph nodes or MALT via high endothelial venules (HEV)

MALT tissue lining mucosal surfaces

Part 2: Defence against disease

upregulation of endothelial adhesion molecules and the activation of lymphoid tissue in the lymph nodes draining an acutely inflamed site lead to their proliferation. Foreign antigens and particles can also enter the lymph, sometimes loose but often having been engulfed by specialised cells of the monocyte/macrophage lineage. A particularly important member of this lineage is the dendritic cell, an antigen-presenting cell (APC), of which each tissue with potential exposure to the outside world has its own representative – the first discovered was the Langerhans' cell of the epidermis (see Table 4.2 and Fig. 6.9).

DENDRITIC CELLS: DENDRITIC APC AND FOLLICULAR DENDRITIC CELLS

The response of T and B cells requires help from other cells because they are incapable of (T cells) or inefficient at (B cells) recognising native foreign antigens such as bacterial products. These helper cells are called dendritic cells – formerly called accessory cells – and fall into two broad categories.

The first group belongs to the macrophage/monocyte lineage and are termed dendritic cells. These are APCs. Dendritic cells are widely distributed in the epithelium of the skin and the mucosa, and are originally formed in the bone marrow, from where they migrate to their stations in the body's perimeter defence system. Dendritic cells are therefore well placed to encounter any invading pathogens. They can take up antigen in a variety of ways, including via immunoglobulin (Fc) or complement (C3b) receptors. Dendritic cells are so called because they have processes to increase their cell surface contact area (Fig. 6.9); three-dimensional electron microscopy shows that they have a peony-like, complex, folded appearance. The dendrites and surface folds are constantly moving, sampling the environment, and can insinuate themselves between adjacent cells and engulf particles and microbes. Their processes can even reach beyond the epithelium, e.g. into the lumen of the gut, to grab microbes – dendritic cells can form tight bonds with epithelial cells so that the integrity of the epithelial lining is not breached.

The foreign antigen sampled by these cells at the site of entry (the tonsils in the case of 10-year-old Bertie) is carried to the MALT and the draining lymph nodes, where an army of T and B cells resides.

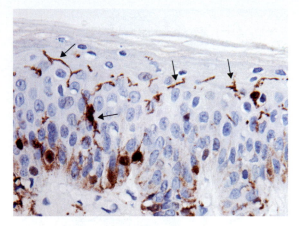

Figure 6.9 Langerhans' cells (stained here for S100 protein) are dendritic antigen-presenting cells (APCs) found in the skin and other squamous epithelium, such as that found in the oropharynx.

After engulfing a microbe or antigenic particle, the dendritic cell can break it down to smaller fragments (epitopes) and present selected antigens to T cells in a suitable form on type II major histocompatibility complex (MHC) receptors.

Dendritic APCs express high levels of co-stimulatory molecules on their surfaces and are the most potent and effective cells in the stimulation of the immune response. Although dendritic APCs are the most efficient cells for the presentation of antigen to T cells, B cells and macrophages also express type II MHC molecules and are useful APCs. There are many other cells that can express type II MHC molecules, but lack the co-stimulatory molecules essential for T- and B-cell activation (see later for a fuller discussion of MHC class I and II molecules and the way in which they work).

At the time of microbe recognition, the activation of other inflammatory cells due to the recognition of PAMPs (pathogen-associated molecular patterns, see Chapter 4) in the organisms by PRRs (pattern recognition receptors) leads to the secretion of cytokines which instruct the dendritic cell to pull in its processes, detach any bonds that it has formed with adjacent epithelial cells and migrate to the lymph nodes. Chemokines secreted by the lymph nodes attract the dendritic cell, and during the migration it processes the antigen in its phagolysosome ready to display selected polypeptide segments via the MHC II molecules once it is in the T-cell zone (paracortex) of the lymph node (Fig. 6.10).

Chapter 6: Chronic inflammation and the adaptive immune response

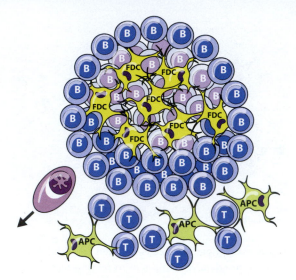

Figure 6.10 Diagram of a secondary follicle and adjacent paracortex to illustrate the anatomical relationship between B cells and follicular dendritic cells (FDCs), and T cells and antigen-presenting cells (APCs).

For T-cell activation to take place, both antigen presentation and engagement of co-stimulatory molecules must occur. T cells can unleash such a deadly and destructive inflammatory response that it is a bit like launching a guided missile – it will not happen unless a defined sequence of checks and codes has been correctly applied.

The dendritic APCs should not be confused with follicular dendritic cells (FDCs), the second type of dendritic cell. FDCs are confined to the lymphoid follicles, and can be found in lymph nodes during embryonal development. They are thought to derive from embryonal mesenchymal tissue, and are not of macrophage/monocyte lineage, but nevertheless express Fc and complement receptors. They do not express MHC II molecules and thus cannot interact with T cells. FDCs cannot phagocytose or process antigen, but can trap and display loose antigen or whole organisms removed from the lymph on their surfaces as unprocessed molecules and they are potent stimulators of B-cell differentiation. Similarly B cells can come to the follicles loaded with antigen from the tissues, which they pass to the FDCs. Activation of FDCs leads to the development of germinal centres (secondary follicles) in which B cells are retained to test out the fit of their antigen receptors. Those B cells that have a reasonable fit with the antigens on the FDCs' surface are retained and induced to proliferate, at which point hypermutation occurs and a generation of daughter cells is produced with alterations in their receptors. Those with tightly fitting receptors for the antigen are selected for further proliferation.

APCs carrying the processed antigen reach the lymph node through afferent lymphatics and filter through the paracortex, where they meet an army of T cells ready to respond. The T cells recognise the antigenic epitopes presented by APCs, proliferate and differentiate into helper T cells (CD4 Th1 or Th2) which coordinate key aspects of the immune response and initiate the B-cell response.

FDCs stimulate germinal centre formation and B cells to proliferate, differentiate and produce antibodies tailored to fit a particular antigen. The paracortex of the node enlarges in response to T-cell activation by dendritic APCs. The lymph node enlarges, and can be palpated – this occurred in the neck of the patient in this case.

ORIGIN OF OPSONINS USED IN THE ADAPTIVE IMMUNE RESPONSE

As we discussed, APCs and FDCs can trap organisms (in this case, streptococci) to prevent their dissemination into the blood; they do this by grabbing attached portions of immunoglobulin (Fc component) or complement (C3b), attached to the surface of the bacterium. These, of course, are opsonins (see Chapter 4). The opsonins are fixed to the surface of the organism itself or to any soluble antigenic fragments that it may have released. The difference is that APCs engulf the antigen or organism and process it, presenting short polypeptide molecules via their MHC II surface molecules, whereas FDCs grab on to C3b or Fc and hold the attached organism or antigen on the cell surface.

From where do these opsonins, C3b and immunoglobulin appear? You already know that bacterial wall fragments can directly stimulate the complement cascade to produce C3b (see Fig. 4.21), and that macrophages in the tissues constitutively secrete C3. But if antibody production is a function of the adaptive immune response, which takes several days to happen, how can there be any immunoglobulin for the phagocyte to adhere to?

Similar to any good general anticipating a possible attack, the body prepares itself during the early months of a baby's life. For the first weeks and months, protection by passive immunisation is given by the mother's IgG antibody, which can transfer via the placenta, conferring protection for 4–6 months, or IgA via the mother's milk bathes the gut lining and traps organisms. Breast milk contains minimal IgG. A child's own immune resistance begins to develop at around 2–3 months, but is not complete until about the age of 5 years.

At the same time the bone marrow generates millions of different B-cell clones, each capable of secreting an antibody likely to be of use in combating infection. In early life, so-called 'early endogenous antibody production' occurs and a variety of antibodies are produced and secreted into the blood and in tissue fluid. A population of natural antibodies is thus present within human (and higher primate) tissues from an early age. 'Natural' antibodies are often reactive with the disaccharide galactose, which is a key component of many bacterial glycosylated cell surface proteins, and may have been generated in response to bacteria resident in the gut.

If a natural antibody is a reasonable 'fit' with an antigen, it can stick to it well enough to opsonise the particle for phagocytosis, or to set off the complement cascade. Low-affinity antibodies can attach loosely to more than one type of antigenic epitope, whereas high-affinity antibodies produced in the lymph nodes after antigen presentation are highly specific, bind tightly and react not just with one antigen, but with one epitope on an antigen.

The type of microbial agent that initiates an inflammatory reaction determines the most appropriate host immune response, e.g. parasitic infection leads to the release of cytokines such as interleukin (IL)-5, with the attraction of eosinophils.

THE ADAPTIVE IMMUNE RESPONSE

Now it is time to think about the adaptive immune response in a bit more detail – hopefully your appetite has been whetted and you wonder how on earth these amazing defences are brought about.

The adaptive response can be split into humoral immunity and cell-mediated immunity. Humoral immunity is due to antibody production by B cells, which transform to plasma cells and produce immunoglobulin (antibodies). Cell-mediated immunity is mediated by T cells. To understand T-cell function we must briefly go into what MHC molecules are and how they function. There are several subsets of T cells, but the two most important are the T-suppressor/cytotoxic cells, capable of direct attack and lymphokine production, and the T-helper cells, which act to regulate the immune response and interact widely with other immune-reactive cells. A third important type of lymphocyte, probably a T subtype, is the natural killer (NK) cell (see page 200). This cell acts independently and is important in the innate immune response and in cancer surveillance and killing.

We start life with an army of B and T cells, raw recruits perhaps, but willing to go into action at the press of a button (not for nothing are these called 'naïve' cells!). Both originate from stem cells in the bone marrow, but early on the T cells are lured away to the thymus; the mechanisms governing how they get there are still not clear (see Fig. 6.8).

The 'action' buttons, or cell surface receptors, on the B and T cells have been prepared by an intelligence service that has attempted to anticipate every eventuality. Each cell has a unique receptor for antigen on its surface, generated after the B cells have undergone a series of rearrangements in their heavy chain genes while in the bone marrow. The T cells have similarly rearranged their T-cell receptor genes during their maturation process in the thymus.

HUMORAL IMMUNITY

How does a B cell produce the correct antibody for a new antigen?

Surprisingly, it is the antigen that chooses the B cell best equipped to fight it, rather than the other way round! The B cell has no choice in the matter. Mother nature equips the body with an enormous number of B cells, each armed with a receptor for a unique antigen.

Imagine the B cells, lined up around the wall of a dance hall, waiting for the right antigen to ask them to dance! Each B cell has the genetic code for a single antibody and it displays the antigen-binding portion of the antibody on its own cell surface – this is the 'B-cell receptor complex' (BCR) (Fig. 6.11). Like secreted antibody, the BCR has both heavy and light chain components. The Fc end (see Fig. 6.12) of the antibody is anchored to the surface of the B cell by a transmembrane component. Other than this, the BCR is identical to the antibody that the B cell will produce when the BCR binds the matching antigen.

Of course, the body requires an enormous number of B cells with different genetic codes (and thus BCRs) to ensure that it has a B cell equipped to fight any new foreign antigen. Several B cells may have receptors that can bind to different parts of the antigen (epitopes); only those that fit closely are used. (There are at least 10^8 different immunoglobulin molecules in the serum.) Nature discovered a brilliant way of producing this variety of codes, which we discussed in Chapter 1, and then used a similar approach for T-cell receptor molecules (TCRs) (Fig. 6.13).

While we are discussing the receptors displayed on the surface of a B cell, we should mention CD40, which binds to the CD40 ligand present on helper T cells (Fig. 6.13). Without this, B cells cannot mature to form plasma cells or 'class switch' (see page 197) to secrete IgG, IgA or IgE antibodies. B cells also express a number of other 'co-stimulatory' cell surface molecules, such as CD80 and CD86, which interact with CD28 on T cells and are critical in the development of the T-cell-B-cell interaction.

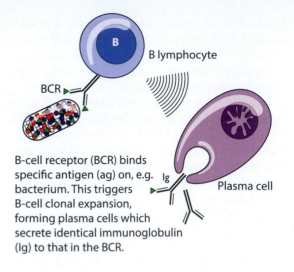

B-cell receptor (BCR) binds specific antigen (ag) on, e.g. bacterium. This triggers B-cell clonal expansion, forming plasma cells which secrete identical immunoglobulin (Ig) to that in the BCR.

Figure 6.11 Each B lymphocyte bears a unique B-cell receptor (BCR) which is tethered to the membrane but is otherwise identical to the antibody that it will secrete. Antibody secretion occurs when B cells are activated by antigen, either by cross-linkage of BCRs on the surface or by Th cells, with which they interact via MHC II and co-stimulatory molecules.

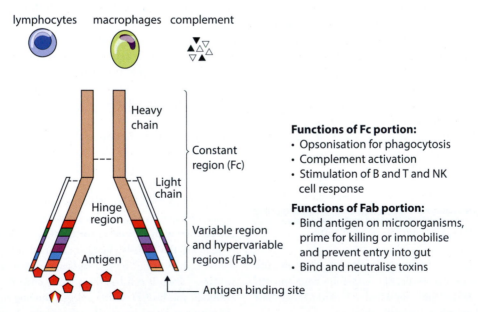

Functions of Fc portion:
- Opsonisation for phagocytosis
- Complement activation
- Stimulation of B and T and NK cell response

Functions of Fab portion:
- Bind antigen on microorganisms, prime for killing or immobilise and prevent entry into gut
- Bind and neutralise toxins

Figure 6.12 The immunoglobulin molecule is active at both ends! The antigen-binding site is a three-dimensional structure, with three key sites at which bonds are made. Through a process of hypermutation, low-affinity binding sites can mutate to show high affinity for antigen.

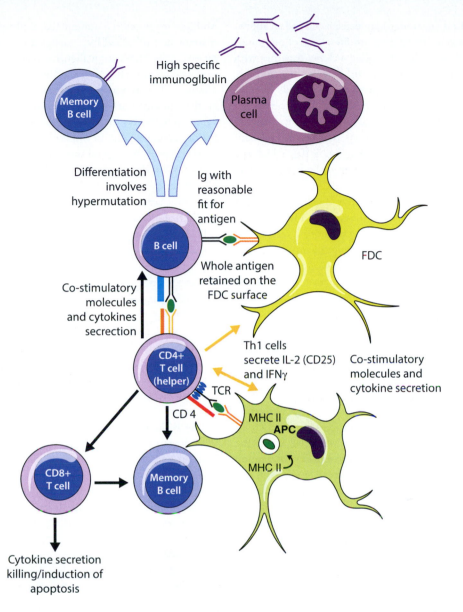

Figure 6.13 Interactions between antigen-presenting cells and T and B lymphocytes; arrows indicate direction of stimulation. APC (antigen presenting cell), FDC (follicle dendritic cell), Ig (immunoglobulin = antibody), TCR (T cell receptor), Th1 (Th1 = subtype of CD4+ helper T cell; Th2 subtype not shown here; other Th subtypes exist. All modulate the immune response according to the type of stimulus – see page 197).

B cells, similar to macrophages, and dendritic cells, express MHC class II molecules, and carry receptors for complement (CD21) and the Fc component of immunoglobulin molecules, although they do not phagocytose. For a B cell to be activated, there must be multiple binding of the BCR to antigenic sites. B cells can recognise complement that is bound to the surface of a microbe (and complement binds only to foreign cells), and the BCR is much more easily activated if the B cell also binds to complement. The FDCs in the lymph node's germinal centre can bind complement and use it to gather complement-bound antigen for presentation to the B cell.

At the first encounter between a B cell and an appropriate antigen, there is a proliferation of immunologically identical B cells, called a clone, together with memory cells. However the 'fit' between the antibody and the antigen can be improved considerably by a bit of adjustment; the difference, you might say, between an off-the-peg and a tailored garment. The tailoring is achieved through interactions between B cells and antigen presented by FDCs (see Fig. 6.10), which is a potent inducer of hypermutation in the hypervariable region of the antibody.

A large population of daughter B cells is produced through cloning and these differentiate to form plasma cells; a population of circulating memory B cells is also generated. Plasma cells secrete huge quantities of antibody into the bloodstream and tissues, neutralising free antigen and marking up whole organisms for attack by complement, phagocytes and NK cells. Initially, the antibody is of IgM type and later the B-cell clone switches to produce IgG.

The 'heavy chain switch' is useful clinically; the presence of IgM antibody in the blood of a patient indicates a recent infection, of no more than a few weeks' duration. Finding IgG antibody is less useful, as it can indicate that the patient has been exposed to the antigen at almost any time in the past, from weeks to years.

If the particular antigen is encountered again, the memory cells will quickly undergo clonal proliferation and swamp it with specific antibody. A memory B cell requires far less stimulation and co-stimulation than an unprimed ('naïve') B cell.

It is worth stating that a B-cell inflammatory reaction will lead to the generation of numerous different antibodies, all directed at different antigenic sites (epitopes). Different B cells, each with a different BCR and therefore each with unique antibody properties, if stimulated by the same antigen, will produce a variety of different antibodies against its various epitopes. This is a 'polyclonal' response, i.e. several different B-cell clones are stimulated and each will produce its own particular antibody.

This is very different from what happens in multiple myeloma, a malignant disease of plasma cells. All malignancies originate from a single mutated cell, so it follows that all the antibodies secreted by a malignant proliferation of plasma cells will be identical. Compare the electrophoretic strips shown in Fig. 6.14.

This is an important concept because certain cancers that affect the lymphoid system may be difficult to distinguish from a reactive inflammatory proliferation.

Multiple myeloma is a tumour that infiltrates the bone marrow and is caused by a malignant proliferation of plasma cells. Certain malignant lymphomas, representing cells at earlier stages in the path from B cell to plasma cell, may secrete antibodies. The finding that all the antibodies are exactly the same is of diagnostic value, because this is virtually unheard of in an inflammatory response, e.g. to a microbe.

This is probably a good moment to think about antibodies! What do they look like and exactly how do they bind to antigen? Why are there antibody subtypes? We have come across IgM and IgG so far, and we have mentioned that IgD is encountered on the surface of some B cells. How can such a large number of diverse antibodies have been generated from cells with a common ancestor?

What are antibodies and how do they work?

Antibodies come in two forms: secreted, soluble antibody (Ig molecules) and membrane-bound antibody, with just the component, which recognises microbes and other antigens on the surface – this is known as the BCR. B-cell maturation precedes that of T cells in intrauterine fetal life, and both somatic mutation and class switching (see below) have been identified before 26 weeks' gestation.

The BCR is unique to that B cell. It is complexed with other proteins that traverse the cell membrane. This protein anchors the BCR in the membrane and is essential for signal transduction after antigen binding. Only a B cell with a BCR that binds the antigen will be stimulated to proliferate and secrete antibody with an identical structure to the BCR, lacking only

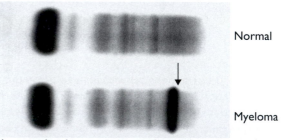

Normal

Myeloma

An extra band is present in an electrophoretic strip from a patient with myeloma, due to very large quantities of a single type of immunoglobulin secreted by a malignant clone of plasma cells. The control strip shows no significant bands in the gamma region

Figure 6.14 Myeloma: M band (arrowed), compared with normal plasma electrophoresis. (Courtesy of Dr Jo Sheldon, St George's Hospital, London.)

the transmembrane component. The type of immuno-globulin molecule in the BCR is usually an IgM mono-mer or sometimes IgD (Table 6.1).

Immunoglobulins are collectively known as 'γ-globulins' because of their motility on electrophor-esis – although this is a historical term you may still hear it used, e.g. an immunodeficient patient is likely to be treated with 'pooled γ-globulin', i.e. concen-trated immunoglobulins derived from blood from a number of different donors.

Each immunoglobulin molecule (monomer) is formed from two identical heavy chains and two iden-tical light chains joined by interchain disulphide links. There are two types of light chain (κ and λ) and five types of heavy chain (G, A, M, D, E). The light chains can combine with any type of heavy chain and do not influence biological function whereas each heavy chain type supports different biological functions (Fig. 6.12 and Table 6.1).

IgG is the most prevalent antibody in the blood and exhibits the most basic immunoglobulin structure, referred to as a monomer. There are four subtypes, the constant regions of which are slightly different. IgG3, for instance, is best at fixing complement, whereas IgG1 is the best at opsonising for phagocytosis, because phagocytes such as macrophages and neutrophils bear specific receptors for FcIgG1. In general, IgG is the best antibody for attacking bacterial pathogens. IgG can cross the placenta, carrying passive immunity from mother to fetus to 'tide it over' until the newborn child can produce its own antibody about 3 months later. Different subtypes of IgG exist, each with fascinating properties: the precise subtype produced is governed by the cytokine balance prevalent at the time, which is determined by the innate immune system. Thus the innate immune system can influence the adaptive immune system. Between them these subclasses of IgG can neutralise toxin, opsonise microbes for macrophage and neutrophil ingestion (and destruction), and activate complement. IgG can also cross the placenta to confer passive immunity. The transplacental IgG molecule's half-life is approximately 3 weeks, compared with the 3–5 days that is normal for the other antibodies.

IgM is the largest antibody molecule: five mono-meric units are joined by a J-chain to make a mas-sive pentamer. By clustering antigen together in one spot (10 antigen molecules can be bound by one IgM pentamer), it is good at mopping up large quantities of invading pathogens at the start of an infection. IgM

Dictionary

Transduction: the translation of a signal into an action, usually involving messengers that refer the signal to the nucleus. Here particular genes are switched on and translated into proteins with actions appropriate to the initial stimulus.

can also activate the C1 fragment of the complement pathway. A key feature of this molecule is that it is too large to cross the placenta (see Fig. 2.35).

In the blood, IgA exists in a monomeric form, but IgA is most important as a mucosal protector, in gut secretions, for example. At these sites it exists as a dimer, linked by a J-chain and a 'secretory compon-ent' derived from the gut epithelial cell. Its constant (Fc) region ends are linked by a chain, which prevents mucosal enzymes from digesting and destroying the antibody. Also, the dumbbell-shaped dimer can bind antigen at either end, clumping bugs together and making it easier for them to be caught up in mucin and carried out of the body. In the gut-associated lymph-oid tissue (GALT, e.g. Peyer's patches in the terminal ileum) there are 20–30 times more IgA-producing than IgG-producing plasma cells. This means that IgA is the most plentiful antibody in the body but most of it is on the mucosal surfaces, not in the blood. (Look out for trick questions in exams!)

IgA defends both the luminal and the subepithelial zones of mucosa-lined surfaces. About 1:600 people are IgA deficient. They may be asymptomatic or present with recurrent ear or sinus infections. An interesting finding is that patients with coeliac disease (see page 23) are 10–15 times more likely to be IgA deficient (2–3% of coeliac patients compared with 0.2% of the general population). In a clever exten-sion of mucosal immunity, intestinal B lymphocytes can be stimulated by antigens and then migrate via the lymphatics and blood to localise in other areas, such as breast or salivary glands, so that specific IgA also defends these sites.

IgE binds mast cells and has a major role in the defence against parasites. When IgE is made, its Fc end is bound by mast cells, which are long-lived cells in the tissues. Here the antibody is ideally placed to recognise pathogens such as worms or flukes that invade the tissues.

Table 6.1 Immunoglobulin types

Immunoglobulin	Complement fixation by		Macrophage/ polymorph binding	Mast cell/ basophil binding	Cross placenta	Function
	Classic pathway	Alternative pathway				
IgG	++	−	++	−	++	Combats microorganisms and toxins. Most abundant immunoglobulin in blood and extravascular fluid
IgA	−	±	±	−	−	Most important immunoglobulin for protecting mucosal surfaces. Combines with secretory component to avoid being digested
IgM	+	−	−	−	−	Important in early response to infection because it is a powerful agglutinator
IgD	−	−	−	−	−	? Function. Present on the surface of some lymphocytes and may control lymphocyte activation/suppression
IgE	−	−	±	±	−	Involved in mast cell degranulation, thereby protecting body surfaces. Important in allergy and parasitic infections

A curious feature of IgE is that, once secreted by a plasma cell, it becomes bound to the surface of a tissue mast cell. When two or more IgE molecules, which are bound to the same mast cell, recognise and bind to epitopes on a marauding parasite, the IgE molecules can cross-link.

Cross-linkage of IgE molecules triggers the immediate release of toxic granules from the mast cell on to the parasite. This is clever, because most parasites are too large to be phagocytosed by the usual defenders, the neutrophils and macrophages. However, the idea backfires in 'atopic' individuals (those people who are 'hypersensitive' to certain allergens), who produce IgE under circumstances in which most people would produce IgM or IgG. A common feature of allergens, such as peanut, pollen or bee sting, is that they bear repeated sequences of the same epitope. Exposure of an individual allergic to, already sensitised to, the allergen leads to cross-linkage of mast-cell-bound IgE, and the subsequent degranulation sets off a local allergic reaction or an anaphylactic (body-wide allergic) response. Bee sting is a particularly potent allergen and each year causes fatalities.

IgD is a slightly mysterious monomeric immunoglobulin: it is found in MALT, particularly in the wall of the gut, but its precise role is not clear. It is present on the surface of B cells, where it seems to be very important in the primary activation of a naïve B cell, i.e. a B cell that has never before encountered its unique antigen. Such 'virgin' B cells have more surface IgD (sIgD) than IgM on their surfaces, whereas those that are 'experienced' bear more sIgM. Whether they have sIgD or sIgM on their surfaces, the first Ig produced and secreted into the blood is almost always IgM. There is very little IgD in circulating blood.

Most importantly, immunoglobulin *recognises antigen* using the variable regions on the molecule. After that, the constant regions *initiate the biological functions*, such as complement fixation and opsonisation, appropriate to the immunoglobulin class. Although diagrams generally depict antibody molecules as simple bent dinner forks (see Fig. 6.12), they actually have a complex three-dimensional structure with many 'folds' (Fig. 6.15).

Antigen binds to the variable end of the antibody molecule (Fab), in which there are three hypervariable regions forming a potential 'pocket' for antigen attachment. The shape of the pocket depends on the outer electron clouds of its atoms, which of course determines the antigen shape that it recognises. The important point is that this interaction depends on the antigen and antibody having complementary profiles; no covalent bonding is involved, so the chemical composition is not crucial. The shapes don't have to be a perfect fit but a close fit gives the strongest binding. Fine-tuning can be undertaken by a B cell together with FDCs in the germinal centre and Th cells in the paracortex of the lymph node.

The genetic make-up governing the three hypervariable regions in each pocket can be adjusted for a better antigen 'fit' before the clonal expansion of the B cell is undertaken. This requires new instructions from the cell's genome and the alterations within the genome to generate better-fitting BCR proteins produce the phenomenon of hypermutation. The B cell is the only cell in the body in which it is normal for the body to play about with its genetic make-up! This generates a population of antibodies with excellent binding properties.

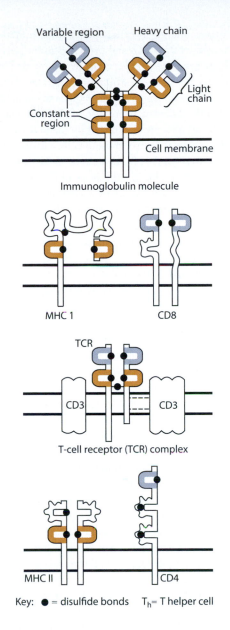

Figure 6.15 The immunoglobulin homology unit is the basic building block for a range of molecules involved in cell–cell recognition, the 'immunoglobulin gene superfamily'. These often interact with other Ig superfamily members, e.g. the T-cell receptor complex/CD4 or -8 and MHC molecules. Several other molecules play an important role in immune reactions, such as intercellular adhesion molecules (e.g. ICAM-1) and lymphocyte function antigen (e.g. LFA-3).

Chapter 6: Chronic inflammation and the adaptive immune response

In hypermutating B cells, spontaneous mutation is encouraged by the FDCs and cytokines in the germinal centre to generate a hypervariable region that more closely fits the antigen. Mutation generates daughter B cells with new hypervariable regions in their BCRs, each slightly different from that of the first cell that bound the antigen presented to it by the FDCs.

In the germinal centre, those B cells with new BCRs are tested to see if they fit the antigen more snugly than the original. It is a bit like the story of Cinderella! The failures (the ugly sisters) die by apoptosis. The best fitting are rescued by binding with CD40 ligand on Th cells (Prince Charming) and the debris from the dead B cells is hoovered up by macrophages; their clear cytoplasm, speckled with debris, makes the germinal centre appear rather like a 'starry sky'.

Binding of antigen to antibody can lead to a variety of effects. Cross-linking of antigenic particles or cells will produce precipitation or agglutination whereas binding of antibody to an active site on a virus or toxin can result in neutralisation. Other components of the immune response can become involved, as when antibody fixes complement to produce lysis or enhanced phagocytosis. Antibody can also promote cell-mediated cytotoxicity involving NK cells, so-called antibody-dependent cell-mediated cytotoxicity (ADCC).

Nature has found this approach so useful that a common structure, the immunoglobulin homology unit, is the basic building block for a range of molecules involved in cell–cell recognition, the so-called immunoglobulin gene superfamily (see Fig. 6.15). Its members include the MHC I and II receptors, CD4 and CD8, among others: CD4, expressed on T-helper cells, binds to MHC II, and CD8 on T-suppressor/cytotoxic cells binds to MHC I.

These interactions are essential; in fact T-cell activation requires there to be both antigen binding by the TCR *and* T-cell stimulation by the binding of either CD4 or CD8 to their respective MHC receptors. There are also several co-stimulatory molecules that are very important in cell–cell interactions.

THE MHC SYSTEM

The human MHC system is also called the HLA (human leukocyte antigen) system and is encoded on chromosome 6 where there are six loci: three for class I antigens (A, B and C) and three for class II antigens (DP, DQ and DR).

Each person has a unique set (almost) of MHC molecules that are present on most cells and are inherited: three class I MHC and three class II MHC from each parent. The A, B and C HLA proteins encoded by the class I MHC loci are paired with their other halves and linked to β_2-microglobulin to make up the whole class I MHC molecule. The class II MHC molecules are made up of DP, DQ and DR proteins. All the genes for MHC molecules received from both parents are fully represented.

MHC class I molecules are expressed on virtually all nucleated cells (with the exception of glial cells and platelets), whereas class II molecules are normally restricted to the APCs, which comprise macrophages and B cells. T cells cannot react with free, native antigens (that is a job for the Ig molecules). T cells can bind only to peptide antigens that have been processed by APCs or macrophages and are then displayed on the cell surface *alongside* MHC molecules.

Under some circumstances MHC class II molecules can be expressed on many other cell types, including intestinal, epidermal and respiratory epithelial cells, vascular endothelial and stromal cells, particularly stimulated by cytokines, e.g. interferon-γ. It is thought that most of these cell types are unable to present antigens to Th cells; the 'professional' APCs have co-stimulatory molecules that bind CD4 to 'verify' the activation stimulus given when MHC class II molecules bearing polypeptide fragments bind to CD4+ Th cells.

This is relevant to transplantation where it is essential to have a good 'match' between the donor and the recipient to minimise the risk of rejection. An identical twin provides an excellent match. Some siblings are a good match but other people's organs carry MHC antigens that will be identified by the recipient's immune system as 'foreign' and the tissue will be rejected.

Key facts

Examples of the immunoglobulin superfamily

- Immunoglobulins
- MHC class I and II antigens
- T-cell receptors
- CD2, -3, -4 and -8 antigens
- Adhesion molecules, e.g. intercellular adhesion molecule (ICAM) and vascular cell adhesion molecule (VCAM)

The class I molecules on all nucleated cells present antigens on the cell surfaces that reflect the protein content within the cell. Intracellular protein is processed through an organelle called a proteasome, cut into small polypeptide chunks and placed in a lysosome along with an MHC I molecule. The polypeptide molecule fits into a groove on the MHC I molecule and is transported to the cell surface for display. This mechanism is brilliant for declaring to the world that a virus has infected the cell, and attracts CD8+ Ts/c cells (cytotoxic) or NK cells to come and kill the cell by inducing it to undergo apoptosis. Class I MHC antigens bind CD8+ T cells (suppressor/cytotoxic) (Fig. 6.16).

Class II molecules are very different. For a start they are normally expressed only by APCs and their job is to reflect what is happening in the world around them. They engulf microbial particles, digest them into small fragments and place selected small peptides in lysosomes, which fuse with lysosomes containing class II MHC molecules. The class II molecule has a 'clip' mechanism that protects it from binding intracellularly derived protein. The clip is released only by a 'CLIP' protein, allowing the binding of MHC II with extra-cellularly derived protein. This is then moved to the APC surface for presentation to CD4+ Th cells, and the generation of an immune response.

Thus, class I antigens show what is happening within the cell and class II antigens show what is happening in the tissues outside the cell.

GENERATION OF DIVERSITY IN THE IMMUNE SYSTEM

If you have forgotten what genes are made of and how they can be translated to form proteins, refer to Fig. 2.34. You will be aware that much of our DNA appears to code for nothing, and that the 'important' sections that encode genes are called *exons* and the in-between stuff consists of *introns*. For many years the introns were largely disregarded as 'silent', but recently it has been appreciated that the non-encoding segments of DNA exert an influence. It is a little like the startling realisation that the fatty tissue in between islands of haematopoietic cells in the bone marrow, previously thought to be mere packing material, is highly active and secretes cytokines which play a large role in the inflammatory responses.

T and B cells both use a similar mechanism to generate multiple unique antigen recognition sites in the chains that make up the TCR and those that make up the BCR/Ig molecule, respectively. We considered this in Chapter 2 (page 49), when we discussed the processes involved in generating diversity in the B-cell heavy chains (Fig. 6.17) from four building blocks, the variable (V), diversity (D), joining (J) and constant (C) regions, which can be recombined in a variety of ways by looping, joining, excision and splicing, to create an almost infinite number of different antigen recognition sites. The process is rather like cutting and pasting using a computer. Immunoglobulin heavy chain gene rearrangement is a two-stage process.

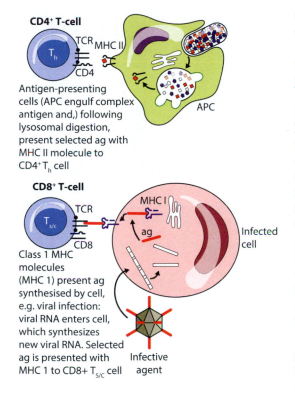

CD4⁺ T-cell

TCR MHC II
CD4

Antigen-presenting cells (APC engulf complex antigen and,) following lysosomal digestion, present selected ag with MHC II molecule to CD4⁺ Tₕ cell

APC

CD8⁺ T-cell

TCR
CD8

MHC I

Class 1 MHC molecules (MHC 1) present ag synthesised by cell, e.g. viral infection: viral RNA enters cell, which synthesizes new viral RNA. Selected ag is presented with MHC 1 to CD8+ Tₛ/ᴄ cell

ag

Infected cell

Infective agent

Figure 6.16 T-helper (Th) cells recognise processed antigen presented to their unique T-cell receptor (TCR) by MHC II molecules on the surface of antigen-presenting cells (APCs). CD4 is their co-stimulatory molecule. Their role is to activate CD8⁺ T-suppressor/cytotoxic (Ts/c) and B cells and stimulate them to proliferate and form circulating memory cells. Without T-cell help, class switching of antibody and the formation of memory B cells do not occur.

Chapter 6: Chronic inflammation and the adaptive immune response

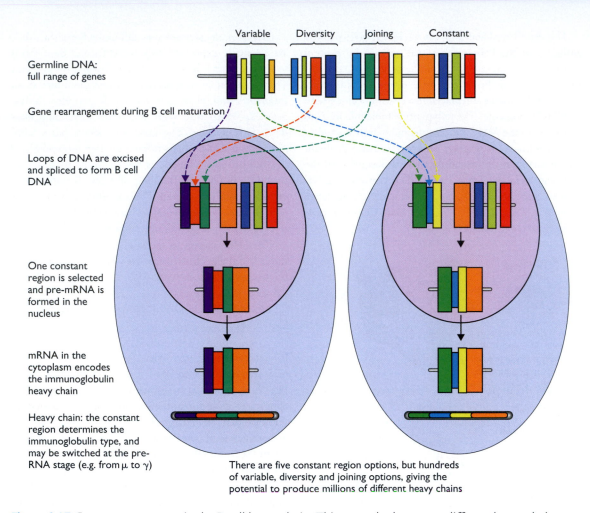

Figure 6.17 Gene rearrangement in the B-cell heavy chain. This example shows two different heavy chains being produced in each cell.

Having rearranged the V, D and J regions satisfactorily, within the nucleus, the cell must make its choice of heavy chain. Later on, the daughter cells can alter the heavy chain by altering the gene encoding the C region.

Why is the C region treated differently? Probably because a lymphocyte that has recognised an antigen with the variable region on its surface molecule may need to produce molecules with different constant regions to best address the needs of a particular challenge. Each immunoglobulin molecule has slightly different functions (see Table 6.1); IgG is better at opsonising bacteria for phagocytosis or NK cell killing, whereas IgM is best at activating the complement complex, IgE is best at eliminating parasites and IgA is designed to neutralise microbes in the mucosal surfaces. The switch of constant region occurs during the germinal centre reaction before differentiation to plasma cells. The antigen recognition site and the variable region remain the same; only the constant region has changed. It is easy to see how an enormous variety of molecules can be produced in this way.

Although exactly the same type of process is undertaken for the generation of diversity in the TCR and other similar molecules, it must be emphasised that the V, D, J and C groups of genes are different for each type of molecule (i.e. heavy chain, light chain, $TCR\alpha$, $TCR\beta$, etc.). Also, the genes for each chain are rearranged independently. Once the genes on one of a pair of chromosomes are rearranged, the locus on the paired chromosome is inhibited (allelic exclusion).

Unlike B cells, T cells do not undergo hypermaturation with alteration of the binding site.

CELL-MEDIATED IMMUNITY

Cell-mediated immunity is achieved by T cells. There are several types of T cell, which differ in their biological roles (Table 6.2). Some act as helper cells, some as cytotoxic cells, some as killer cells, some as regulator cells and some as suppressor cells. And of course memory T (and B) cells are generated after an initial response. A population of null lymphocytes, which are neither T nor B, are called the natural killer (NK) cells. The NK cell is a strange cell – technically a lymphocyte, but with actions akin to the innate immune system (see page 199). It is particularly important in killing tumour cells. Most T cells have an α and β chain in their receptor, but a small proportion bear receptors made of γ and δ chains and are known as γ–δ cells. These may be quite significant in some types of autoimmune disease and in coeliac disease.

T cells bear a variety of surface molecules, some of which are common to most T-cell types (e.g. CD3), whereas others are individual to the various subtypes (e.g. CD4 Th1 or Th2 cells, CD8 Ts/c cells). These molecules are essential in antigen recognition by combining with the T-cell receptor (TCR) and linking with MHC molecules (a little like shaking hands) to give a confirmatory signal. As the response unleashed by T cells has such profound effects, there are other molecules on the surfaces of T cells and APCs that must also link to 'confirm' the veracity of the signal (one imagines this is bit like launching a nuclear strike – the order to do so must include code entries and simultaneous turning of keys before the countdown starts.).

How is antigen recognised by T cells?

We have mentioned already that T-cell recognition of antigen is similar to that of B cells, though it is a bit more complicated. The TCR on the surface of the majority (about 95%) of T cells is composed of an α chain and a β chain with constant and variable regions analogous to those of immunoglobulins. The remaining minority of T cells have a receptor made up of γ and δ chains. These T cells tend to migrate to epithelial surfaces, such as the gut or respiratory epithelial lining cells, and are thought to act independently of APCs.

Antigen binds to a receptor site on the TCR. The TCR is linked to a cluster of polypeptide chains, collectively named the CD3 molecules; together, TCR and CD3 make up the T-cell receptor complex. We have already said that the TCR and immunoglobulin molecule belong to the same family. In fact, the similarities between the TCR and immunoglobulin structure go even deeper, as the genetic mechanisms for producing the necessary enormous diversity are almost identical. As in the immunoglobulin molecule, rearrangement of germline DNA during the T-cell maturation process (in the thymus, rather than the bone marrow, where B cells mature) generates a series of somatic mutations such that every T cell has a unique receptor.

Mode of action	CD4$^+$			CD8$^+$	
	Suppressor/inducer	Helper/Th1	Inducer Th2	Suppressor cells	Cytotoxic cells
Genetic restriction (MHC)	II	II	II	I	I
Suppressor activity	++ (provide help)	–	–	++	–
Cytotoxic activity	–	+	–	–	++
Help for immunoglobulin	–	+	+++	–	–

Table 6.2 T-lymphocyte subsets

MHC, major histocompatibility complex.
T cells are important against infections, in graft rejection, in graft-versus-host disease, in some hypersensitivity reactions and in tumour immunity. Th1 cells assist macrophages in stimulating cell-mediated immunity. Th2 cells assist B cells by stimulating immunoglobulin production and regulating immunoglobulin class.

Chapter 6: Chronic inflammation and the adaptive immune response

Small print

Cluster designations

CD (cluster designation) numbers indicate that a particular cell has a surface antigen that can be detected with a specific antibody. This has proved very useful for identifying leukocytes.

Some CD antigens useful in leukocyte identification

Antigen	Principally expressed on
2	Most T cells
3	Mature T cells
4	'Helper/inducer' T cells
8	'Suppressor/cytotoxic' T cells
15	Monocytes and granulocytes
16 and 56	Natural killer cells
20	Most B cells
21	Complement receptors
68	Monocytes/macrophages
79a	Most B cells, including plasma cells

Self-reacting cells are deleted at an early stage of development. The mechanism is known as 'tolerance'. T-cell tolerance is achieved in the thymus.

Like the immunoglobulin molecule, the T-cell antigen receptor site depends on a three-dimensional fit to form non-covalent bonds with antigen. However, the hypermutation mechanism, so important in the generation of high-affinity immunoglobulin molecules, is not used for TCR genes.

What is the difference between Th and Ts/c cells and NK cells?

Th cells, which have the molecule CD4 included in the TCR complex, make up about 60% of the body's mature T cells (Fig. 6.13). They will recognise processed antigen if the CD4 receptor combines with class II MHC molecules. By a combination of releasing interferon and interleukins, and binding co-stimulatory molecules, Th cells stimulate other T cells, NK cells, macrophages and B cells. CD4+ T helper cells co-ordinate the immune response (fig 6.13). The two main CD4+ helper T cell subsets are Th1 and Th2. Th1 is useful against pathogens which have invaded host cells,

such as viruses and some bacteria; Th2 is useful against extracellular pathogens, especially parasites such as helminths, also bacteria. Th1 is involved in autoimmune responses against tissues, whereas Th2 is involved in allergic responses such as asthma.

In HIV/ AIDS the immune system is devastated because HIV has a receptor for CD4, and thus enters and destroys the function of the cell most centrally placed in the coordination of cell-mediated immunity and much of humoral (antibody-mediated) immunity. Macrophages and dendritic APCs express low levels of CD4 on their surfaces and are also target cells for HIV infection.

Ts/c cells, with CD8 included in the TCR complex, make up about 30% of mature T cells. They bind to processed antigen associated with class I molecules (see Figs 6.15, 6.16). In fact, they cannot recognise any antigen that is not presented together with MHC class I molecules, which are expressed by every nucleated cell in the body. One of their most important roles is that of recognising virally infected cells. When viruses replicate within cells, viral antigen is expressed on the cell surface. When viral antigen is presented, together with MHC I, to a cytotoxic T cell with an appropriately matched receptor, it springs into action and lyses the cell membrane, destroying the infected cell. CD8+ T cells are also thought to be pivotal in the generation of peripheral tolerance (see later) and in tumour cells; both the Tc cells, which can lyse other cells, and the Ts cells, which induce anergy, are important in this process.

NK cells are neither T nor B cells and make up 10–15% of the circulating blood lymphocytes (Fig. 6.18). They contain cytoplasmic granules and have three important membrane receptors: one for the Fc component of IgG, one antigen receptor that can recognise particular molecular signatures on the surfaces of some organisms or tumour cells, and a non-CD8 type receptor for MHC I molecules. Cells coated with antibody will be recognised via the Fc receptor and lysed by NK cells (ADCC, antibody-dependent cellular cytotoxicity). The lysis is achieved by molecules such as perforin and granzyme contained within the cytoplasmic granules. Other cells, the antigens of which bind with the NK cell antigen receptor, are saved from death by possession of their MHC I molecule, which represses NK cell lysis. If the NK cell cannot 'see' the MHC I receptor due to its alteration (e.g. by tumour, viral infection or drug binding) the target will be lysed.

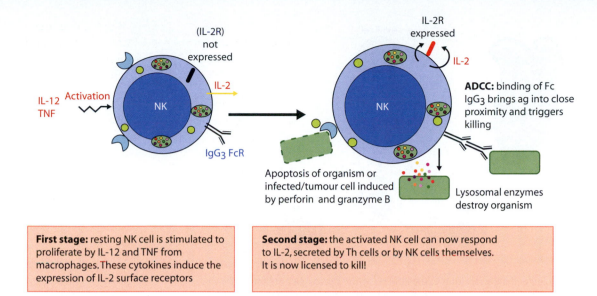

First stage: resting NK cell is stimulated to proliferate by IL-12 and TNF from macrophages. These cytokines induce the expression of IL-2 surface receptors

Second stage: the activated NK cell can now respond to IL-2, secreted by Th cells or by NK cells themselves. It is now licensed to kill!

Figure 6.18 Natural killer (NK) cells act independently of MHC molecules and may directly cause apoptosis of cells displaying stress antigens or indirectly induce apoptosis when stimulated by binding antibody attached to antigens on a cell surface. They are particularly useful in defence against intracellular particles such as viruses, or altered host cells, e.g. tumour cells. ADCC = antibody-dependent cellular cytotoxicity.

TOLERANCE

The ability of the body to mount a major immune response against foreign antigens raises a very important question. How does the immune system distinguish between an antigen that is foreign and one that is normally present on the cells in the body?

This process, known as tolerance, can be adaptive at two stages. The first occurs during lymphopoiesis (lymphoid cell maturation) and is known as central tolerance. The second occurs once the mature B and T cells have been released into the peripheral tissues, and is referred to as peripheral tolerance. Tolerance is achieved by a mixture of cell deletion and generation of anergy (non-reactivity) in self-reacting cells (Fig. 6.19). These mechanisms induce tolerance such that antigens that are exposed to the immune system during fetal life are not capable of eliciting a response in later life. One means of inducing tolerance is to present antigen on MHC molecules without co-stimulatory molecules.

In the thymus, for example, T-cell tolerance is determined by not selecting (to stay alive) any cells that do not bind MHC antigens at all (because MHC I and II molecules are the means by which lymphocytes

know that they are dealing with 'self' cells). However, equally the body does not want cells that react by binding very strongly to host MHC molecules, because these may stimulate activation against self. So both extremes are eliminated, by either forced apoptosis or anergy (the antigen is recognised but no reaction takes place).

So, nature has devised a neat system of differentiating self from non-self. Or has it?

AUTOIMMUNE DISEASE

As the heading suggests, the body can on occasion turn on itself and begin to react with self-antigens, to destructive purpose. Autoimmune diseases are estimated to affect up to 7% of the population. Mostly this is due to a breakdown in tolerance, but occasionally new antigens are generated that cross-react with the body's cells. For reasons that are, as yet, unknown, middle-aged women seem to be the most likely to develop autoimmune disease.

It is thought that a loss of tolerance can develop in several ways: exposure of the immune system to

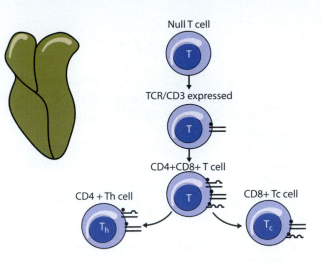

Null T cell

TCR/CD3 expressed

CD4+CD8+ T cell

CD4 + Th cell

CD8+ Tc cell

- T cells enter the thymus without the T-cell receptor/CD3 complex or CD4 or CD8. Care is taken that they cannot encounter antigen en route–macrophages phagocytose antigen and apototic debris.

- They mature and proliferate, going through a stage of being double-labelled as CD4+8+ cells, becomimg either CD4+ or CD8+ cells within the cortex. Cells which aid to later develop CD4 or 8 are deleted.

- 'Positive selection' for cells which can recognise MHC molecules presented on thymic epithelial cells occurs. Reacting cells survive, the rest undergo apoptosis.

- 'Negative selection' against cells which bind self MHC molecules takes place in the medulla. The antigen is carried to the thymic medulla by bone marrow-derived dendritic cells. Those cells which bind are self-reactive and are deleted.

- Non-autoreactive cells which can interact with antigen displayed on an appropriate MHC molecule move to the medulla and are secreted into the blood.

Figure 6.19 Mechanisms of acquiring tolerance: these may break down because of cross-reaction with molecules on microbes or drugs (molecular mimicry), exposure to previously sequestered antigen (e.g. lens protein or spermatozoa) and 'switching on' of previously anergic cells.

previously sequestered (hidden) antigens is the most easily understood. Parts of the body that are not exposed to the immune system during fetal life can produce a response later on; lens protein and spermatozoa are just two examples. A person who has had severe trauma to one eye, with the release of lens protein into the blood, runs the risk of forming antibodies to the protein. A few weeks later, the other eye may be severely damaged by an antibody-mediated inflammatory reaction (a condition known as sympathetic ophthalmitis). Mumps may give rise to inflammation of the testis (orchitis), which causes sperm antigens to be released into the blood circulation. The anti-sperm antibodies, which may develop as a result, can cause infertility. Similarly, mumps may inflame the islets of Langerhans in the pancreas and has been implicated in the causation of type 1 diabetes mellitus.

Occasionally there is cross-reaction between a microbial antigen and normal body cells, due to a similarity in shape or structure between the self and microbial antigens; unwittingly, the body develops an antibody or cell-mediated reaction against itself as it expunges the microbe. A person at risk of such a complication would be our 10-year-old patient, who presented at the start of this chapter with a sore throat due to a Lancefield group A streptococcal infection. This bacterium is notorious for carrying antigenic determinants that mimic the endocardium of the heart so closely that the patient develops inflammatory foci in the wall of the heart. This life-threatening condition, which can lead to long-term cardiac damage, is known as rheumatic fever. The incidence of rheumatic fever has plummeted in recent years, possibly due to early eradication of the infection by prompt antibiotic therapy.

Following on from this point it is obvious that defects in any of the immune system's regulatory molecules may lead to aberrations in tolerance. It is thought that much of the peripheral tolerance mechanism relies on anergy on the part of self-reacting immune cells, easily lost if the system controls are tampered with.

Lastly, even anergic T or B cells may be induced to behave against their will by outside forces. An example of this is infection by Epstein–Barr virus (EBV), which can infect B cells directly, entering via their CD21 complement receptors. The presence of EBV can stimulate an infected B cell to produce antibodies. If an anergic self-reacting B cell happens to become infected, it will be switched on, however inappropriately.

The autoimmune diseases include many clinically important and potentially life-threatening conditions. They are generally divided up into those diseases that affect a single organ (organ specific) and those that affect several organs or tissues (non-organ specific) (Table 6.3).

Table 6.3 Autoimmune diseases and autoantigens

Disease	Antigen(s)
Organ-specific diseases	
Hashimoto thyroiditis	Thyroid peroxidase, thyroglobulin/thyroxine (T_4)
Pernicious anaemia	Intrinsic factor
Type 1 diabetes mellitus	Cells in the pancreas (tyrosine phosphatase)
Addison's disease	Adrenocortical cells (ACTH receptor and microsomes)
Autoimmune haemolytic anaemia	RBC membrane antigens
Graves' disease	TSH receptor on thyroid cells
Pemphigus	Epidermal keratinocytes
Bullous pemphigoid	Basal keratinocytes
Guillain–Barré syndrome	Peripheral nerves (gangliosides)
Polymyositis	Muscle (histidine tRNA synthetase)
Non-organ-specific diseases	
Systemic lupus erythematosus	Double-stranded DNA, nuclear antigens
Chronic active hepatitis	Nuclei, DNA
Scleroderma	Nuclei, elastin, nucleoli, centromeres, topoisomerase 1
Primary biliary cirrhosis	Mitochondria (pyruvate dehydrogenase complex E2)
Rheumatoid arthritis	IgG (rheumatoid factor, connective tissues, collagen)
Multiple sclerosis	Brain/myelin basic protein
Sjögren's syndrome	Exocrine glands, kidney, liver, thyroid
Several organs affected	
Goodpasture's syndrome	Basement membrane of kidney and lung (type IV collagen)
Polyendocrine	Multiple endocrine organs (hepatic – cytochrome P450; intestinal – tryptophan hydroxylase)

ACTH, adrenocorticotrophic hormone; RBC, red blood cell; TSH, thyroid-stimulating hormone.
From Lydyard *et al.* (2000) *Pathology Integrated: An A–Z of disease and its pathogenesis*. London: Arnold.

Chapter 6: Chronic inflammation and the adaptive immune response

You will come across some of these diseases elsewhere, e.g. the antibodies in Graves' disease stimulate the thyroid-stimulating hormone receptor to secrete thyroxine and thus cause thyrotoxicosis, an example of a type II hypersensitivity reaction.

We will briefly describe systemic lupus erythematosus (SLE) to illustrate the wide-ranging effects of a non-organ-specific autoimmune disease. Systemic lupus erythematosus is a systemic disorder in which there is chronic, relapsing and remitting damage to the skin, joints, kidneys and almost any organ. Similar to most immune disorders, it has a higher incidence in women. In America, it is also more common in black people and it tends to occur in the second and third decades. Patients may present with a characteristic 'butterfly' rash on the face, or with more subtle symptoms. Many present after their kidneys have been damaged beyond repair, in chronic renal failure. This is an example of type III hypersensitivity, in which antigen–antibody complexes lodge in blood vessels or tissues and stimulate acute inflammation due to their presence and actions.

The fundamental feature of the disease is inflammation of the small arterioles and arteries, i.e. a vasculitis, often related to the deposition of antigen–antibody complexes in the vessel walls. Involvement of the glomerular capillaries in the kidney produces a variety of types of glomerulonephritis. The other sites of involvement are joints (synovitis), heart (non-infectious endocarditis – named after Libman and Sacks – and pericarditis), lungs (pleuritis and effusions) and central nervous system (focal neurological symptoms due to vasculitis). This is an example of type II hypersensitivity, in which anti-DNA and other antibodies formed by the body attack its own tissues.

The course of the illness is extremely variable and unpredictable and may range from mild skin involvement to severe renal disease leading to death.

In the autoimmune disorders, the immune response is well controlled but the initiating event of antigen recognition is wrong. There is another group of disorders in which antigen recognition proceeds normally but the body's response is exaggerated. These are called hypersensitivity reactions.

HYPERSENSITIVITY REACTIONS

We discussed types of hypersensitivity in Chapter 2 (see page 80). Just as a reminder, here are the types and some examples. You will notice that all but type IV involve antibodies in one shape or form, and so are driven by B cells; the only T-cell-driven response is the type IV 'delayed-type' hypersensitivity. The difference between autoimmune and hypersensitivity disease is that, in autoimmune disease, the problem is one of loss of tolerance of self-antigens, whereas in hypersensitivity the problem is that of an unnecessarily excessive immune response to a foreign antigen. In the case of type IV hypersensitivity, this division is artificial.

In a nutshell, the following are the hypersensitivity reactions.

Type I hypersensitivity

This affects about 20% of the general population. On first exposure there is an inappropriately strong Th2 response to an allergen, e.g. pollen, in people predisposed towards this. IL-3 and IL-4 are secreted by Th2 cells and induce a class switch in plasma cells, so that IgE is generated and becomes attached to mast cells. The person is now sensitised. Activation of the antibody by binding the same antigen on subsequent exposure causes degranulation and the release of histamine and other acute inflammatory mediators, e.g. asthma, anaphylaxis.

 Read more about asthma in Pathology in Clinical Practice Case 2

Type II hypersensitivity

Antibody forms against cells/entire tissues or cell receptors, with a variety of effects depending on whether the antigen is 'fixed', as in thyroid tissue in thyrotoxicosis, or 'circulating' as in platelets in ITP. Circulating cells, coated in antibody, will be taken up in the spleen as happens to platelets in idiopathic thrombocytopenic purpura (ITP).

 Read more about idiopathic thrombocytopenic purpura in Pathology in Clinical Practice Case 5

Antibodies against the thyroid-stimulating hormone receptor lead to permanent activation, without negative feedback causing thyrotoxicosis and Graves' disease.

Anti-gastric parietal cell and intrinsic factor antibodies in pernicious anaemia cause absorptive problems and increase the risk of gastric cancer.

 Read more about pernicious anaemia in Pathology in Clinical Practice Case 22

History

Robert Graves (1797–1853)

Figure 6.20 Robert Graves. (Reproduced with permission from the Wellcome Library, London.)

Robert Graves (Fig. 6.20) was an Irish physician renowned for his teaching, who introduced bedside teaching and student clerking, and taught in English rather than in Latin, which was the custom at that time. His non-medical exploits included being imprisoned in Austria as a suspected German spy (because he spoke such fluent German), travelling and painting with J.M.W. Turner, and taking command of a sinking ship in the Mediterranean, repairing the pumps with leather from his own shoes and taking an axe to the lifeboat to keep the crew safely on board!

Type III hypersensitivity

IgG or IgM antibody-antigen complexes, which form against previously soluble antigen in the blood, circulate and lodge in capillaries, sparking off an acute inflammatory response because of Fc receptors on the antibody. The complexes tend to lodge in so-called 'high-pressure vascular beds,' such as kidney, joints and other small vessel beds. Larger complexes are easier to clear. The faster the deposition rate and the greater the affinity between antigen and antibody, the less likely it is to be cleared. Chronic inflammation supervenes and causes fibrosis and tissue destruction. Examples are membranous glomerulonephritis, SLE, and rheumatoid arthritis.

 Read more about glomerulonephritis in Pathology in Clinical Practice Case 34

Type IV hypersensitivity

This is a T-cell response against the body's own components (and therefore a crossover between hypersensitivity and autoimmune disease). The helper CD4 Th cells and cytotoxic CD8 Tc cells coordinate in an excessive T-cell reaction to attack the body's own cells.

Excessive macrophage activation by IFN-γ secreted by Th1 cells stimulates fibroblast activation and fibrosis, and sometimes granulomatous inflammation. Examples are Crohn's disease, multiple sclerosis, contact dermatitis, type I diabetes mellitus and tuberculosis (TB). The reaction is the basis of the tuberculin test for TB.

 Read more about examples of Type IV hypersensitivity in Pathology in Clinical Practice Case 10 (tuberculosis), Case 11 (multiple sclerosis) and Case 38 (rheumatoid arthritis).

TRANSPLANTATION

We have already mentioned the role played by the MHC class I and II antigens, displayed on cell membranes, in governing the body's response to transplanted tissue or cells.

Host rejection of a tissue or organ graft may occur at various stages:

- **Hyperacute rejection** is mediated by pre-existing antibody or complement. Complement damage can be a particular problem in xenografts (grafts between species). For reasons that are unclear, some organs need far more precise tissue matching than others. The most stringent requirements are for kidney and bone marrow grafts. Surprisingly, the heart and liver require little more than ABO compatibility (i.e. matching blood types).
- **Acute rejection** is usually mediated by antibodies. The host may already carry antibodies against the tissue transplanted into him or her, hence the requirement for cross-matching red blood cells and checking that the host's serum does not carry antibodies that can react with the donor's T lymphocytes. Thus, type II hypersensitivity reactions are also important in tissue transplantation.

Chapter 6: Chronic inflammation and the adaptive immune response

- Chronic rejection: occurs where T cells identify as foreign the endothelial cells of the blood vessels in the donor organ. By damaging the blood supply they cause gradually more severe ischaemic damage to the transplanted organ.

You will realise from reading the hypersensitivity sections that the body may generate either a type IV cell-mediated response or a type II antibody response, against any tissue that fails to give the right MHC signals to the T cells.

GRAFT-VERSUS-HOST DISEASE

It is worth pointing out that, for patients who require a bone marrow transplant, it is the *donor* immune cells that react against the host, causing graft-versus-host disease (GvHD) if there is a mismatch. GvHD particularly involves the skin, gastrointestinal tract and liver, and it has been suggested that dendritic cells as well as donor T cells are involved in GvHD. The symptoms can be extremely severe, and include blistering and sloughing of the skin, ulceration and sloughing of the intestines, and intense jaundice due to the disappearance of the intra-hepatic bile ducts. Treatment is by immunosuppression.

IMMUNISATION

The immune system is a natural defence mechanism but it can also be manipulated so that it will respond more quickly to an antigen and so, hopefully, reduce the impact of the infection. This is called immunisation.

History

Edward Jenner (1749–1823)

It would be a crime to consider immunisation without pausing for a moment to think about its history and the man responsible for developing its use. The man is Edward Jenner, a pupil of John Hunter (Fig. 6.21). Jenner lived with Hunter for the first 2 years after coming to London and the friendship they developed continued after Jenner left London to start general practice in Berkeley, Gloucestershire. Jenner was profoundly influenced by Hunter's interest in natural history and in his methods of scientific investigation. To one of Jenner's questions, Hunter is said to have replied, 'I think your solution is just; but why think? Why not try the experiment?'

Even before Jenner, it had been noticed that an attack of smallpox protected against further disease. It was known that the epidemics varied in severity and that it was best to contract a mild form of smallpox as this resulted in life-long protection. This knowledge was widespread; in India, children were wrapped in clothing from patients with smallpox; in China, scabs from smallpox patients were ground and the powder was blown into the nostrils; in Turkey, female slaves were injected under the skin with dried preparations of pus from smallpox patients. Inoculated slaves fetched a high price whereas pock-marked slaves were worth nothing! Lady Mary Wortley Montagu, the wife of the British Ambassador in Constantinople, was aware of these techniques and she took the risk of having her own children inoculated. When she returned to England in 1718, she tried to convince her friend, the Prince of Wales, that he should do the same. He was worried about experimenting on the royal children but, when six orphan children

Figure 6.21 Edward Jenner. (Reproduced with permission from the Wellcome Library, London.)

were successfully immunised against smallpox, he consented and the royal children were inoculated. Medical ethics have made some advances since those days!

Jenner and others had noticed that cows suffered from a pustular disease resembling smallpox called 'variolae vaccinae' – cowpox. It was known that it could be transmitted to humans and that, apart from local symptoms, there were no ill effects. There was a widespread belief that those who had suffered from cowpox became immune to smallpox and a farmer in Dorset, Trevor Jesty, tried it out on his own children. The idea of using the cowpox virus to induce immunity to smallpox thrilled Jenner but, rather than jumping to conclusions, he followed Hunter's example and experimented. On 14 May 1796, Jenner inoculated a boy of 8 years named James Phipps with cowpox. The boy's illness took a predictable course and he recovered. On 1 July, he inoculated the boy with smallpox and no reaction occurred, either on this occasion or on a subsequent occasion a few months later (Fig. 6.22). Jenner described this experiment to the Royal Society but it was rejected. He continued to make his observa-

Figure 6.22 Jenner successfully inoculated against smallpox with cowpox. (From Lakhani S (1992) Early clinical pathologists: Edward Jenner. *J Clin Pathol* 45. Reproduced with permission from the BMJ Publishing Group.)

tions and, in 1798, published his work entitled *An Inquiry into the Causes and Effects of the Variolae Vaccinae*. Hence, inoculation with smallpox was replaced by inoculation with cowpox. The word **'vaccination'** came into use and smallpox cases dropped in the UK as a series of laws (Vaccination Acts of 1840, 1841, 1853, 1861, 1867 and 1871) made vaccination free and compulsory, with parents liable to repeated fines until their children were vaccinated. Compulsion was withdrawn in 1948 and smallpox was eradicated globally by 1980 with the last naturally occurring case being in Somalia in October 1977.

Immunisation may be active or passive. Active immunity involves using inactivated or attenuated live organisms or their products. The effect is reasonably long lasting and calls up an adaptive immune response. The duration of cover varies with each type of vaccine. Passive immunity results from injecting human immunoglobulin and the effect is immediate, but only lasts 1–2 weeks.

ACTIVE IMMUNISATION

First exposure to an antigen provokes a primary response in which IgM is the major antibody. Further exposure to the antigen produces a secondary response, which occurs faster and produces higher levels of antibody, and the class is predominantly IgG. This is because memory B cells circulate after the first exposure and it takes far less cross-linkage of antigen to stimulate the extensive B-cell proliferation and plasma cell formation, with copious antibody production, than it did the first time that antigen was encountered.

In addition to humoral (B-cell-mediated) immunity, cell-mediated (T-cell) immunity is also induced by vaccination.

Dictionary

Antibody is the same as **immunoglobulin** and makes up most of the γ-globulin fraction of the plasma proteins
Vaccination and **immunisation** are the same thing. It all began with inoculation of vaccinia virus (coxpox to protect against smallpox). Vaccination generates immunity against particular antigens. Immunisation may generate either humoral or cellular immunity, often both.

Chapter 6: Chronic inflammation and the adaptive immune response

The aim of active immunisation is to give sufficient doses of antigen to ensure that, after completing the course of immunisation, the person can mount a rapid effective response if exposed to the disease. The number of doses, time intervals and need for booster doses varies with the vaccine, the natural history

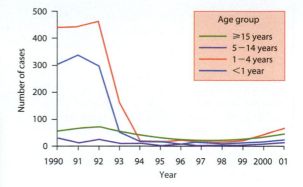

Figure 6.23 The effect of immunisation against Hib has been dramatic: *Haemophilus influenzae* type b disease in England and Wales by age group (PHLS and CDSC data). (From Public Health Laboratory Service (2002) Continuing surveillance of invasive *Haemophilus influenzae* disease. *CDR Weekly* **12**(26).)

Table 6.4 NHS immunisation schedule for children in the UK, obtained in 2015 from www.immunisation.nhs.uk

Vaccines only for 'at-risk' groups:

- BCG (TB) vaccine (birth to age 35)
- Hepatitis B vaccine (birth onwards)

2 months:

- 5-in-1 vaccine (diphtheria, tetanus, pertussis, polio and Hib)
- Pneumococcal vaccine
- Rotavirus vaccine
- Meningitis B vaccine

3 months:

- Meningitis C vaccine
- 5-in-1 vaccine (second dose)
- Rotavirus vaccine (second dose)

4 months:

- 5-in-1 vaccine (third dose)
- Pneumococcal vaccine (second dose)
- Meningitis B vaccine (second dose)

12–13 months:

- MMR (measles, mumps, rubella) vaccine
- Hib/meningitis C booster vaccine
- Pneumococcal vaccine (third dose)
- Meningitis B vaccine (third dose)

2, 3 and 4 years plus school years 1 and 2:

- Children's annual flu vaccine

3 years and 4 months:

- 4-in-1 pre-school booster (diphtheria, tetanus, pertussis and polio)
- MMR vaccine (second dose)

12–13 years:

- HPV vaccine – girls only currently in the UK

13–18 years:

- 3-in-1 teenage booster vaccine (tetanus, diphtheria and polio)
- Men ACWY (meningitis W)

Table 6.5 NHS adult immunisation schedule

≥65 years:

- Pneumococcal vaccine
- Annual flu vaccine

70 years:

- Shingles vaccine

Vaccines only for 'at-risk' groups:

- BCG (TB) vaccine (birth to age 35)
- Chickenpox vaccine (any age)
- Flu vaccine (adults)
- Flu vaccine (children)
- Pneumococcal vaccine (2–65 years)
- Hepatitis B vaccine (birth onwards)

of the disease and the likelihood of encountering the infection.

In the UK, the present immunisation schedule for children is shown in Table 6.4 and the recommendation for adults who are unimmunised or in a high-risk group is shown in Table 6.5.

Live attenuated viral vaccines, such as polio, and measles, mumps and rubella, generally produce the most long-lasting immune responses and oral polio vaccine has the advantage of having maximal effect on local gut immunity, the natural portal of entry for wild polio. Non-live vaccines may be more effective when combined with adjuvants to enhance the immune

The common types of vaccine

Live attenuated organisms

- Polio (oral: Sabin)
- Measles
- Mumps
- Rubella
- Tuberculosis (BCG)

Inactivated organisms

- Polio (subcutaneous: Salk)
- Pertussis
- Typhoid
- Hepatitis A
- Influenza (variable subtype depending on prevailing strain)

Toxoid (toxin inactivated by formaldehyde)

- Tetanus
- Diphtheria

Components of the organism

- *Haemophilus influenzae* type b (Hib) (capsular polysaccharide)
- Hepatitis B (surface protein (recombinant))
- Pneumococci (capsular polysaccharide)

Important pathogens for which there are no vaccines

- Rhinoviruses (colds)
- HIV/AIDS*
- Cytomegalovirus (CMV)
- Epstein–Barr virus (EBV)
- Hepatitis C virus
- Gonorrhoea
- Syphilis
- Leprosy
- Trachoma
- Malaria*
- All parasitic and protozoal infections

*Trials of possible vaccines are in progress

response, e.g. aluminium phosphate and aluminium hydroxide are used in DTP vaccine.

New vaccines are being developed all the time and create interesting decisions about the best use of health funding. Recently, it has become possible to immunise against papillomaviruses that cause cervical cancer and genital warts. Clearly, prevention should be better than cure. The cervical cancer screening programmes have been so effective in reducing morbidity and mortality that, in countries where global screening systems are established, it can be difficult to justify vaccination because this will only protect those receiving the vaccine before they are infected and it will still be necessary to keep screening the majority of people who are past that age. Anti-meningitis type B strain is a vaccine that has only recently been approved in the UK. A malaria vaccine and an Ebola vaccine are about to be launched.

PASSIVE IMMUNISATION

Passive immunity relies on using either pooled plasma containing a variety of immunoglobulins to local infectious agents or specific immunoglobulin obtained from convalescent patients or recently immunised donors. Specific immunoglobulins are available for tetanus, hepatitis B, rabies and varicella-zoster. They are most commonly used for post-exposure prophylaxis.

Anti-toxins/anti-venoms are particularly helpful in high-risk countries such as Australia.

PART 3

FEATURES OF CARDIOVASCULAR DISORDERS

INTRODUCTION
FEATURES OF CARDIOVASCULAR DISORDERS

If I have seen further it is by standing on the shoulders of giants

Isaac Newton (1642–1727)

It is extraordinary to think that diseases with effects as diverse as those of gangrene, strokes, heart attacks and divers' 'bends' are all disorders of the circulatory system. The general features of circulatory disorders are almost the opposite of the cardinal features of inflammation, which are covered in Part 1: for 'calor (heat), rubor (redness), tumor (swelling) and dolor (pain)' read 'coldness, pallor, cyanosis, pain and loss of sensation'.

Why is this? The drop in temperature and change in colour are easily understood, because blood carries body heat from the core and dissipates it to the extremities, and it is the red colour of the oxygenated haemoglobin pigment in the red blood cells that makes pale-skinned persons look pink. Anything decreasing blood flow to a finger or toe will decrease the tissue perfusion by warm blood, making it cold and pale, and any delay in delivery of red blood cells to the affected digit will mean that more of the haemoglobin will have given up its oxygen load, leaving blue-coloured deoxyhaemoglobin (cyanosis). Pain is a variable phenomenon, depending on the tissue affected and the type of injury, e.g. a 'heart attack', or myocardial infarction, caused by sudden blockage of a coronary artery, is usually associated with intense central chest pain, often radiating down the left arm, whereas a gradual 'furring up' of the arteries supplying the legs causes severe pain on walking, which disappears when the demand for oxygen by the leg muscles is removed by rest (intermittent claudication). By comparison, a 'stroke', in which the blood supply to part of the brain is suddenly interrupted, will generally cause weakness or paralysis, but no pain. Loss of sensation also varies according to the type of vascular disease and the tissue or organ affected; a stroke may destroy a sensory pathway to the brain, leading to a large area of numbness, which may involve half the body, whereas blockage of the blood

flow to a toe would cause numbness in just the area supplied by the vessel because of ischaemic damage to the local sensory nerves.

Cardiovascular disorders include disorders in the vessels, the blood or the heart. They may be 'local' or 'systemic' and may gradually develop over months or years, or strike suddenly and catastrophically. Perhaps the easiest way to look at these diseases is to relate them to a domestic plumbing system, the main components of which are the pipes and the pump (Fig. 1). Pipes may gradually 'fur up' (atherosclerosis), or become blocked (vascular occlusion). The blockage is most commonly blood clot, i.e. thrombosis. Sometimes small fragments of thrombus may break off and be carried around the system until they lodge in a pipe with a diameter too small to let them through (embolism). Burst pipes cause haemorrhage but, thankfully, the vessels can constrict and the blood constituents can clot to limit the leak. Sometimes the pipe may bulge alarmingly before it bursts (aneurysm) and this may be detected and repaired, e.g. in the abdominal aorta. Pump failure for whatever reason is fairly disastrous, and in the heart this may be due to valve disease, myocardial infarction, infection, congenital abnormality, etc.

Some solutions to these problems have been found; thus affected segments of piping can be replaced (arterial bypass grafts), pumps can be tinkered with (valve grafts) or replaced (heart transplants), high pressure causing strain on the system can be relieved (antihypertensive drugs), and the narrowed vessels may be stretched with balloon catheters and then stented to keep them open. Of course these are usually only partial solutions and there is no doubt that prevention is the best medicine.

This moves us naturally from the mechanistic, reductionist approach to a more holistic one that takes account of our genetic background, environment and lifestyle. Consider Fig. 2; this is an integrated gene by environment (GxE) report card for a personal genome.

Dictionary

Common terms in cardiovascular disorders

Embolus: intravascular solid, liquid or gaseous mass carried in the blood from its origin and lodged in another site

Thrombus: solid mass of blood formed within the cardiovascular system involving the interaction of endothelial cells, platelets and the coagulation cascade

Blood clot: solid mass of blood formed by the action of the coagulation cascade

Infarct: localised area of ischaemic tissue necrosis generally caused by an impaired blood supply

Haematoma: extravascular accumulation of clotted blood

Haemorrhage: discharge of blood from the vascular compartment into the extravascular body spaces or to the exterior

Petechiae, **purpura**, **ecchymoses**: small haemorrhages

Hyperaemia: an increased volume of intravascular blood in an affected tissue which may result from increased flow (active hyperaemia) or reduced drainage (passive hyperaemia = congestion)

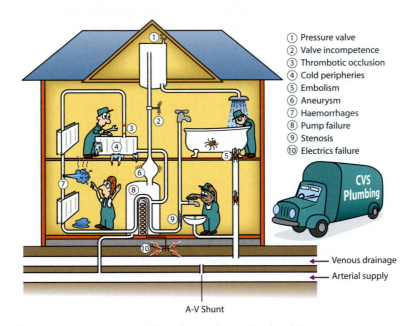

1. Pressure valve
2. Valve incompetence
3. Thrombotic occlusion
4. Cold peripheries
5. Embolism
6. Aneurysm
7. Haemorrhages
8. Pump failure
9. Stenosis
10. Electrics failure

Figure 1 The cardiovascular system can be likened to a domestic plumbing system.

The environmental risk factors are around the circumference and the lines show their links to specific diseases. The diseases themselves however also interact so, for example, obesity increases the risk of hypertension and coronary artery disease. This graphic is about an individual's risk of diseases based on analysis of their genome. The size and colour of the risk factors shows which changes in lifestyle would be likely to have greatest impact. For this person, it is smoking and exercise. That would also be true for most of the population but, if he had a different genome, the importance of the risk factors could be different; this is how personalised medicine could evolve over the next decade. In the following chapters, we will highlight a few of the most important genetic and environmental factors but remember that there are actually hundreds involved and full assessment of individual risk will require large databases and powerful computers.

So let's move from the future to the past. If you had been alive in the sixteenth century, you would have been taught that blood was produced by the liver and then carried in the veins to the organs where it was consumed; this was Galen's theory of the regeneration of the blood. He believed that the portion of blood from the liver that entered the right side of the heart divided into two streams. One route was through the pulmonary artery to

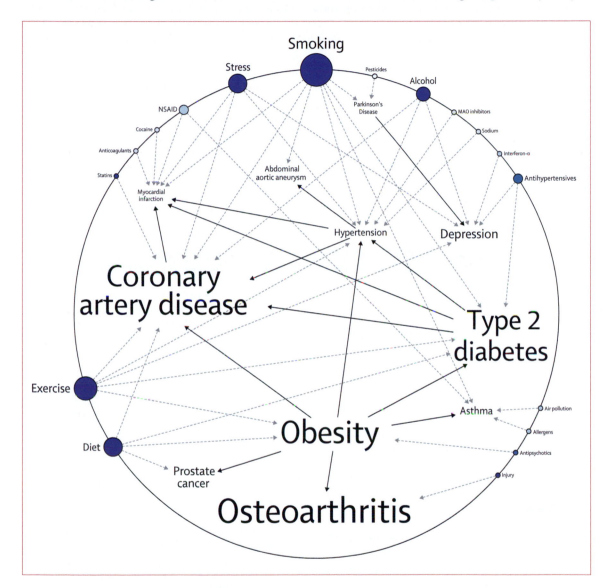

Figure 2 Gene–environment interactions for a specific 40-year-old man with a family history of coronary artery disease and sudden death. His genome was compared with disease-specific mutation databases and pharmacogenomics databases to estimate his clinical risk. For each environmental factor on the circumference of the chart, the text and circle size are proportional to the number of diseases associated with that factor. The colour intensity of the circle represents the risk probability. For diseases, text size is proportional to clinical risk. (Reproduced with permission from Ashley EA, Butte AJ, Wheeler MT *et al.* Clinical assessment incorporating a personal genome. *The Lancet* 2010; **375**: 1525–1535.)

bathe the lungs and the other route was across the heart through 'interseptal pores'. The left ventricle received this blood, which mixed with the 'pneuma' (air) coming to the heart through the pulmonary veins. The blood, fortified by the 'pneuma', was then ejected via the aorta towards the peripheral organs.

There were several problems with the theory. First, nobody had managed to identify 'interseptal' pores, so Michael Servetius, the Spanish theologian and physician, suggested that blood travelled from the right to the left ventricle by circulating through the lungs – an idea for which he died a martyr's death after being denounced by John Calvin for holding heretical opinions! Second, Galen's theory proposed a mixture of air and blood in the left side of the heart, which was a difficult concept

to accept once the structure of the heart valves had been established. Leonardo da Vinci had drawn these accurately but it was Andrea Caselapino, in 1571, who described the valves' actions correctly and went on to use the term 'circulatio'. Thus William Harvey, who studied in Padua from 1600 to 1602, would have been familiar with the Italians' ideas and was able to reach his own conclusions by standing on the shoulders of these giants. He published his famous work, *De Motu Cordis*, describing the dual circulation of the blood in 1628. Even Harvey was left with a problem: he could not demonstrate the connections between the arterial and venous sides of the circulation. The discovery of the capillaries had to wait for Marcello Malpighi's microscopical analysis of frog lung in 1661.

History

William Harvey (1578–1657)

William Harvey was born in Folkestone on 1 April 1578. It may have been April Fools day, but this man provided medicine with the boost that it needed to get it out of stagnation. The value of his work is put into perspective when you realise that to be honoured as a Harveian Orator by the Royal College of Physicians is the greatest distinction that one can aspire to.

Harvey did his medical training at Caius College, Cambridge, and later at Padua, Italy. In Padua, Harvey studied with Fabricius who had succeeded Fallopio (of fallopian tube fame). Galileo was the Professor of Mathematics at Padua at the time, but does not appear to have been influential in Harvey's development.

Harvey was elected a full fellow of the College of Physicians in 1607 and soon afterwards became

Assistant Physician to St Bartholomew's Hospital. This helped to establish his private practice and his famous patients included King James I, King Charles I and the Lord Chancellor, Sir Francis Bacon. Although Bacon is given the credit for inductive thinking, it was Harvey who applied it to his investigations of the heart. Harvey, in fact, had very little respect for Bacon and had stated that Bacon 'writes philosophy (science) like a Lord Chancellor; I have cured him of it'.

In his famous book *De Motu Cordis*, he described the challenge of working out the sequence of the heart's movements. *De Motu Cordis* evolved in two stages: initially it was an investigation into the heart beat and the arterial pulse, and only later did he include the investigation of the circulation. The book is quite small, only 72 pages, but is one of the most important pieces of scientific work.

CHAPTER 7

VASCULAR OCCLUSION AND THROMBOSIS

VASCULAR OCCLUSION

We start our consideration of cardiovascular disorders with the problems of vascular occlusion.

Vascular occlusion may be arterial or venous and the effect of any occlusion will depend on the following:

- The type of tissue involved
- How quickly the occlusion develops
- Whether there is collateral circulation.

Collateral vessels provide an alternative route for the blood and they are sometimes able to compensate completely, especially if the occlusion develops slowly. The venous system has more collaterals than the arterial system, e.g. there are anastomoses between the portal and systemic veins, around the lower end of the oesophagus, and also linking veins between the deep and superficial venous plexus in the leg. This means that occlusion of a deep vein in the calf does not produce haemorrhagic infarction of the foot, but just a mild oedema of the tissues and congestion of the superficial veins because of their increased flow. Unfortunately, not all veins have a collateral system. If the central vein of the retina is occluded, as may happen in thrombosis of the cavernous sinus due to local infection, the tissue of the orbit becomes oedematous and congested so that the eye is pushed forward (proptosis), and there may be local haemorrhage as the small vessels rupture because of the increased pressure. In the worst cases the venous pressure rises until it exceeds the arterial pressure and prevents arterial flow. This produces infarction, i.e. death of the tissue, and the infarcted tissue is red or purple and swollen because of the haemorrhagic oedema. The word 'infarction' actually comes from the Latin '*farcire*' meaning to stuff, and it is thought to have originally been used for the appearance of venous infarcts stuffed with blood.

Arterial collaterals exist in various areas such as the gut, the circle of Willis and, to some extent, the heart (see Fig. 9.5). Arterial occlusion without the benefit of collaterals will produce ischaemic infarction where the tissue is pale without any swelling. Occasionally, arterial infarcts are haemorrhagic because there is reperfusion or some limited arterial flow leading to leakage of blood from necrotic small vessels. In incomplete arterial occlusion, the effects depend on the tissue's demand for metabolites. Brain and heart tissue are highly susceptible to ischaemic injury whereas bone and skeletal muscle are quite resistant. It is possible to reduce a tissue's demand by cooling the tissue as is done in some types of surgery.

Vascular occlusion can be due to:

- thrombosis
- embolism
- atherosclerosis
- spasm
- external compression.

We discuss the first two of these causes in some depth in this chapter, starting with thrombosis.

THROMBOSIS

Patients presenting with an arterial thrombus are generally middle-aged or elderly and may have circulatory problems due to atherosclerosis. Many will be smokers and some may have diabetes. Their symptoms and signs will depend entirely on which vessel is affected. In contrast, a patient with venous thrombosis may be any age but generally will be rather immobile or forced to be immobile, such as after an operation. Such patients frequently complain of pain in a calf muscle and often swelling of the foot and ankle. But why should such people suddenly develop a thrombus? Much is known now about normal haemostatic mechanisms but the most important factors influencing thrombus formation were described more than a century ago by Rudolf Ludwig Karl Virchow (Fig. 7.1).

Virchow was born on 13 October 1821. As a child he excelled at school and his examination reports were rather monotonous because they contained only three terms: 'excellent', 'very good' and 'most satisfactory'! Virchow attended medical school in Berlin in 1839 and, even before the existence of platelets and clotting factors was known, had suggested that the development of a thrombus depended on the following:

- Alteration to the constituents of the blood
- Damage to the endothelial layer of the blood vessel
- Changes in the normal flow of blood.

These three factors are known as Virchow's triad and they are the clues that allow us to understand what has happened to our patients with venous and arterial thrombosis. But first we must revise the body's normal haemostatic mechanisms.

NORMAL HAEMOSTATIC MECHANISMS

Blood and the blood vessels perform an incredibly complex set of tasks. The blood vessels provide a conduit and the blood transports nutrients, waste products and components to defend against infection. However, the vessels are not simple tubes because it is crucial that substances can diffuse between the blood in the vessel and the surrounding tissues. This means that the small vessels (capillaries) must be permeable, i.e. they are designed to leak. The trick is to ensure that the leakage is under control. For larger vessels that do not need to be permeable, the potential problem occurs because the vascular system is a closed, pressurised system and so at risk of rupture. We must remember that the network of blood vessels is vast and many are superficial, and so susceptible to traumatic damage.

All of this means that the normal haemostatic mechanisms are crucial for stopping blood from leaking but they must also be finely controlled so that thrombus does not form under normal circumstances. There are three main components:

1 Platelets
2 Vessel wall endothelium
3 Soluble blood proteins involved in haemostasis.

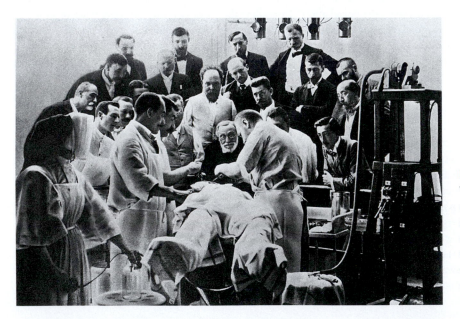

Figure 7.1 Rudolf Virchow (1821–1902) observing an operation in Berlin: he is in the dark suit at the head of the table. (Reproduced with permission from the Wellcome Library, London.)

The haemostatic blood proteins comprise the following:

- Coagulation factors
- Coagulation inhibitors
- Fibrinolytic components.

Briefly, the sequence of events is as follows. Injury to the vessels causes an initial *vasoconstriction* which helps to slow the blood flow. The damaged endothelium of the vessel exposes the subendothelial connective tissue, which attracts platelets and causes them to adhere to the damaged area. The activation of the platelets causes them to release soluble factors, resulting in platelet aggregation called the primary haemostatic plug. Tissue factor and platelet phospholipid cause the formation of fibrin via the coagulation pathway. The fibrin acts to stabilise the platelet plug and the process is termed 'secondary haemostasis'.

If this was all that was involved in maintaining haemostasis, could we really survive the assault on our circulation? Of course not! If the above system were set in motion with nothing to check its progress, soon the whole circulation would come to a standstill and become one big mass of thrombus. This is avoided by clearance, inhibition and inactivation of the coagulation factors as well as by digestion of fibrin.

Now, if we return to Virchow's triad, we can consider both the normal physiology and pathology of each component in more detail.

Endothelial cells

Endothelial cells are dynamically managing haemostasis all the time through expression of anticoagulant factors, which inhibit platelet aggregation and activation, fibrinolytic activity to remove thrombus, and their ability to switch to procoagulant actions if required. The details are covered in Figs 7.2 and 7.3 (and see Fig. 8.10).

Blood constituents in normal haemostasis

The most important blood constituents involved in normal haemostasis and thrombosis are platelets and the numerous components of the coagulation pathway.

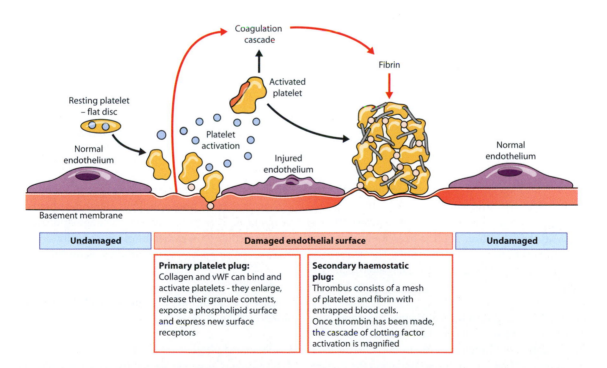

Figure 7.2 Platelet adhesion, activation and aggregation: platelet aggregates ('primary platelet plugs') can plug small breaches in the endothelium, but serious tears require the addition of a fibrin mesh (generated by the coagulation cascade) to stabilise the platelets and form thrombus while healing takes place.

Chapter 7: Vascular occlusion and thrombosis

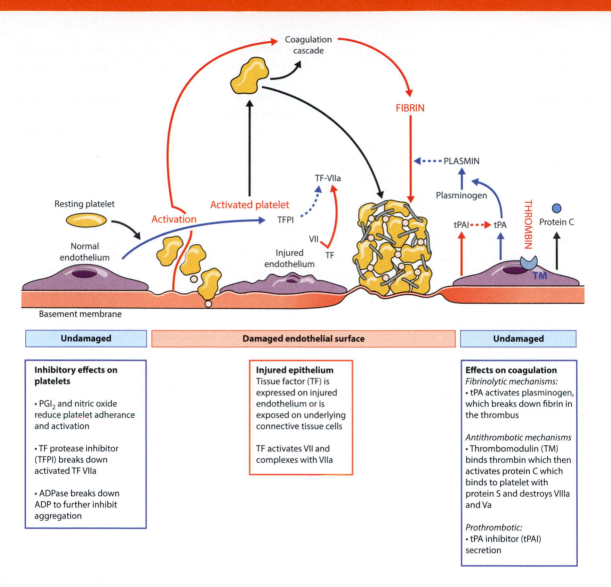

Figure 7.3 Coagulation is an energy-dependent process, usually initiated by the activation of factor VII by contact with tissue factor (TF) (extrinsic or tissue factor pathway). Platelet phospholipid membrane binds clotting factors so that a cascade of sequential activation can occur. Thrombin is crucial to coagulation: it catalyses the formation of fibrin from fibrinogen, exerts positive feedback on earlier stages of coagulation and initiates fibrinolytic mechanisms. The Intrinsic or contact activation pathway initiated by factor XII activation (Hageman factor) is closely inter-related.

Platelets

Platelets are small (3 μm) cytoplasmic fragments produced by megakaryocytes in the bone marrow under the regulation of thrombopoietin produced in the liver and kidneys. They survive for 7–10 days in the peripheral circulation and contain a variety of active components in α and dense granules.

Their role in thrombosis can be divided into three phases (see Fig. 7.2):

1 Adhesion of platelets to vessel wall
2 Activation of platelets leading to change in shape and secretion of granules
3 Aggregation of platelets and amplification of processes.

Basic biology of platelets

Some diseases caused by platelet abnormalities

- Von Willebrand's disease: lack of vWF
- Bernard–Soulier syndrome: lack of gpIb receptor
- 'Grey platelet' syndrome: lack of α granules
- Wiskott–Aldrich syndrome: lack of dense granules

All these conditions are rare but they illustrate the consequences of platelet receptor and granule abnormalities.

Platelet plasma membrane receptors

- GpIb complex: binds vWF and thrombin; initiates aggregation and activation
- GpIa/IIa (integrin collagen receptor $\alpha_2\beta_1$): binds type I, II, IV and IV collagen
- GpIIb/IIIa (platelet integrin $\alpha_{IIb}\beta_3$): binds fibrinogen

Platelet granule contents

- α *Granules*
  - Platelet factor 4 (an anti-heparin)
  - β-Thromboglobulin
  - Platelet-derived growth factor (PDGF)
  - vWF
  - P-selectin
  - Thrombospondin
  - Vascular endothelial growth factor (VEGF)
  - Fibroblast growth factor (FGF)
  - Insulin-like growth factor (IGF)
  - Transforming growth factor α (TGF-α)
  - Factors V and XIII
  - Fibrinogen
  - Fibronectin
  - Albumin
  - Proteoglycans
- Dense granules
  - CD63
  - ADP/ATP
  - Calcium
  - Magnesium
  - Histamine
  - Adrenaline
  - Serotonin
- Lysosomes
  - Lysosomal membrane proteins (LAMPs)
  - Acid hydrolases

When the endothelium is damaged and collagen is exposed, the first event is adhesion of platelets. This is achieved via platelet surface membrane receptors:

- gpIa, which binds to collagen
- gpIb, which binds to von Willebrand's factor (vWF)
- gpIIb/IIIa, which binds to fibrinogen and vWF.

After adhesion, the platelets are activated and release the contents of their granules. There are two main types of granules: α granules and dense granules. The most important secretory products are calcium, which is needed for the coagulation pathway, and adenosine diphosphate (ADP), which induces platelet activation. Activated platelets change shape and also synthesise thromboxane (TxA$_2$), a prostaglandin. Prostacyclin (PGI$_2$) (produced by endothelial cells) and thromboxane (produced by platelets) interact with endothelium and other platelets to influence clot formation in opposing ways, but both are affected by aspirin (see below).

Platelet aggregation involves the gpIIb/IIIa receptor complex mentioned above. This is expressed after activation and is most important in binding fibrinogen, which acts as a bridge to the adjacent platelet. Not surprisingly, there are 'loops' in this process to amplify the reaction. Most importantly, activated platelets express membrane phospholipid (formerly known as platelet factor 3), which stimulates the intrinsic (also called contact activation) pathway of the coagulation cascade (see below), resulting in the production of thrombin. Thrombin acts to stimulate platelets and so enhances the reaction.

The platelet has another important facet to its character; it has mechanical properties. An unstimulated platelet has a disc shape maintained by microtubules, and actin and myosin filaments at the periphery. On activation, the platelet is transformed into a sphere with long pseudopods which spread over the damaged surface and then, after aggregation, the internal filaments slide so that the platelet plug contracts to stabilise and anchor it.

The most common cause of defective platelet function is aspirin therapy due to inhibition of cyclooxygenase, resulting in impaired thromboxane synthesis. After a single dose of aspirin, the defect lasts 7–10 days, i.e. the life of the platelet. Aspirin also inhibits PGI_2 production in endothelial cells and you might think this would increase clotting; however, endothelial cells have nuclei and full protein synthetic abilities, so more PGI_2 can be produced.

> 🔗 Read more about easy bruising in Pathology in Clinical Practice Case 5

Coagulation components

The components and pathway involved in coagulation are shown in Fig. 7.3.

This is the same system as the one we mentioned in Chapter 4 when discussing inflammatory mediators. Then we were particularly interested in fibrin degradation products; now our interest focuses on fibrin, which is the final product of the pathway and acts to stabilise the plug of aggregated platelets.

The process achieves a massive amplification effect such that one molecule of activated factor XI creates up to 2×10^8 molecules of fibrin. The common

pathway begins at factor X, which acts on prothrombin to produce thrombin; this itself has a variety of actions but most importantly converts fibrinogen to fibrin. Generally, each step in this cascade involves the following:

- Activated enzyme
- Substrate for a coagulation factor
- Co-factor
- Calcium ions
- Phospholipid surface.

Some feedback loops are included in Fig. 7.2 but, for simplicity, the control mechanisms that inhibit or inactivate reactions have been omitted. The following are the mechanisms included:

- Depletion of local clotting factors
- Clearance of activated clotting factors by the liver and mononuclear phagocyte system
- Neutralisation of activated coagulation factors by forming a complex, e.g. anti-thrombin, α_2-macroglobulin
- Proteolytic degradation of active coagulation factors, e.g. protein C
- Fibrinolysis: this is of major importance.

The most important enzyme capable of digesting fibrin is plasmin. This is produced from plasminogen by a factor XII-dependent pathway, therapeutic agents such as streptokinase, or tissue-derived plasminogen activators. Plasminogen activators (PAs) fall into two classes:

1 Urokinase-like PA (uPA)
2 Tissue-type PA (tPA).

They differ in that uPA activates plasminogen in the fluid phase, whereas tPA (principally produced by endothelial cells) is active only when attached to fibrin. Conveniently, some plasminogen is bound to fibrin as a thrombus is formed and so is perfectly situated for conversion by the tPA to plasmin, which can then digest the thrombus. Compounds capable of breaking down thrombi have enormous therapeutic potential for restoring blood flow before significant myocardial or cerebral infarction has occurred.

VIRCHOW'S POINT 1: ALTERATION IN THE CONSTITUENTS OF THE BLOOD

Blood that clots more readily than usual is termed hypercoagulable. This may be caused by a variety of different mechanisms including the following:

- An increase in blood cells (polycythaemia)
- Loss of the plasma fraction of the blood (severe burns)
- Increased numbers of platelets (thrombocythaemia)
- Increased amount or aggregation of plasma proteins (myeloma, cryoglobulinaemia)
- Severe trauma
- Disseminated cancer
- Pregnancy and contraceptive pill use
- Thrombophilia (enhanced coagulation).

Thrombophilia is the increased tendency to have arterial or venous thrombi because of increased coagulation; it can be inherited or acquired. Patients generally either have recurrent thrombi or start having them at an early age. There is also an association with pregnancy problems such as recurrent fetal death, growth restriction and prematurity. The most common form of *acquired* thrombophilia is anti-phospholipid antibody syndrome where a patient has antibodies and clinical thrombosis or pregnancy problems. Around 2% of the population have anti-phospholipid antibodies and half will also have autoimmune systemic lupus erythematosus (SLE) or a lupus-like disease. It is not known whether these antibodies act through endothelial damage or by activating complement, platelets or the coagulation cascade directly. The *inherited* thrombophilias result in venous thrombosis and are caused by a shortage or inactivity of circulating inhibitors of coagulation.

Figure 7.4 shows how protein C and protein S normally form a complex with thrombin and the endothelial surface membrane components, thrombomodulin (TM) and phospholipid (PL), which deactivates the normal activated co-factors Va and VIIIa. This is a natural anticoagulant mechanism. *Protein C deficiency* is an autosomal dominant disorder or can be required in liver disease and disseminated intravascular coagulation (DIC) (see later page 236). *Anti-thrombin deficiency* can also be inherited as an autosomal dominant but has several subtypes depending on which binding sites are affected. It

naturally inhibits thrombin and clotting factors IXa, Xa, XIa and XIIa. Up to 80% of patients with the abnormal gene develop venous thrombosis by the age of 55 years.

The most common inherited thrombophilia is factor V Leiden disease, which is generally due to a single point mutation that results in glycine replacing arginine at the position on the protein factor V Leiden, where protein C would normally bind and cleave the molecule to activate it. This is an autosomal dominant pattern of inheritance and affected heterozygotes have a four- to eightfold increase of venous thrombosis and homozygotes have a 50- to 100-fold increase. The gene prevalence is around 3–7% in western Europe.

Small print

Main inherited causes of thrombophilia

- Factor V Leiden: point mutation results in protein that is resistant to cleavage
- Prothrombin gene polymorphism: point mutation in non-translated region results in increased transcription and raised prothrombin levels
- Protein C and S deficiencies
- Anti-thrombin deficiency

At the end of this chapter we discuss disseminated intravascular coagulation (DIC), which is a derangement of normal human stasis that can result in bleeding or increased clotting. The causes of DIC (see Fig. 7.11) can induce a hypercoagulable state.

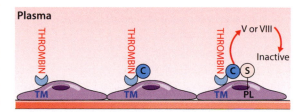

Figure 7.4 Protein C and S deactivate Factors V and VIII.

Chapter 7: Vascular occlusion and thrombosis

VIRCHOW'S POINT 2: CHANGES IN THE ENDOTHELIUM

The endothelium is quite remarkable, because it is capable of initiating both thrombogenic and anti-thrombogenic stimuli (see figures 7.2 and 7.3). Normally, these two groups of actions are finely balanced in favour of preventing thrombus formation. Damage to the endothelium will, however, tip the balance towards thrombosis. The endothelium also has another very important role; this is to prevent the elements of blood from coming into contact with the subendothelial connective tissue, which is highly thrombogenic. In vessels affected by atheroma, not only is the endothelium more readily damaged but the subendothelial tissue also consists of the components of atheroma that are extremely thrombogenic.

Endothelial damage is of most significance in arterial thrombosis. Endothelial cells may be lost where an atheromatous plaque has ulcerated or when vessels are damaged by surgery, infection, immune-mediated damage (arteritis) or indwelling vascular catheters. More subtle metabolic damage occurs in areas where there is turbulent flow, creating haemodynamic stress, or in patients with prolonged high blood pressure. The important damaging agents are listed in the Key facts box below.

🔑 Key facts

Key causes of endothelial cell damage

Physical

- Haemodynamic forces
- Trauma
- Irradiation

Chemical

- Advanced glycation end-products generated in diabetes and ageing
- Hypercholesterolaemia
- Oxidised lipoprotein products
- Cigarette smoke

Inflammatory/Infective

- Proinflammatory cytokines
- Bacterial products
- Viruses

In the heart, the endocardial surface is covered by endothelium, which can be damaged in a myocardial infarction. Also, the valve surface endothelium may be damaged by inflammatory endocarditis, which promotes thrombus formation on valve leaflets, resulting in altered function, a variety of heart murmurs and the danger of throwing emboli into the systemic circulation.

Clinically, the most important change is the endothelial damage related to atherosclerosis, which is discussed in Chapter 8.

VIRCHOW'S POINT 3: CHANGES IN THE NORMAL FLOW OF BLOOD

There are two principal ways in which the normal flow can be disturbed: the normal lamellar flow pattern can be altered (turbulence) or the speed may be reduced (stasis) but both lead to similar changes.

During normal flow, red and white blood cells concentrate in the central, fast-moving stream whereas platelets flow nearer to the periphery, and the layer closest to the endothelium is usually devoid of cells and platelets. If the blood flow slows down or turbulence produces local countercurrents, several factors increase the likelihood of thrombus formation:

- Platelets come into contact with the endothelium.
- Turbulence may damage endothelial cells.
- There is no inflow of fresh blood containing clotting factor inhibitors.
- There is no clearance of blood containing activated coagulation factors.

As you see, both turbulence and stasis operate in thrombosis but turbulence is most important in arteries whereas stasis is more important in veins.

Arterial thrombosis

Turbulence tends to occur where arteries branch and over the irregular surface of an atheromatous plaque. It also occurs when cardiac valves have been damaged by inflammation, as may occur with rheumatic fever and infective endocarditis, or have been replaced by artificial valves. This topic is covered in more detail in the section on atherosclerosis on page 235. Briefly, it is now apparent that the endothelial cells can develop different phenotypes in response to biomechanical stimuli related to blood flow patterns. This involves at least

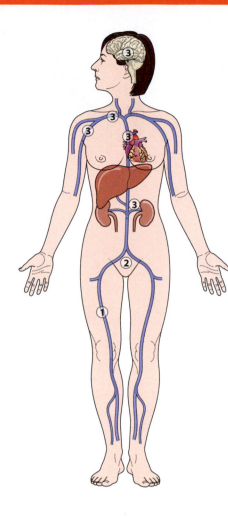

In order of frequency	Clinical setting
1 Leg veins	Immobility, post-surgery, and hyper-coagulability states
2 Pelvic veins	Post-childbirth, puerperal sepsis pelvic surgery, and tumours
3 Others:	
Inferior vena cava	Extrinsic compresion by tumour, extension from leg or iliac veins
Renal vein	Tumour extension from kidney
Portal/hepatic veins	Local sepsis, tumour compression
Cavernous sinus	Facial sepsis
Superior sagittal sinus	Hemorrhagic infarction
Superior vena cava	Extrinsic compression by mediastinal tumour
Axillary	Trauma from rucksack, local surgery

Other clinical settings which can predispose to thrombosis:
Patients with malignant tumours, especially carcinoma of ovary, brain
    and pancreas
Inflammatory disorders due to down-regulation of protein C
    e.g. inflammatory bowel disease, TB, SLE
Blood disorders, e.g. polycythaemia, sickle cell, PNH
High-dose oestrogen therapy in contraceptive pills and HRT
Anti-phospholipid syndrome

Figure 7.5 Sites and clinical setting of venous thrombosis.

48 distinct transcription factors. Evolving knowledge similar to this helps improve our understanding of the mechanisms operating and the development of potential therapeutic interventions.

Stasis is generally important in arterial thrombosis only if the heart or arteries have been damaged. Abnormal dilatations of large vessels (aneurysms) will produce pockets of stagnant blood that will thrombose, and myocardial infarction may result in a localised area of damaged heart muscle, which does not move, or in an arrhythmia that will affect the contraction of a whole chamber.

Venous thrombosis

Thrombus formation, related to stasis of blood, is more common in the venous circulation and occurs particularly in the legs or pelvic veins of immobile individuals (Fig. 7.5). Why is stasis common in the leg vessels when

Key facts

Key factors influencing thrombus formation

Virchow's triad
    Altered blood
        ↑ cells
        ↑ platelets
        ↑ protein
        ↓ fluid
    Altered wall
        endothelial loss (atheroma)
        endothelial damage (smoking)
    Altered flow
        stasis
        turbulence

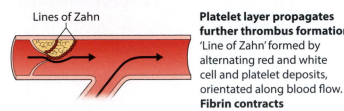

Decreased blood flow or increased coagulability
Thrombus forms in valve pocket. Platelets adhere to surface of thrombus

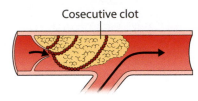

Lines of Zahn

Platelet layer propagates further thrombus formation
'Line of Zahn' formed by alternating red and white cell and platelet deposits, orientated along blood flow. **Fibrin contracts**

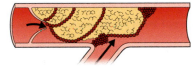

Cosecutive clot

Once lumen occluded, 'consecutive' unstable clot forms
No lines of Zahn, slow flow and no new platelets. Weakly attached to wall and easily dislodged

Entry of tributary
i   may stabilise thrombus by re-attaching to wall

ii   permits further propagation

iii   may carry fragments of thrombus into general circulation: embolization

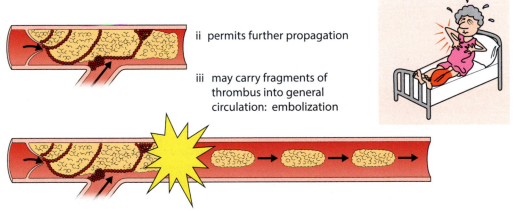

Figure 7.6 Venous thrombosis: development, propagation and embolization.

the patient is immobilised? If you remember the physiology of venous return from the legs, you will recall that it is contraction of skeletal muscles which pushes blood along the veins and it is the presence of valves which ensures the direction of flow. Understanding this has influenced patient management. Patients are

encouraged to move their legs regularly when confined to bed, wear compression stockings on long air flights and leg muscles are stimulated to contract during long operations.

Thrombus formation often begins within the venous valve pockets (Fig. 7.6). The initial cluster of platelets

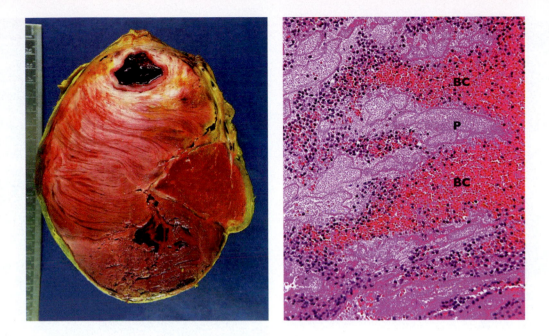

Figure 7.7 Splenic artery aneurysm containing thrombus showing lines of Zahn. Photomicrograph shows alternating layers of platelets (P) and blood cells (BC) trapped in a fibrin mesh.

activates the clotting cascade to produce a small thrombus. A second phase of platelet aggregation then occurs to cover the original thrombus and promote a further wave of coagulation. This process is repeated again and again to extend the thrombus, so-called propagation. The resultant thrombus has alternate layers of platelets and a red cell–white cell–fibrin mixture, which produces a rippled effect apparent to the naked eye, termed 'lines of Zahn' (Fig. 7.7). The direction of the lines relates to the pattern of blood flow in the vessel. These platelet layers anchor the thrombus to the adjacent endothelium, helping to stabilise it.

Once a vessel is completely occluded by thrombus, blood flow ceases and the stagnant column of blood clots without the production of any 'lines of Zahn'. This is called 'consecutive' clot, and it is particularly dangerous because it is adherent to the vessel wall through only its attachment to the original focus of thrombus. This makes it especially likely to break off and embolise to another area. The radiological appearances of vein thrombus are shown in Figure 7.8.

It is also worth emphasising at this point that, similar to most phenomena in the body, the three major factors of Virchow's triad rarely work in isolation. In myocardial infarction, ischaemia damages the endocardium but the affected myocardium also fails to move normally, hence causing local stasis of blood, which is also important in formation of the thrombus within the ventricle. So, although it is imperative that one knows the basis for Virchow's triad, it is also important to remember that many factors interact to produce the final picture in any one patient.

Key facts

Arterial and venous thrombosis

	Arterial	Venous
Clinical setting	Person with atheroma	Immobile person
Pathogenesis	Turbulent flow	Stasis
	Damaged endothelium	Hypercoagulable blood
	Platelet rich	Clotting factor rich
Symptoms	Sudden onset	Slow onset
Complications	Infarction	Pulmonary embolus

Chapter 7: Vascular occlusion and thrombosis

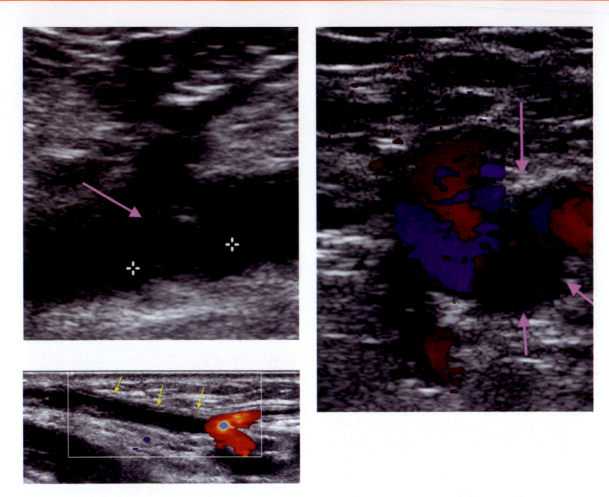

Figure 7.8 Top left: ultrasound scan of the common femoral vein showing a filling defect (purple arrow) due to an acute deep vein thrombus. Left: ultrasound scan showing a thrombus occluding a calf vein with colour flow up to but not through the vessel (yellow arrows). Top right: ultrasound scan with colour Doppler showing colour flow in collateral veins but no colour flow in the thrombosed femoral vein (purple arrows).

NATURAL HISTORY AND COMPLICATIONS OF THROMBOSIS

Once a thrombus has formed, what are the possible outcomes? As you know, the body possesses many effective systems for regulating thrombus formation during normal haemostasis. The ideal solution is that these systems halt the thrombotic process and remove the debris to leave a normal blood vessel. This process is termed 'dissolution'. If the thrombus cannot be removed, it may be organised or recanalised (Fig. 7.9).

Alternatively, it may be cast off into the circulation, i.e. it may embolise (Fig. 7.10).

Dissolution is thought to occur commonly in the small veins of the lower limb. Interestingly, venous intima contains more plasminogen activator than arterial intima, which may be the reason. Drugs with a thrombolytic action, such as tPA, can be given to patients early after thrombosis to promote dissolution of the clot and, hence, resolution. It is important that this drug be given within hours because the drug has much less effect on polymerised fibrin, which predominates later.

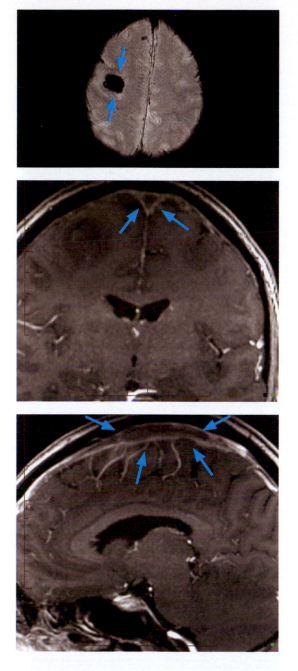

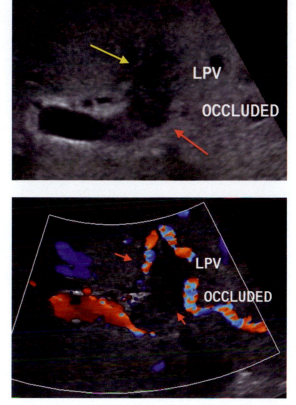

Figure 7.10 Ultrasound scans of the liver showing acute occlusion of the left main portal vein by thrombus, with collateral veins filling around the thrombosed vein. The main risk factors for portal vein thrombosis are a prothrombotic disorder, cirrhosis, abdominal inflammation and infection, and hepatocellular carcinoma with tumour invading the portal vein. If the thrombosis is not adequately treated, the patient may develop devastating portal hypertension with splenomegaly, ascites and varices at the site of portosystemic venous shunts.

Figure 7.9 Top: MRI of the brain using a susceptibility-weighted sequence designed to show haemorrhage. This axial section shows a haemorrhagic infarct (blue arrows) of the right frontal lobe secondary to venous congestion and ischaemia related to a sagittal sinus thrombosis. Middle and bottom: MRI of sagittal and coronal post-intravenous gadolinium sequences showing the filling defect in the superior sagittal sinus due to dural venous thrombosis (blue arrows).

Organisation of a thrombus involves similar processes to the organisation of inflammation described in Chapter 4. When the thrombus has formed, polymorphs and macrophages begin to degrade and digest the fibrin and cell debris. Later, granulation tissue grows into the base of the thrombus so that the thrombus is converted into a mass of small vessels separated by connective tissue. These vessels originate from the vasa vasorum of the adventitia of the blood

Chapter 7: Vascular occlusion and thrombosis

vessel, and it is unlikely that the blood flowing through these is of much clinical importance (but see below for collateral circulation).

'Recanalisation' is a term used by clinicians to indicate that there is useful flow through a previously occluded vessel. Obviously, if thrombolytic treatment has been successful, the thrombus will be dissolved, the original intimal lining will still exist and the clinician will see flow on the arteriogram. A similar situation occurs if the clot retracts so that it is obstructing only part of the flow and the blood flow

is, at least partially, restored through the original lumen (Fig. 7.11).

It is also possible to have *new* endothelium-lined channels through the occlusive thrombus. This is thought to occur by the production of clefts within the thrombus, resulting from a combination of local digestion and shrinkage. The clefts extend through the clot and become lined by endothelial cells derived from the adjacent intima. The amount of flow through such a segment will depend on the number and size of the conduits, but the vessel will not be 'as good as new'.

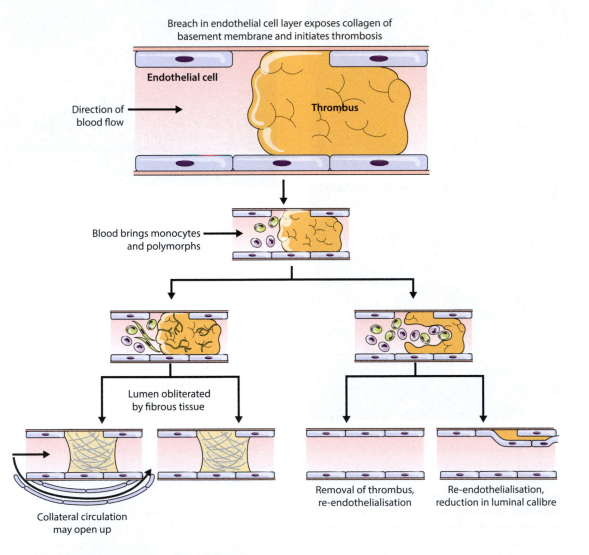

Figure 7.11 Possible outcomes of vessel thrombus

Part 3: Features of cardiovascular disorders

EMBOLISM

One of the complications of thrombosis is embolism and we now go on to consider the different types of emboli and their complications.

An embolus is solid, liquid or gaseous material, which is carried in the blood from one area of the circulatory system to another.

Almost all (about 99%) emboli arise from thrombi and, thus, there is a tendency to use the term 'thromboembolism' as synonymous with embolism. This is not strictly true because there are many other, though admittedly rarer, causes of emboli, including the following:

- Fragments of atheromatous plaques
- Bone marrow
- Fat
- Air or nitrogen
- Amniotic fluid
- Tumour
- Foreign material, e.g. intravenous catheter.

As most emboli come from thrombi, we start our discussion with this particular type.

Where emboli lodge depends on their size, origin and the relevant cardiovascular anatomy. Those that arise in the venous system can travel through the right side of the heart to end up in the pulmonary circulation. Those that arise in the left side of the circulation will block systemic arteries, and the clinical effect will depend on the organ involved, be it brain, kidneys, spleen or the periphery of limbs.

Emboli to the lungs from venous thrombosis represent an important preventable cause of morbidity in hospitalised patients and we consider these first.

PULMONARY EMBOLISM

The lungs are very interesting organs because they have a dual blood supply. The lung receives not only deoxygenated blood via the pulmonary arteries but also oxygenated blood from bronchial arteries feeding directly from the aorta. Hence, the lungs have an established collateral arterial circulation. This means that occlusion of a branch of the pulmonary artery rarely causes infarction of the lung parenchyma and,

as the alveolar walls are intact, resolution is possible. The effects of a pulmonary embolus will depend on three factors:

1 The size of the occluded vessel
2 The number of emboli
3 The adequacy of the bronchial blood supply.

The size of the occluded vessel (Fig. 7.12)

If a large embolus occludes a main pulmonary artery or even sits astride the bifurcation of the pulmonary trunk, a so-called saddle embolus, the patient's blood pressure may suddenly drop (shock) or there may be acute right-sided heart failure (cor pulmonale). Either can cause immediate death if at least 60% of the pulmonary blood flow is occluded. If the patient survives to reach hospital, it may be possible to lyse the embolus.

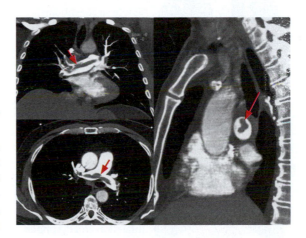

Figure 7.12 CT pulmonary angiography has become the investigation of choice for detecting pulmonary embolism. The technique relies on a bolus of intravenous iodinated contrast given through a large cannula in a proximal arm vein with the scan being timed to start during maximal opacification of the pulmonary arterial tree. CT pulmonary angiography became possible because of the advent of the helical or spiral CT scanners which appeared in the 1990s and which are capable of acquiring very thin slices very quickly, allowing an image to be produced in any anatomical plane. The CTPA shows a saddle embolus (red arrow) from a deep vein thrombosis of the leg lying across the bifurcation of the pulmonary outflow tract.

Chapter 7: Vascular occlusion and thrombosis

Around 95% of emboli originate in the iliofemoral venous system, with a small number coming from the pelvic veins, calf muscle veins and superficial veins of the legs. Obviously, the diameter of these emboli will correspond with the diameter of the vessel of origin, which is less than the size of the major pulmonary arteries. So how does an embolus block a vessel larger than itself? It becomes coiled, as illustrated in Fig. 7.13.

Not infrequently, a long single embolus may fragment in the circulation to produce numerous small emboli. These may reach the small pulmonary arteries as a 'shower' to occlude several vessels at the same time, producing similar, sudden, severe clinical effects to a single large embolus.

If a medium-sized pulmonary artery becomes blocked, this may produce no clinical effect because the bronchial circulation is able to supply the lung parenchyma. Generally, there will be local haemorrhage but no damage to the framework of the lung and so complete resolution can occur. If the haemorrhage is small, the patient may be asymptomatic but, if large, the patient may have some dyspnoea or haemoptysis.

If the small peripheral pulmonary arteries are involved, there may be infarction because the area is beyond the territory of the bronchial collateral supply so the pulmonary arteries are, in effect, end-arteries. Generally the area affected will be quite small but may produce symptoms, especially if there are multiple emboli.

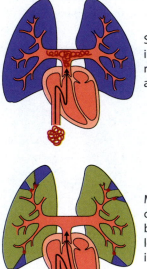

Single large embolus in major artery causing right sided heart failure and circulatory collapse

Multiple emboli combined with poor bronchial blood supply leads to pulmonary infarction and dyspnoea

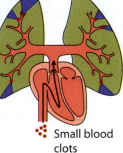

Small blood clots

Figure 7.13 The effects of pulmonary embolism range from catastrophic shock with circulatory collapse, through moderate or mild respiratory impairment with pleuritic chest pain, to virtually no symptoms at all. The size, number and timing of the emboli and the quality of the bronchial artery circulation all influence the outcome. If small embolic events are sufficiently separated by time, thrombolytic mechanisms may clear them and restore normal circulation. The type of embolus also influences the sites affected and the clinical outcome. Here we illustrate two patterns of pulmonary thromboembolism.

Dictionary

Dyspnoea: sensation of shortness of breath. When associated with cardiac failure, it may be due to pulmonary oedema interfering with gaseous exchange and lung stretch reflexes. If worse on lying flat, it is called orthopnoea. If it is fast, it is tachypnoea.

Haemoptysis: coughing up blood from the respiratory tract.

The number of emboli

Multiple emboli may be thrown into the lungs as a single event or there may be successive embolic episodes. The first situation occurs when a single large embolus fragments into smaller emboli before reaching the lungs. The second scenario happens when initially only part of the thrombus breaks off but, hours or days later, a second piece follows. It also occurs because the patient may produce additional thrombi. If a patient survives the initial pulmonary embolus, there is a 30% risk of having a further embolus. This makes it extremely important that the patient receive prompt and effective anticoagulant therapy to reduce the risk. However, the anticoagulant therapy will not remove the existing embolus; this requires fibrinolytic treatment as described earlier. Sometimes a patient will remain in 'shock' despite complete lysis of the embolus, and this is possibly due to intense vasoconstriction of the peripheral pulmonary vessels.

The adequacy of the bronchial blood supply

If a patient has heart failure or pre-existing pulmonary disease, the bronchial blood supply will be impaired and emboli lodging in medium-sized pulmonary arteries will result in infarction. As the blockage is relatively proximal, the infarct will be large, extending as a cone with the apex at the blocked vessel and the base on the pleura (Fig. 7.10). Initially, the area will be firm and purple because of the haemorrhage and congestion but later it will be replaced by pale fibrous tissue and the area will shrink. Infarcts are most common in the lower lobes of the lungs and are multiple in 50% of cases.

These patients tend to get chest pain related to inflammation of adjacent pleura and shortness of breath due to both a reduction in lung volume and humoral and neural factors leading to vasoconstriction and bronchoconstriction.

A typical clinical scenario is that of an elderly patient in hospital who has cardiac failure and a fractured neck of femur after a fall. The combination of recumbency, cardiac failure and postoperative dehydration creates an ideal situation for the formation of a deep vein thrombosis (DVT) in the leg veins. A moderate-sized embolus, over a background of an inadequate collateral supply due to cardiac failure, results in significant ischaemia of the lung parenchyma and infarction.

THE FATE OF THE EMBOLUS

In some ways this is similar to that of a thrombus. Ideally, it will be lysed by the fibrinolytic system to restore patency of the vessel. If not, organisation takes place and the mass will be incorporated into the wall with possible recanalisation of the vessel. Spontaneous lysis is often very good so it is important to support the patient to allow 'nature' to do the healing.

If there are multiple emboli or repeated episodes of embolisation and organisation, the pulmonary vessel

---

Case study: pulmonary embolism

A 51-year-old woman notices swelling of her right calf in the taxi ride home from the airport after a trans-Atlantic flight. It is 1 o'clock in the morning and she is very tired and keen to go straight to bed so she does not pay close attention to her leg. Shortly after waking the following morning, she develops a sudden onset of shortness of breath that is associated with central chest pain. The pain has no radiation and is not associated with nausea or vomiting. The woman calls an ambulance and is taken to the accident and emergency department. On arrival she is tachypnoeic but does not have central cyanosis. Her pulse is 108 beats/min and regular. The blood pressure and apex beat are normal. Examination of the lungs and abdomen is normal. The patient's right calf is red, swollen and tender. It has a circumference 5 cm greater than the other side.

What is the diagnosis?
The patient has a pulmonary embolism secondary to lower limb DVT.

What factors would cause this person to be hypercoagulable?

There are various conditions that can make a person more prone to thrombosis. The following are those that are particularly associated with a DVT:

- Immobility
- Flight in a pressurised aircraft
- Obstruction to the venous outflow.

Why is examination and imaging of the pelvis important?
Some lower limb DVTs are secondary to compression of the iliac vessels by a tumour mass in the pelvis. If this possibility is overlooked, an opportunity to diagnose a malignant neoplasm may be missed.

Other than the tachypnoea, examination of the lungs was normal. Is that the expected?
Yes. Ventilation of the lungs is not disturbed and therefore parameters such as expansion, percussion and auscultation will be unaffected.

What is the reason for the appearance of the right calf?
The DVT in the right lower limb hampers venous drainage of the limb. This results in congestion and accounts for the clinical features of being red, swollen and tender.

Read more in Pathology in Clinical Practice Case 18

Chapter 7: Vascular occlusion and thrombosis

wall will thicken, resulting in a rise in pulmonary arterial pressure (pulmonary hypertension). This in turn means an increased workload for the right ventricle, which tries to compensate by becoming thicker (hypertrophy). Eventually, the right ventricle may not be able to compensate and cardiac failure will ensue. Right ventricular enlargement due to pulmonary disease is called cor pulmonale.

SYSTEMIC EMBOLISM

Systemic emboli travel in the internal circulation and commonly originate in the left side of the heart from thrombi forming on areas of myocardial infarction, abnormal cardiac valves or thrombus forming on atheromatous plaques in the aorta or carotid arteries. Other causes include fragments of atheromatous plaques that result from fissuring or ulceration of a plaque, and release lipid and cholesterol mixture into the circulation.

Arterial emboli, unless very small, nearly always cause infarction. Emboli to the lower limbs may produce gangrene of a few toes or of the entire limb. Cerebral emboli cause death or infarction unless the embolus lodges in an area that receives adequate collateral supply through the circle of Willis. Alternative sites are the upper limb and the vessels supplying the gut, kidney and spleen.

OTHER TYPES OF EMBOLI

The other types of emboli generally enter veins rather than arteries because veins have thinner walls and a lower pressure. Therefore, most are venous emboli that lodge in the lungs.

Bone marrow emboli

Bone marrow emboli are occasionally seen in histological sections of lungs *post mortem*. This is especially likely if the patient has suffered major trauma, such as a road traffic accident, but can even occur with the 'trauma' of attempted cardiac resuscitation, particularly in elderly people whose costal cartilages have ossified. Anything that fractures bone can release bone and bone marrow into the venous circulation, with resultant pulmonary emboli, but the clinical significance of this type of embolisation is unclear.

Fat emboli

Fat from the marrow cavities of long bones or from soft tissue can also enter the circulation as a result of severe trauma. However, they even form without any trauma and so alternative mechanisms must operate. Fortunately, although fat globules are found in the lungs of most victims of severe trauma, less than 5% will suffer from the 'fat embolism syndrome', which is characterised by respiratory problems, a haemorrhagic skin rash and mental deterioration 24–72 hours after the injury. The syndrome is unlikely to result merely from mechanical blockage of vessels but probably involves chemical injury to the small vessels of the lungs, producing pulmonary oedema and activation of the coagulation pathway to cause disseminated intravascular coagulation (DIC). However, the exact mediators have not been identified. The origin of the fat in the non-trauma cases may be chylomicrons and fatty acids in the circulation coalescing to form droplets: the emulsion instability theory.

Air and nitrogen emboli

Large quantities of air within the circulation can act as emboli by forming a frothy mass that can block vessels or become trapped in the right heart chambers to impede its pumping. Air can either enter the circulation from the atmosphere or be produced within the circulation by alteration of pressure.

Severe trauma to the thorax may open large vessels (e.g. internal jugular veins), allowing air to be sucked in during inspiration, or air may be forced into the uterine vessels during badly performed abortions or deliveries. Fortunately, small quantities of air, as may be introduced during venesection, dissolve in the plasma and it probably takes about 100 mL to produce problems. Air in the arterial system is more damaging.

A special type of air embolism occurs in deep-sea divers. Normally insoluble gases, such as nitrogen or helium in the diver's breathing mixture, will dissolve in the blood and tissues at the high pressures that occur deep beneath the sea surface. As the diver surfaces, the pressure is reduced and the gas begins to come out of solution as minute bubbles. If the reduction of pressure is rapid then these bubbles form emboli, which are particularly likely to lodge in the skeletal and cerebral circulation. The situation is slightly more complicated because platelets adhere to the nitrogen bubbles, activate the coagulation system and produce DIC (see below). The acute form of decompression sickness or 'bends' involves pain around joints and in skeletal muscle, respiratory distress and, sometimes, coma and death. In the early stages, it can be treated by putting the victim in a 'decompression' chamber where the high pressure will redissolve the bubbles and allow a slow, controlled decompression. The chronic form or Caisson's disease produces multiple areas of ischaemic necrosis in the long bones. (Caissons are high-pressure underwater chambers.)

Amniotic fluid emboli

This is an uncommon but life-threatening form of embolisation with a mortality rate of around 80%. Basically, amniotic fluid is forced into the circulation due to tearing of the placental membranes and rupture of the uterine or cervical veins. These emboli are a mixture of amniotic fluid, fat, hair, mucus, meconium and squamous cells from the fetus, and they most commonly lodge in the mother's alveolar capillaries. Clinically, there is sudden onset of respiratory failure often followed by cerebral convulsions and coma. There is also excessive bleeding as a result of DIC and the consumption of clotting factors (Fig. 7.14). The exact mechanism is still unclear but it is not due simply to blockage of the pulmonary vasculature; it is postulated that some factor, such as prostaglandin $F_2\alpha$, in the amniotic fluid may be involved.

Tumour emboli

Embolisation of tumour is an important mechanism of tumour spread but it is unlikely to have any immediate cardiovascular effects. The mechanisms involved in this process are discussed in Chapter 14.

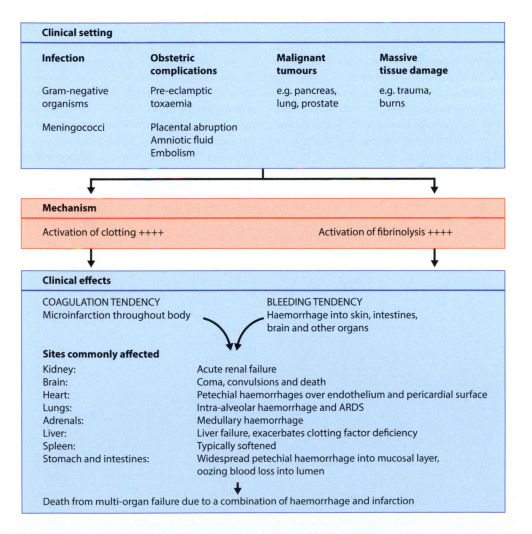

Figure 7.14 Disseminated intravascular coagulation: clinical settings and effects.

Chapter 7: Vascular occlusion and thrombosis

Foreign material

This commonly occurs in intravenous drug users where magnesium trisilicate (talc) is deposited in small lung vessels and provokes a granulomatous reaction.

DISSEMINATED INTRAVASCULAR COAGULATION

This is a convenient moment to discuss DIC. We have just mentioned amniotic fluid embolism and we will soon move on to 'shock' (Chapter 9) – both conditions that can produce DIC. Furthermore, DIC results from a loss of control in the clotting and fibrinolytic systems, which should still be fresh in your memory!

There is no typical clinical presentation because any organ may be affected and the major problem may be excessive clotting, which blocks numerous vessels, or inadequate clotting resulting in haemorrhage. As a general rule, sudden-onset DIC presents with bleeding problems, is particularly associated with obstetric complications, and may resolve once the obstetric situation has improved. In contrast, chronic DIC is more common in patients with carcinomatosis and the thrombotic manifestations dominate.

Small thrombi form anywhere in the circulation and produce microinfarcts. In the brain this may result in convulsions and coma, lung damage produces dyspnoea, and cyanosis and renal changes cause oliguria and acute renal failure. Fibrin deposition not only produces thrombi but also results in a haemolytic anaemia as the red cells fragment while squeezing through the narrowed vasculature (microangiopathic haemolytic anaemia).

The fundamental problem is that there is excessive activation of coagulation, which ultimately is complicated by consumption of the coagulation factors and overactivity of the fibrinolytic system. Clotting activation occurs through increased activity of tissue factor, either released from damaged tissues or upregulated on circulating monocytes or endothelial cells in response to proinflammatory cytokines. It also occurs in widespread endothelial damage (e.g. by endotoxaemia), which exposes underlying collagen. Ultimately, intense fibrinolysis results in excess fibrin-degradation products that inhibit fibrin production. These latter mechanisms predominate in haemorrhagic DIC.

Not surprisingly, the prognosis is very variable and the management extremely difficult because you are trying to balance a see-saw that is out of control. If you inhibit the clotting system too much, the patient will bleed, but any bleeding tendency may require fresh frozen plasma, which may contribute to microthrombus formation. Generally, the most important thing is to treat the underlying cause. DIC is one of the complications of the systemic inflammatory response syndrome described on pages 139 and 142.

CHAPTER 8
ATHEROSCLEROSIS AND HYPERTENSION

So far we have concentrated on thrombosis and embolism, conditions that may suddenly affect healthy people of any age. Now we move on to the major cardiovascular problems of later life, namely hypertension, myocardial infarction and strokes. These are overwhelming causes of morbidity and mortality in developed countries and are associated with atherosclerosis of arteries (Fig. 8.1), so it is of great importance to try to identify the causes and mechanisms of atherosclerosis.

CLINICAL SCENARIO: ATHEROSCLEROSIS

Let us briefly consider a possible clinical picture in a patient debilitated by vascular problems.

A 48-year-old man complains of blurring of his vision. He is known to have diabetes, which is a complex metabolic disorder characterised by hyperglycaemia (raised blood glucose). At the age of 14 years, he presented with the typical diabetic symptoms of tiredness, weight loss, polyuria (increased urine production) and polydypsia (increased thirst). His blood glucose was raised, glucose was found in his urine and he has been on insulin therapy since that time.

His present complaint of blurred vision started 3 months ago. On questioning, he also complained of shortness of breath, especially on exertion, and cramps in his calf muscles on exercise. On examination, he was found to have raised blood pressure, a mild degree of cardiac failure with pulmonary oedema, small haemorrhages and small blood vessel proliferation in his retina, and systolic bruits in his neck (abnormal sounds, heard through the stethoscope, caused by turbulent

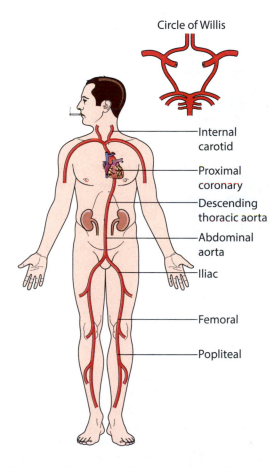

Figure 8.1 The arteries most commonly affected by atheroma.

blood flow). His blood tests showed a small rise in urea and creatinine, indicating a degree of renal impairment.

235

This unfortunate man has widespread disease related to arterial pathology. The arterial pathology comes under the general heading of arteriosclerosis, commonly referred to as 'hardening of the arteries', although this does not relate to a specific pathologically recognised entity. His large and medium-sized arteries are likely to be narrowed by fibrolipid atherosclerotic lesions (see next section) and his small arteries and arterioles will show the proliferative or hyaline changes of arteriolosclerosis. Atherosclerosis is principally a disease of the intima and may result in narrowing of the vessel, obstruction or thrombosis. Arteriosclerosis, on the other hand, affects the media with a resultant increase in wall thickness and decreased elasticity, which may lead to hypertension (see page 250 and Fig. 8.11).

Let's look at his symptoms to see if we can suggest a cause for each problem:

- His long-standing diabetes makes him much more likely to develop atherosclerosis than non-diabetic people of the same age.
- Fibrolipid atheromatous plaques in his coronary arteries will reduce the perfusion of the cardiac muscle, resulting in chronic ischaemia, which damages the heart muscle so that it pumps less efficiently. As the left side of the heart generally fails first this will result in pulmonary oedema.
- Atheroma in the carotid arteries produces the bruit heard on auscultation and may lead to cerebral infarction.
- The combination of poor cardiac function and atheromatous plaques in the abdominal aorta and femoral vessels will explain the pain and cramp in his calf muscle, which is secondary to poor perfusion.
- Hyaline arteriolosclerosis will affect small renal vessels, leading to glomerular damage that will induce hypertension through a mechanism involving the hormones renin and angiotensin. This exacerbates the atheroma and worsens the cardiac failure.
- The cause of his blurred vision may be of vascular origin because the retina is frequently damaged by small haemorrhages, microaneurysms and new vessel formation, although diabetes can also produce a host of other ocular changes.

The next stage in understanding this man's disease is to consider the actual appearance of his vessels.

WHAT DOES THE VESSEL LOOK LIKE?

ATHEROMATOUS PLAQUES

His main problem is atherosclerosis, which is a spectrum of arterial changes focused on the intima. These range from an asymptomatic small fatty streak, through atheromatous lesions to plaques complicated by thrombosis and rupture.

An atheromatous plaque is whitish-yellow in colour and varies in size from 0.5 cm to 1.5 cm. It is raised above the surrounding intima and protrudes into the lumen. Adjacent plaques can coalesce to produce larger lesions. On slicing, the plaque is composed of a fibrous cap covering a soft yellow lipid centre, which reminded the early pathologists of porridge or gruel and so was termed 'atheroma'. The intima is greatly thickened by the fibro-fatty deposition and the media may be thinned due to a loss of smooth muscle cells, resulting in both a loss of elasticity and a weakening of the wall. The adventitia may show new vessels budding off the vasa vasorum and providing the potential for a collateral circulation if obstruction occurs.

Generally, the fibrous cap is composed of smooth muscle cells, collagen, elastin and proteoglycans. Beneath this there is a more cellular region of macrophages, T lymphocytes and smooth muscle cells covering the soft gruel-like mass of lipid, cellular debris, cholesterol clefts, plasma proteins and lipid-laden cells (foam cells) derived from macrophages and smooth muscle cells. At the edges of the lesion, there may be new vessel formation (Fig. 8.2 and 8.3).

These plaques are more common in the aorta, and the femoral, carotid and coronary arteries, where they may produce clinical problems by causing partial or complete occlusion, thrombosis, embolism or aneurysm formation (see later). Areas of turbulent flow are worst affected, so that lesions often occur around the ostia of vessels. However, the abdominal aorta is more liable to atheroma than the thoracic aorta but the explanation for this is unknown.

From the clinical standpoint, it would be nice to know which lesions are fairly stable and which are liable to cause problems. This led to classifying plaques as lipid or fibrous rich depending on the predominant feature and, in the American Heart Association classification, are 'type IV atheromatous lesions' and 'type V fibro-fatty lesions'. Also the plaque can be concentric or eccentric,

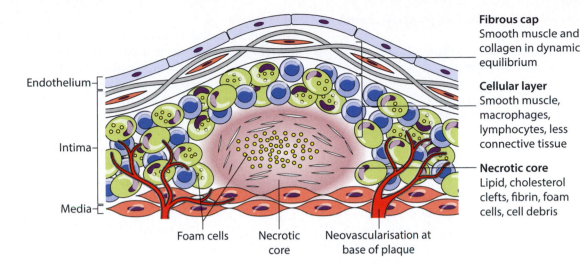

Endothelium

Intima

Media

Fibrous cap
Smooth muscle and collagen in dynamic equilibrium

Cellular layer
Smooth muscle, macrophages, lymphocytes, less connective tissue

Necrotic core
Lipid, cholesterol clefts, fibrin, foam cells, cell debris

Foam cells   Necrotic core   Neovascularisation at base of plaque

Figure 8.2 The key components of the atheromatous plaque.

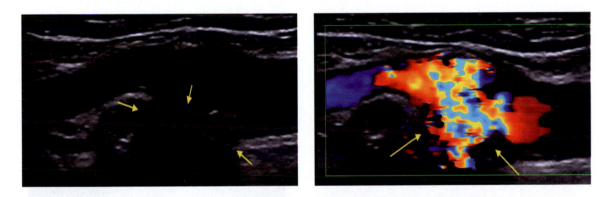

Figure 8.3 An ultrasound scan of the internal carotid artery showing calcified plaque at the origin of the artery (arrows). Colour flow Doppler shows turbulent flow in an ulcer crater (arrows) that has formed in the plaque. This patient is at risk of thrombosis and embolism.

with the eccentric plaque retaining some active media capable of responding to vasomotor signals. About 12% of atherosclerotic plaques in coronary arteries are both eccentric and lipid rich, and it is these that are most likely to undergo acute thrombosis because of plaque injury. In around 25% of cases of plaque injury, the damage is superficial, involving only surface endothelium and superficial collagen. Some believe that these result from vasospasm and are more common in cigarette smokers. In the remainder, there is deep plaque fissure leading down to the lipid pool, which is a powerful stimulant for platelets and the clotting cascade.

To summarise, plaques are not static lesions but can change in a gradual or abrupt way and can also be influenced by changes in other layers of the vessel wall. In the coronary arteries, if there is a slow increase in plaque volume, it is likely to lead to stable angina. Deep plaque fissuring will produce acute thrombosis and episodes of unstable angina, myocardial infarction or sudden death (Fig. 8.4). Endothelial dysfunction or superficial plaque injury may lead to inappropriate vasoconstriction. Atrophy of the media can result in aneurysm formation, and vessel proliferation in the adventitia can provide a network of collateral vessels.

Chapter 8: Atherosclerosis and hypertension

Ruptured plaques can heal and plaques, in general, may remodel or calcify.

Clinicians will refer to *acute coronary syndromes*. Stable angina is chest pain related to the heart that develops during exertion and resolves on rest and is fairly predictable. It is due to myocardial hypoxia because the delivery of oxygen is insufficient, most commonly as a result of coronary artery narrowing. Unstable angina occurs suddenly and may be unrelated to exercise. It overlaps with new-onset angina and crescendo angina, and is one of the acute coronary syndromes. The other two types of acute coronary syndrome are ST-elevation myocardial infarction (STEMI) and non-STEMI (see page 261). Initially unstable angina and NSTEMI are indistinguishable but are later differentiated by the rise in serum markers of muscle necrosis (e.g. troponin) in NSTEMI. All are associated with progressive coronary artery changes and require urgent medical attention.

Fatty streaks

Fatty streaks, similar to atheromatous plaques, occur in large muscular and elastic arteries but often differ in the regions affected. They do not affect blood flow but could represent a precursor lesion for atheromatous plaques. They first appear as tiny, round or oval, flat yellow dots, which become arranged in rows and finally coalesce to form a streak. The population distribution is very different as fatty streaks are found from a very early age and are independent of sex, race or geography whereas fibrolipid plaques are more common in males in developed countries.

HOW IS THE ATHEROMATOUS PLAQUE PRODUCED?

It is now believed that many of the changes are a consequence of repeated damage and attempts at

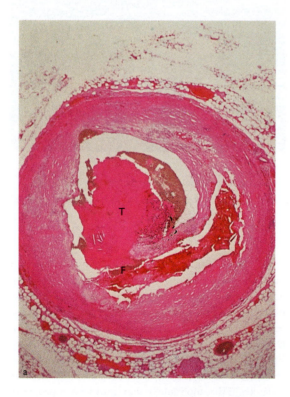
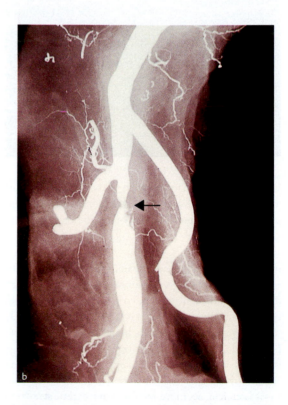

Figure 8.4 (a) Atheromatous plaque with fissuring (F) and thrombus (T) formation. (b) Angiogram showing narrowing (arrow) of the coronary artery corresponding to (a).

healing and repair. This 'reaction to injury' theory has its origin deep within the history books because it incorporates the ideas of Virchow, Duguid and Rokitansky. It involves repeated endothelial damage, provoking a chronic inflammatory reaction that alters the vessel wall. Virchow believed that leakage of plasma proteins and lipid from the blood to the subendothelial tissue stimulated intimal cell proliferation. He regarded the cell proliferation as a form of low-grade inflammation and termed it the 'imbibition hypothesis', later often called the 'insudation' or 'infiltration' hypothesis. Rokitansky is credited with the 'encrustation' theory, which suggested that thrombi forming on damaged endothelium could become organised to form a plaque. The modern 'reaction to injury' theory was proposed by Ross and Glomset in 1976 and modified in 1993.

The *reaction to injury theory* (Fig. 8.5) involves some change or damage to the vascular endothelium that causes increased permeability to proteins and lipid, and also leads to the aggregation of platelets and monocytes. Monocytes migrate from the blood into the subendothelial layers, where they become macrophages and ingest the lipid. These leukocytes release various enzymes and growth factors, which promote recruitment and proliferation of smooth muscle cells and production of extracellular matrix (ECM). A short, sharp injury can be almost completely repaired but chronic repeated injury leads to the formation of an atheromatous plaque.

Stable atheromatous plaques may not produce any clinical effect but problems occur if a lipid lesion covered by a thin fibrous capsule is disrupted, releasing material that promotes local thrombosis. This may become incorporated into the plaque to increase its size. The factors involved can be thought of as atherogenic factors, important in producing the early plaque, and thrombotic factors, important in its progression. Thrombosis and the role of the endothelium are covered on page 217.

Central to all of this is the role of the endothelial cell (EC) (see also page 250), which normally maintains a non-thrombogenic and non-adhesive surface by not expressing surface molecules involved in EC–leukocyte interactions.

Although a continuous sequence, the stages of atherogenesis can be divided into the following.

Endothelial activation

Endothelial cells can be activated by a variety of factors, the most important of which are blood flow-generated endothelial sheer stress (ESS), dyslipidaemia and proinflammatory cytokines. Components of cigarette smoke, products of glycoxidation associated with diabetes/metabolic syndrome and hypertension are also likely to be important.

The initial reaction of the endothelial cell aims to maintain homeostasis. ECs have different phenotypes, which occur because of differential expression of atheroprotective and proatherogenic genes, often initially in response to their location in the vascular tree and the blood flow and blood pressure-related stresses, e.g. the transcription factor, nuclear factor κB (NF-κB), upregulates genes coding for low-density lipoprotein (LDL) receptors. This can occur as a homeostatic response in an area of haemodynamic stress but it then makes the endothelial cell more vulnerable to any dyslipidaemia.

Endothelial cell dysfunction

High concentrations of plasma LDL increase its influx and accumulation in the intima, where it can be modified to produce two pathogenic substances: oxidised LDL and cholesterol crystals.

Oxidised LDL is thought to be capable of causing:

- EC damage
- Smooth muscle cell injury, leading to central necrosis of the plaque
- Foam cell formation, as the oxidised lipoproteins are taken up by the receptor for modified LDL
- Stimulation of local release of growth factors and cytokines, leading to the recruitment and retention of macrophages.

For the ECs, this is the stage when there is a vicious cycle because their initial defence reaction starts to trigger a multipart inflammatory process with expression of cell adhesion molecules (e.g. vascular cell adhesion molecule 1 [VCAM-1], intercellular adhesion molecule 1 [ICAM-1], E-selectin, P-selectin and fractaline [CX3CL1]) and production of cytokines (e.g. monocyte chemotactic protein-1 [MCP-1] and interleukin 8 [IL-8]). This amplifies the damage by recruitment of immune cells.

Chapter 8: Atherosclerosis and hypertension

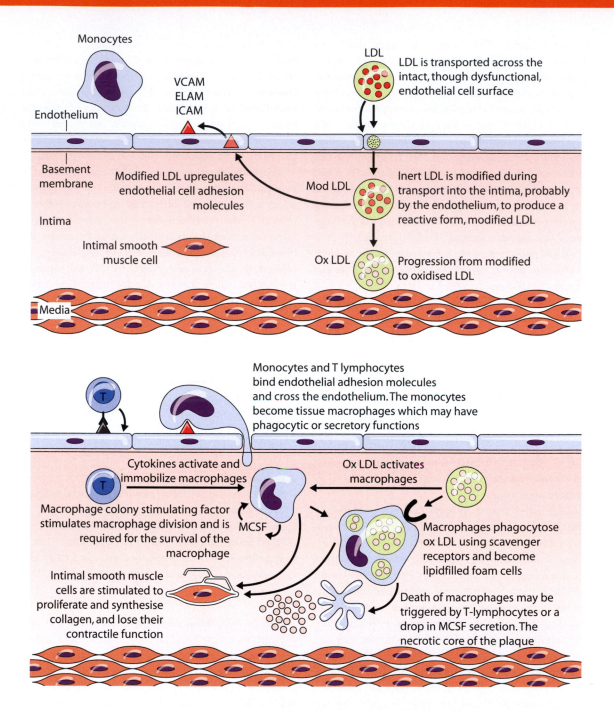

Figure 8.5 The development of the atheromatous plaque. Endothelial cells may be damaged by cigarette smoke, inflammatory mediators, flow disturbances and hyperlipidaemia. This increases their permeability to low-density lipoproteins, which enter the arterial intima by insudation. See Fig. 8.2 for the structure of the established plaque.

Part 3: Features of cardiovascular disorders

Initiation of inflammatory response and recruitment of immune cells

Macrophages have different phenotypes and some are atherogenic whereas others are protective. This is why strategies to remove or destroy macrophages may do more harm than good. It is the 'classic' M1 phenotype that predominates in plaques and produces more inflammatory cytokines. It appears that the altered LDL is recognised by the 'scavenger' receptor on macrophages and phagocytosed so that the macrophage becomes a foam cell. The oxidised lipoprotein may ultimately contribute to the death of the macrophage.

VCAM-1 expression by the ECs facilitates the entry of T lymphocytes, which are predominantly of the T-helper 1 (Th1) phenotype, which interact strongly with M1-type macrophages. Th2 cells stimulate M2 macrophages, which inhibit inflammation. There are also anti-inflammatory T_{reg} cells.

Smooth muscle cell recruitment and activation

Intimal smooth muscle cells (SMCs) proliferate and synthesise a range of ECM proteins, proteases and cytokines. Medial SMCs do not but instead have contractile proteins. The switch between the two phenotypes is related to atherogenic stimuli. There is also a phenotype with osteoblastic features that may be important in the calcification of atheroma and is produced in response to the nuclear factor of activated T cells (NFAT) signalling pathway.

Complications such as plaque rupture

There is much interest in the 'unstable' plaques and whether it is possible to make them more stable. Unstable plaques are known to have fewer smooth muscle cells, less extracellular matrix, more extracellular lipid and many lipid-filled macrophages. Plaque stability is seen as a balance between damage and repair in the intima (Fig. 8.6). The damage is due to inflammation

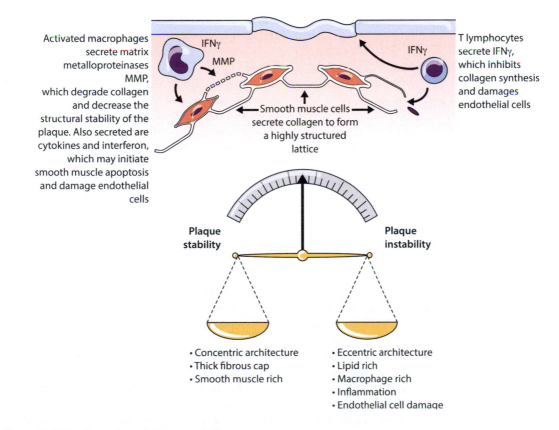

Activated macrophages secrete matrix metalloproteinases MMP, which degrade collagen and decrease the structural stability of the plaque. Also secreted are cytokines and interferon, which may initiate smooth muscle apoptosis and damage endothelial cells

IFNγ
MMP

Smooth muscle cells secrete collagen to form a highly structured lattice

T lymphocytes secrete IFNγ, which inhibits collagen synthesis and damages endothelial cells

IFNγ

Plaque stability

Plaque instability

- Concentric architecture
- Thick fibrous cap
- Smooth muscle rich

- Eccentric architecture
- Lipid rich
- Macrophage rich
- Inflammation
- Endothelial cell damage

Figure 8.6 The dynamics of plaque stability.

Chapter 8: Atherosclerosis and hypertension

mediated by macrophages and T cells, and the repair involves SMCs producing matrix and fibres. It is hoped that influencing the macrophage activity might reduce the incidence of acute coronary events. A number of proteases have been implicated in thinning of the fibrous cap, in particular the matrix metalloproteinases (MMPs). These can be neutralised by tissue inhibitors of metalloproteinases (TIMPs), which are produced by ECs, SMCs and macrophages. Some individuals have genetic abnormalities of genes related to MMPs and TIMPs that can alter their predisposition to atheroma and aneurysms. Recent evidence suggests that statins influence plaque stability through a reduction in their inflammation, as well as their action on lipids.

Basic biology

The manifestations of endothelial dysfunction include:

- Vascular reactivity
- Nitric oxide metabolism
- Lipoprotein permeability and oxidation
- Increased leukocyte adhesion
- Altered extracellular matrix deposition
- Altered haemostasis/thrombosis balance

RISK FACTORS FOR ATHEROSCLEROSIS

Now that we have a mental image of the appearance and distribution of atherosclerosis, we consider the risk factors that are believed to be important. Some of these can be modified but the most significant cannot and these are age and gender.

Age

Deaths from ischaemic heart disease increase with advancing age. How much depends on your gender and other risk factors. There are websites available that allow you to calculate your (or your patient's) cardiovascular risk, e.g. qrisk.org. The results need to be interpreted with caution and in context but will provide some guidance on how each factor influences risk (and it is quite fun to play with!). So try putting in your current values for age, gender, height, weight, etc. It gives you your percentage risk of having a heart attack or stroke in the next 10 years. Then try altering

just the age and go up in steps of 10 years. You'll find that your risk rises dramatically between 40 and 70 years. Then try altering the factors over which you have control, such as weight, smoking, blood pressure and lipid levels. This will give you an idea of the relative importance of the various interventions.

Gender

The death rate from ischaemic heart disease is higher in men than in women up to the age of 75 years, after which the incidence is similar, but the lifetime burden for cardiovascular disease is greater in women because they live longer and have an increased risk of stroke over the age of 75 years. Myocardial infarction is extremely rare in premenopausal women, suggesting that endocrine differences may be important and that the effect of oestrogens on lipid metabolism is a possible mechanism. However, initial evidence suggesting that hormone replacement therapy (HRT) protects against cardiovascular disease was not confirmed in later trials, but it now appears that the timing of providing the HRT may be critical and it needs to be started in the immediate postmenopausal period to be beneficial.

Genes

As our knowledge of the genome increases, we find more associations with particular diseases and can try to unravel the likely mechanism in order to consider possible interventions. Genetic influences on cardiovascular disease (CVD) include the following:

- Genes affecting cholesterol levels through changes to the LDL receptor, apolipoprotein B and apolipoprotein C
- Variants in angiotensinogen associated with hypertension
- Predisposition to type 2 diabetes
- Altered inactivation of nicotine which decreases the likelihood of smoking
- Altered ion channel proteins influencing rare causes of arrhythmias.

Racial differences exist with black people and southeast Asians having a higher risk than white people.

Gene-wide association (GWA) studies are starting to bear fruit and around 34 significant loci have been identified in humans. The really exciting thing is that 75% of them are *not* associated with recognised risk factors or pathways. Some are involved with LDL metabolism or hypertension, but the majority are novel. This raises a whole new area of investigation working out what these genes code for and how they influence cells and tissues.

Smoking

Smoking one packet of cigarettes a day increases the likelihood of having a myocardial infarction by up to 300%. Traditionally, more men than women have smoked but, as women have taken up the habit, their risk has risen. Fortunately, giving up smoking reduces the risk, which means that it is likely that smoking not only promotes atheroma but may also cause occlusion of vessels. This could be due to an increased local clotting tendency because of altered platelet function. Stopping smoking for 1–2 years reduces the risk of myocardial infarction to 'only' twice that of non-smokers. So-called 'safer' cigarettes, which have a lower tar and nicotine content, reduce the risk of bronchial carcinoma, but do not appear to reduce the risk of coronary heart disease. How smoking damages vessels is not known but suggestions include increased free radical activity, raised carbon monoxide levels or a direct effect of nicotine. The recent introduction of e-cigarettes will need to be monitored closely to look for any negative effects.

Hypertension

Hypertension significantly increases the risk of ischaemic heart disease and 'strokes'. The diastolic blood pressure level is considered more important than the systolic level and a diastolic pressure consistently >95 mmHg is deemed harmful. Drug treatment to reduce the blood pressure decreases the risk in patients with moderate-to-severe hypertension, but it is unclear whether it benefits patients with mild hypertension. Factors, such as salt intake, that increase the risk of hypertension also increase the risk of fatal coronary events.

Hyperlipidaemia

Evidence for the role of fats in atheroma comes from a variety of sources and the literature on the role of lipids in atherosclerosis would fill a library, so we just highlight some of the more important points:

- Atheromatous lesions contain far more lipid than the adjacent intima.
- Atheromatous plaques are rich in cholesterol and cholesterol esters (65–80%), derived from blood lipoproteins.

- Intimal lesions can be produced in some animals by increasing the plasma concentration of certain lipids through drug or diet manipulation.
- Macrophages accumulate cholesterol (LDL) and this is increased if there is endothelial damage.
- In families with genetic disorders causing hypercholesterolaemia or in groups with acquired hypercholesterolaemia (e.g. hypothyroidism and nephrotic syndrome), atherosclerosis is increased.
- It used to be thought that cardiovascular mortality could be reduced by lowering the plasma cholesterol with diet or drugs, but it is now suggested that the plasma cholesterol is less crucial than originally thought.

There are a variety of conditions that produce secondary hyperlipidaemia, several of which overlap with the risk factors for atherosclerosis (see Key facts box below).

Key facts

Causes of secondary hyperlipidaemia

Nutritional	Obesity
	Alcohol abuse
Hormonal	Diabetes
	Hypothyroidism
Drugs	β blockers
	High-dose steroids
Miscellaneous	Stress
	Bile duct obstruction and primary biliary cirrhosis
	Nephrotic syndrome and chronic renal failure

Key facts

Risk factors for atherosclerosis

Unmodifiable	Age, gender, personality
Modifiable	Diabetes, metabolic syndrome, hyperlipidaemia
Preventable	Smoking, obesity

Chapter 8: Atherosclerosis and hypertension

Specifically, the risk of coronary heart disease generally increases with:

- Raised serum total cholesterol concentration
- Raised LDL-cholesterol concentration
- Reduced high-density lipoprotein (HDL)-cholesterol concentration.

Figure 8.7 shows the blood cholesterol.

Diabetes

The risk of a myocardial infarction in a patient with diabetes is twice that of a non-diabetic patient and, as our clinical scenario illustrated, their arterial disease is widespread. Arterial disease accounts for around 70% of the deaths of people with diabetes in western countries. As well as frank diabetes, there are also abnormalities of glucose and lipid metabolism which, when combined with obesity and hypertension, are called the 'metabolic syndrome', and this is a recognised risk factor for atherosclerosis. Currently, around 65% of people admitted to hospital with an acute myocardial infarction have metabolic syndrome (Fig. 8.8).

Basic biology

Lipids are an essential part of our metabolism. There are two important pathways for lipid metabolism: exogenous and endogenous. Exogenous (i.e. dietary) lipids are digested to release triacylglycerols (TGs) and cholesterol esters. These combine with phospholipids and specific apoproteins to make them water soluble, and are called chylomicrons. The TG component of the chylomicron can move from the circulation into cells by lipoprotein lipase activity on the endothelial surface of cells. Once inside the cell, it may be converted to glycerol and non-esterified fatty acids which are a major energy source.

Endogenous lipid refers to the various lipids produced by the liver. The building blocks are glycerol and fatty acids, which reach the liver from fat stores or are synthesised from glucose, and cholesterol derived from lipoproteins or synthesised locally from acetate and mevalonic acid using the enzyme hydroxymethylglutaryl-coenzyme A (HMG-CoA). Statins inhibit HMG-CoA. Glycerol and fatty acids combine to produce TGs.

Part 3: Features of cardiovascular disorders

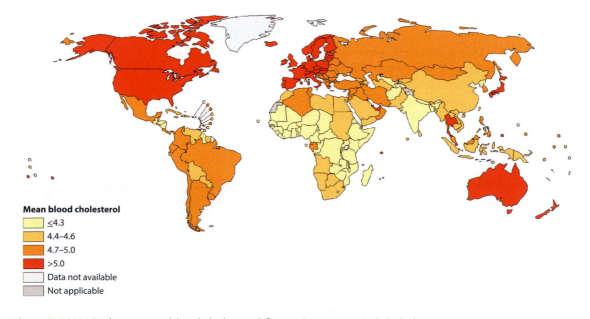

**Mean blood cholesterol (mmol/L), ages 25+, age standardised
Males, 2008**

Mean blood cholesterol
- ≤4.3
- 4.4–4.6
- 4.7–5.0
- >5.0
- Data not available
- Not applicable

Figure 8.7 WHO observatory: blood cholesterol figure. (From WHO global observatory map. © WHO 2011. All rights reserved.)

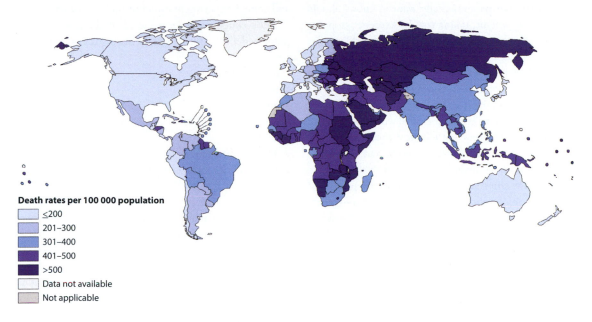

Figure 8.8 WHO observatory: cardiovascular diseases and diabetes. (From WHO global observatory map. © WHO 2011. All rights reserved.)

Other possible risk factors

- Inflammatory conditions: any condition that produces a rise in C-reactive protein (CRP)
- Low socioeconomic group
- Lack of regular exercise
- Obesity: as well as affecting the individual, obesity in pregnancy is associated with an increase in CVS morbidity and mortality in the offspring in midlife
- Stress: increased risk of coronary heart disease (CHD), CVD and strokes with exposure to excessive aircraft or road traffic noise
- Low birthweight or low infant weight: 'fetal origins of adult disease' hypothesis
- Obstetric associations: pre-eclampsia, premature delivery, gestational diabetes, 'type A' personality
- Long-term exposure to particulate air pollution increases acute coronary events.

Any cause of inflammation has the potential to increase the risk of atheroma. This includes both external agents, such as chronic bacterial or virally induced inflammation, and autoimmune conditions. The potential mechanisms could include altering the response of the vascular wall cells to injury and promoting chronic inflammation within the wall. There is also the added complication that steroids used to treat some inflammatory conditions can alter glucose and lipid metabolism.

🔑 Key facts

Metabolic syndrome

Definition	Indicative measurement
Impaired glucose	Fasting glucose tolerance >6.1 mmol/L
Hypertension	BP >130/85
Dyslipidaemia	Triacylglycerols >1.7 mmol/L
	HDL-cholesterol <1.04 mmol/L
Central obesity	BMI >29.4kg/m²
	Waist circumference >102 cm

Chapter 8: Atherosclerosis and hypertension

HOW CAN WE REDUCE OUR RISK FOR CARDIOVASCULAR DISEASE?

This is what your patients really want to know! Should they be losing weight, taking regular exercise, stopping drinking, stopping smoking and altering their diet? The epidemiological studies suggest that these are risk factors, but does intervention make a difference and who should do it? Would they benefit from lipid-lowering drugs or platelet inhibitors?

It is important to remember that you are trying to influence two different pathological mechanisms: the production of an atheromatous lesion and the occlusive event, such as thrombosis or vasoconstriction, so there may be different factors affecting the processes, i.e. 'atherogenic' and 'thrombogenic' factors. In the area between these two are factors that affect the structure and stability of the plaque (the pleiotropic effect) so leading to fissure and thrombosis. It is likely that atherogenic factors operate over decades and starting on a low-fat diet as you retire may be too late, whereas the likelihood of thrombosis could be altered.

The good news used to be that moderate alcohol consumption may actually help reduce strokes but heavy drinking increases the risk by raising blood pressure. Sadly latest UK guidance (2016) no longer suggests that this is true. Also moderate drinking increases high-density lipoprotein (HDL)-cholesterol levels, which are associated with a reduction in CHD. Exercise appears to have a good impact in both primary and secondary prevention provided that it is at a level of around 30 min/day for at least 5 days a week.

What about diet? It is with some trepidation that we write about diet and cardiovascular disease because of the ever-changing dietary fashions, ongoing scientific controversies, and the activities of the food and pharmaceutical industry in influencing government policy and guidance. A little history will help illustrate the problem. The American biologist, Ancel Keys, suggested in the 1950s that diets rich in animal fats led to heart disease. At the same time in the UK, John Yudkin looked at similar epidemiological data and concluded that coronary mortality was more closely associated with sugar consumption than fat consumption, leading him to publish *Pure, White and Deadly* in 1972. Keys also linked the incidence of coronary artery disease to total cholesterol concentration and the intake of saturated fats. By 1988, the first US guidelines were issued that defined the blood lipid levels for starting treatment and the intended target level. Since then there has been a vast range of epidemiological studies, dietary intervention trials and animal experimentation but limited consensus.

Let's concentrate on some of the current topics of popular debate. It is worth discussing the role of:

- *trans*-fatty acids
- Saturated fat
- Sugar
- Polyunsaturated fatty acids: omega-3 and -6 fatty acids
- The 'Mediterranean diet'
- Some factors influencing HDL levels.

Trans-fatty acids are present in margarine and manufactured foods, which use vegetable oil that has been partially hydrogenated to make it semi-solid. They have no known nutritional value and, in 2003, Denmark virtually eliminated them from Danish diets. Australia and Switzerland followed in 2009. In the USA, individual cities or states have also introduced controls and many other countries are considering requiring the labelling of *trans*-fat-containing products. Is it important? In 2006, a US study found that an intake of only 5 g of *trans*-fats per day was associated with a 23% increase in CHD. Sampling of fast food and manufacturing oils around the world has demonstrated how frequently this is exceeded.

Saturated fat consumption in epidemiological studies was associated with CHD and total cholesterol. However, cutting the amount of saturated fat in the diet affects type A LDL, whereas it is type B LDL that is implicated in CHD. Perhaps not surprisingly, prospective studies have not found any significant improvement in CHD risk with reduction in saturated fats, although it is possible that different saturated fats have different effects.

Is *sugar* pure white and deadly? It is looking likely that sugar intake is associated with the metabolic syndrome, i.e. obesity, raised TAGs, low HDL-cholesterol, hypertension and insulin resistance. Just like fats, there will be different ways of consuming sugar and there are some concerns that fructose, because it is metabolised in the liver and can be readily converted to hepatocyte fat, may be more likely to be harmful. This is being consumed in greater quantities because of the use of high-fructose corn syrup in many fizzy drinks. Dietary-added sugars are statistically correlated with blood lipid levels, so that is one way in which they could be atherogenic. They could also act by glycosylating proteins and upsetting normal metabolism. This occurs in diabetes and leads to glycosylated LDL being unable to bind to normal LDL-pathway receptors, ending up with increased uptake into macrophages.

Essential *polyunsaturated fatty acids* include omega-3 and omega-6 fatty acids. Omega-6 fatty acids are plentiful in western diets because they are found in many vegetable oils, but they cannot be converted, by humans, into omega-3 fatty acids, most of which are obtained from fish oils. Until recently it was recommended that dietary changes to increase omega-3 fatty acid intake should be used in secondary prevention after myocardial infarction. Now the Sydney Diet Heart study has demonstrated that this decreases the serum cholesterol but *increases* deaths from all causes, CHD and CVD. There is also greater appreciation that, although consuming 'good' fats as part of a balanced diet is beneficial, adding them as dietary supplements may be ineffective. So you would benefit if you include naturally occurring omega-6 linoleic acid in your diet by eating fish once or twice a week (compared with eating fish less than once a month), but adding pure omega-6 supplements or making major substitution of

omega-6 polyunsaturated fatty acids (PUFAs) for saturated fats is not useful (Fig. 8.9).

Of course what is happening is that the observations from a population with a particular diet (the *Mediterranean diet*) have been used to justify interventions using individual components of that diet, often in excess. Generally that just doesn't seem to be working. So is there still a link between health and the Mediterranean diet? A meta-analysis of 12 prospective studies between 1996 and 2008, involving more than 1.5 million people, concluded that adherence to the diet produced a statistically significant reduction (9%) in cardiovascular mortality. That was about primary prevention. After a heart attack, the Mediterranean diet has been claimed to be better than taking statins for reducing mortality.

Finally, on diet, what about *HDL*? Genes, environment and lifestyle all influence HDL levels, but almost 40% of the variation in HDL-cholesterol levels may be genetic, with polymorphisms of the cholesteryl ester transfer protein (CETP) gene being most important, and roughly half of the environmental variation in men is attributed to increased alcohol consumption, which decreases CETP activity. Hormone replacement therapy (HHRT) raises HDL but is not cardioprotective, whereas exercise and stopping smoking raise HDL and are beneficial, so simply achieving a higher HDL level is

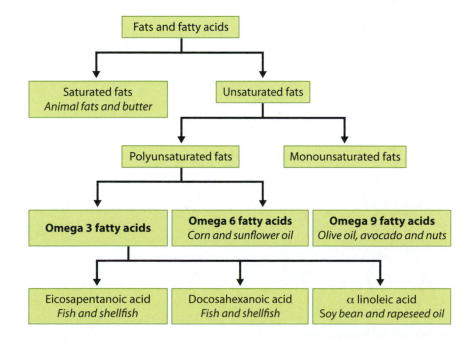

Figure 8.9 Types of fatty acids.

not a guarantee of better cardiac outcomes. Drug interventions to increase HDL levels have not shown any clinical benefit.

If appropriate lifestyle and dietary changes have been made, should drugs also be used? The major class of drugs being used are statins with a fivefold increase in use in the UK between 2001 and 2011. These inhibit the enzyme HMG-CoA reductase and so reduce endogenous production of cholesterol. The statins also produce some lowering of TGs and a rise in HDL levels. However, there has been a recent major shift away from targeting particular levels of cholesterol and instead moving towards an overall assessment of cardiovascular risk. Why has this shift occurred? It is because the accumulating evidence on the effect of statins is that they reduce the risk of cardiovascular events independently of their effect on cholesterol level. Not appreciating this point can lead to under- or over-treatment of patients, e.g. a high-risk patient may be below the cholesterol level trigger for receiving treatment but would have benefited from statin therapy. It is likely that this is because of the statin effect on endothelial cells and

 Read more about intermittent chest pain in Pathology in Clinical Practice Case 1

reducing inflammatory reactions, as evidenced by the drop in CRP levels.

The current recommendation in the UK for the use of statins depends on whether it is primary or secondary prevention (i.e. patients with or without clinically overt atherosclerotic disease). For patients with overt atherosclerotic disease, statin therapy is recommended. For primary prevention, the current approach is to offer statins if the 10-year CHD risk is >20%, although there is a debate about whether that should be extended to lower-risk patients. How effective are statins for primary prevention? Meta-analysis of 65 000 patients in 11 trials demonstrates that 98% achieve no benefit in 5 years, 1.6% avoid a heart attack and 0.4% avoid a stroke. There was no effect on mortality. Reducing thrombus formation is important, so taking low-dose aspirin or platelet inhibitors can be beneficial.

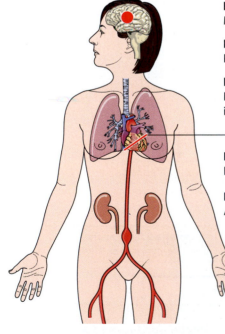

Brain
Microaneurysms and intracerebral haemorrhage

Lungs
Pulmonary oedema due to left ventricular failure

Heart
Left ventricular hypertrophy and failure; myocardial infarction

Kidneys
Ischaemic cortical damage

Blood vessels
Atherosclerosis and aneurysm formation

Figure 8.10 Complications of hypertension.

Part 3: Features of cardiovascular disorders

Key facts

Factors that may raise blood pressure

- High salt intake
- Obesity
- Physical inactivity
- Stress
- High alcohol intake
- Coffee
- Smoking
- Cold environment

HYPERTENSION

Many of the circulatory diseases we have discussed are related to hypertension (Fig. 8.10). Worldwide around 13% of deaths are attributable to hypertension-related problems. Atheroma, arteriolosclerosis, myocardial infarction, left ventricular hypertrophy and aneurysms are linked by their association with hypertension, so we should discuss some of what is known about the aetiology and pathogenesis of hypertension.

Hypertension is extremely common, affecting around 32% of men and 29% of women if a blood pressure >140/90 mmHg is regarded as abnormal. Unfortunately, organ damage may be irreversible by the time a patient presents with symptoms so it is important to screen the people who are most susceptible. So you will need to know about the factors influencing blood pressure. Hypertension is predominantly a condition of middle and later life, and is classified as 'benign' or 'malignant'. Fortunately, 'benign' hypertension is much more common, relatively stable and treatable with long-term anti-hypertensive drugs; however, if untreated, more than 50% will go on to have a stroke or die from heart disease. 'Malignant' hypertension only affects 5% of hypertensive patients but it is more severe and is liable to affect men aged <50 years. It is defined as a diastolic pressure of >120 mmHg or a systolic pressure of >200 mmHg. The major dangers of hypertension are CHD, cerebrovascular accidents, congestive heart failure and chronic renal failure.

In 95% of cases, there is no obvious cause and this is termed 'primary', 'essential' or 'idiopathic' hypertension. Most of the remainder are due to renal disease, with a small number due to endocrine abnormalities (secondary hypertension – see box).

Key facts

Causes of secondary hypertension

Renal disease
  Acute or chronic glomerulonephritis
  Vasculitis, e.g. systemic lupus erythematosus
  Renal artery stenosis
  Chronic pyelonephritis
  Diabetic nephropathy
Cardiovascular disease
  Coarctation of aorta (hypertension in upper half of body)
  High cardiac output states
Hormonal
  Phaeochromocytoma: excess catecholamines from tumour
  Cushing's syndrome: excess corticosteroids
    Primary
      Adrenal cortical adenoma/hyperplasia
    Secondary
      Pituitary basophil adenoma
      Corticosteroid therapy
      Paraneoplastic, e.g. oat cell tumour
  Conn's syndrome: excess aldosterone
    Adrenal cortical adenoma/hyperplasia
    Adrenogenital syndrome
  Acromegaly: excess growth hormone
    Pituitary acidophil adenoma
Neurological
  Raised intracranial pressure
    Haemorrhage
    Tumour
    Abscess
  Hypothalamic or brain-stem lesion
Pre-eclampsia in pregnancy

Key facts

Mediators that influence vascular tone

Constrictors	Angiotensin 1
	Endothelin 1
	Catecholamines
	Thromboxane
	Leukotrienes
Dilators	Prostaglandins
	Kinins
	PAF
	Endothelium-derived relaxing factor (nitrous oxide)

WHAT MECHANISMS OPERATE IN HYPERTENSION?

First, it is useful to review the factors influencing the control of blood pressure. In simple terms, the arterial pressure will depend on the cardiac output and the total peripheral resistance. The cardiac output depends on the heart rate, its contractility and the blood volume. It may increase in the early stages of hypertension but is more usually normal. Thus chronic hypertension appears to be due to increased tone in small arteries and arterioles. The resistance is determined by the arteriolar lumen, which may expand or contract depending on the state of the SMCs in the vessel wall. This is called local vascular tone and is influenced by a variety of mediators (see Key facts box), which may act throughout the body or be produced and have their action locally, i.e. autoregulation.

It is generally thought that reduced sodium excretion by the kidney is a key mechanism in essential hypertension. This would lead to fluid retention, so increasing cardiac output and potentially raising the blood pressure. Most now accept that the salt intake in western countries contributes to hypertension and salt intake should be reduced from a daily average of 9 g to ≤5 g. The evidence for this comes from randomised controlled trials of reducing salt intake, which showed a dose-dependent cause–effect relationship and no threshold effect, i.e. it still has an effect if levels are reduced to 3 g/day. Even more important has been the confirmation that reducing dietary salt intake lowers cardiovascular events by around 30% over the next 10–15 years. This was irrespective of sex, ethnicity, age, body mass or blood pressure. It also applied to fatal stroke and fatal coronary events. Unfortunately, only the minority (up to 20%) of intake is under our control because most is in manufactured products, such as bread and processed food.

The following are other factors in essential hypertension:

- Endothelial dysfunction
- Genetic factors
- Environment (lifestyle) factors.

You will have appreciated by now that the endothelial cell is hugely important in normal and diseased vessels (Fig. 8.11). It produces vasoconstrictors and vasodilators so an imbalance could produce constriction and hypertension. Endothelins predominantly vasoconstrict, are produced by endothelium and primarily act locally on vascular smooth muscle. Production is increased by changes in sheer stress, hypoxia and inflammatory mediators. Nitric oxide vasodilates and reduces proliferation of smooth muscle cells but it is inactivated by reactive oxygen species, such as superoxide.

Genetic factors are generally thought to operate in complex combinations and, except for some rare single-gene disorders, are poorly understood. It is likely that genes involved in vasomotor tone, renal salt handling and smooth-muscle proliferation would be relevant. An example is Liddle's syndrome resulting from mutations in *ENaC*, which produce an increase in aldosterone-mediated salt absorption in the distal renal tubules.

Lifestyle factors are closely similar to those related to atherosclerosis but also include stress, which is believed to act through overstimulation of the sympathetic nervous system, increasing noradrenaline release which leads to vasoconstriction.

Secondary hypertension is most often related to renal disease and results from abnormalities in the renin–angiotensin system, abnormal salt and water balance, and renal vasodepressor substances. Angiotensin II is increased in response to raised renin levels and will increase vascular resistance, by causing vascular smooth muscle contraction, and increase blood volume

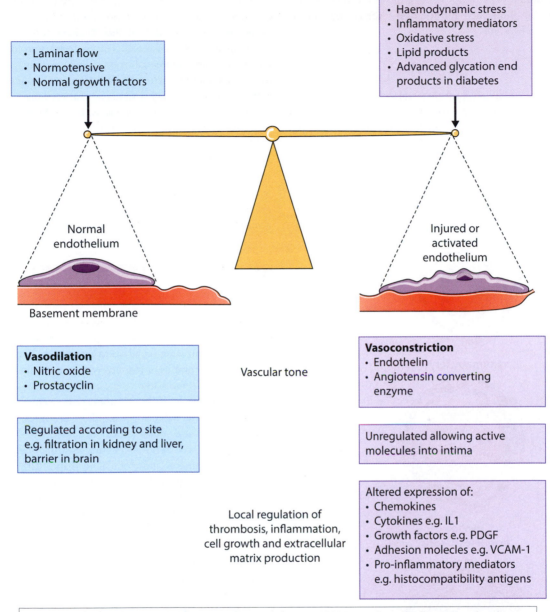

- Laminar flow
- Normotensive
- Normal growth factors

- Haemodynamic stress
- Inflammatory mediators
- Oxidative stress
- Lipid products
- Advanced glycation end products in diabetes

Normal endothelium

Injured or activated endothelium

Basement membrane

Vasodilation
- Nitric oxide
- Prostacyclin

Vascular tone

Vasoconstriction
- Endothelin
- Angiotensin converting enzyme

Regulated according to site e.g. filtration in kidney and liver, barrier in brain

Unregulated allowing active molecules into intima

Local regulation of thrombosis, inflammation, cell growth and extracellular matrix production

Altered expression of:
- Chemokines
- Cytokines e.g. IL1
- Growth factors e.g. PDGF
- Adhesion molecles e.g. VCAM-1
- Pro-inflammatory mediators e.g. histocompatibility antigens

IL = interleukin, PDGF = platelet derived growth factor, VCAM vascular cell adhesion molecule

Figure 8.11 Endothelial function relevant to atherosclerosis and hypertension.

through aldosterone, which promotes distal tubular reabsorption of sodium. Negative feedback is provided through a lowering of renin levels secondary to the increase in angiotensin II, the raised pressure in the glomerular afferent arteriole and decreased proximal tubule sodium reabsorption, which influences the macula densa. Increased renin secretion occurs in all of the renal causes of hypertension listed in the Key facts box on page 249, except for many cases of chronic renal failure. In chronic renal failure, there is sodium and water retention, which is probably related to a reduced glomerular filtration rate influencing tubular sodium handling.

WHAT DO THE VESSELS LOOK LIKE?

Hyaline and hyperplastic arteriolosclerotic changes are very different to atheromatous damage. They affect only small vessels, do not have any increase in lipid and primarily affect the media, whereas atheroma is initially an intimal problem. Both are very important because of their strong association with hypertension (Fig. 8.12).

Hyaline arteriolosclerosis generally occurs in elderly or diabetic patients and involves the deposition of homogeneous pink material that thickens the media, resulting in a narrowed vessel. This material is probably a combination of increased extra cellular matrix, produced by smooth muscle cells, and plasma components that have leaked through a damaged endothelium.

Hyperplastic arteriolosclerosis is found in patients who have a rather sudden or severe prolonged increase in blood pressure. The media of the vessel wall is thickened by a concentric proliferation of smooth muscle cells and an increase in basement membrane material. In the worst cases (malignant hypertension), there may be fibrinoid necrosis of the vessel walls. In diabetes, advanced glycation end-products (AGEs) damage the endothelium.

ANEURYSMS

An aneurysm is a localised dilatation in a blood vessel which may produce no symptoms, cause problems through pressing on adjacent structures, become occluded by thrombus or rupture with potentially devastating effects (Fig. 8.13).

Let us start with berry aneurysms (Figs 8.14 and 8.15). As the name suggests, these are more than just a dilatation and appear similar to a cherry stuck on the side of a vessel. Berry aneurysms are usually small, <1.5 cm in diameter and globular in shape. Although referred to as congenital, they are not present at birth but develop because there is a defect in the media of the blood vessels at sites of bifurcation. These occur most commonly around the circle of Willis (Fig. 8.16). Patients with berry aneurysms generally present with a sudden severe headache and some lose consciousness because the aneurysm has leaked. Occasionally, a patient will have ocular problems or facial pain because of pressure on the cranial nerves by an unruptured aneurysm. Frequently, these patients are young or middle-aged and are not normally hypertensive, but are assumed to have raised their blood pressure by acute exertion.

The other important type of aneurysm affecting cerebral vessels is the microaneurysm or Charcot–Bouchard aneurysm (Fig. 8.17 and see Fig. 8.14). These are generally multiple small aneurysms only a few millimetres in diameter, present on small arteries within the cerebral hemispheres. They occur in older,

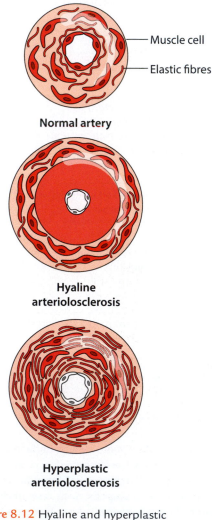

Normal artery

— Muscle cell

— Elastic fibres

Hyaline arteriolosclerosis

Hyperplastic arteriolosclerosis

Figure 8.12 Hyaline and hyperplastic arteriolosclerosis.

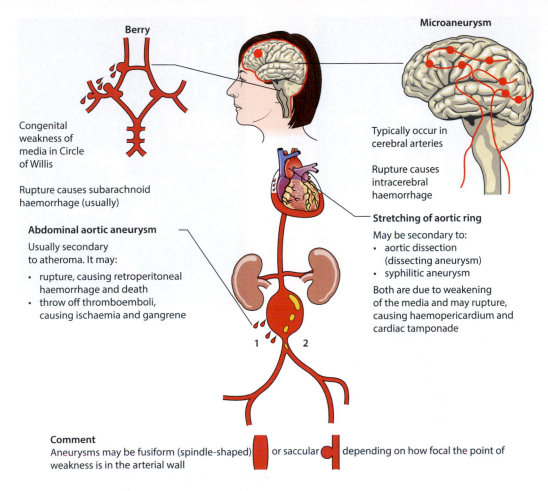

Berry

Congenital
weakness of
media in Circle
of Willis

Rupture causes subarachnoid
haemorrhage (usually)

Abdominal aortic aneurysm

Usually secondary
to atheroma. It may:

- rupture, causing retroperitoneal
  haemorrhage and death
- throw off thromboemboli,
  causing ischaemia and gangrene

Microaneurysm

Typically occur in
cerebral arteries

Rupture causes
intracerebral
haemorrhage

Stretching of aortic ring

May be secondary to:
- aortic dissection
  (dissecting aneurysm)
- syphilitic aneurysm

Both are due to weakening
of the media and may rupture,
causing haemopericardium and
cardiac tamponade

1   2

Comment
Aneurysms may be fusiform (spindle-shaped) or saccular depending on how focal the point of
weakness is in the arterial wall

Figure 8.13 Types of aneurysm and their complications.

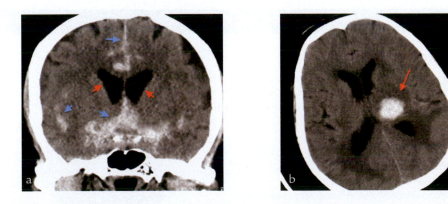

Figure 8.14 (a) Subarachnoid bleed from a berry aneurysm. Coronal CT of the brain. High-density material
in the subarachnoid spaces (blue arrows) is seen in an acute subarachnoid haemorrhage often from a berry
aneurysm. The resulting clot forming in the pathways of cerebrospinal fluid flow leads to obstruction producing
a communicating hydrocephalus — note the dilated frontal horns of the lateral ventricles (red arrows). (b) Axial
CT showing high-density material (red arrow) in the left thalamus in keeping with an acute hypertensive bleed
from a microaneurysm/Charcot–Bouchard aneurysm.

Chapter 8: Atherosclerosis and hypertension

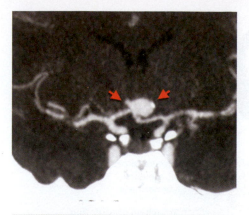

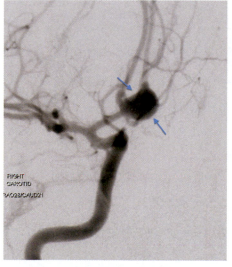

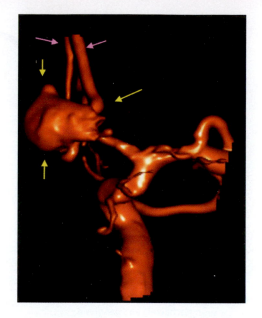

Figure 8.15 Berry aneurysms: CT angiogram (orange arrows) showing a berry aneurysm of the anterior communicating artery with correlative catheter angiogram image (blue arrows) and three-dimensional CT reconstruction (yellow arrows). The aneurysm occurs in an artery that is supplied by both right and left internal carotid arteries, and gives rise to the two anterior cerebral arteries (pink arrows) – a bifurcation point that is subject to turbulent blood flow throughout life.

History

Thomas Willis (1621–75)

Thomas Willis was the son of a Wiltshire farmer who fought with Charles I's army until it was defeated. He then went into medical practice in Oxford where, with others, he formed the Royal Society. Both an anatomist and a clinician, he wrote descriptions of the nervous system that were illustrated by Christopher Wren and made many original observations; but he was not the first to recognise or describe the circle of Willis. Key achievements were his first description of cranial nerve XI, recognising the sweetness of urine in diabetes mellitus but not in diabetes insipidus, and appreciating that hysteria was not a disease of the uterus but a disease of the brain!

Figure 8.16 Thomas Willis. (Reproduced with permission from the Wellcome Library, London.)

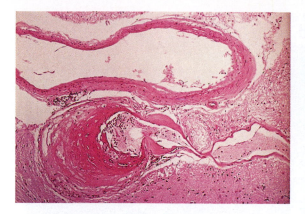

Figure 8.17 Photomicrograph showing a microaneurysm with a thrombus. (Courtesy of Dr S. Edwards, SGHMS.).

hypertensive individuals and are a common cause of intracerebral haemorrhage, a form of 'stroke'.

Atherosclerotic aneurysms are most common in the abdominal portion of the aorta and may present with massive haemorrhage or as a pulsatile mass in the abdomen, which may compress structures such as the ureters. Often they become complicated by thrombosis, with the risk of shedding emboli into lower limb vessels. These aneurysms occur in individuals with risk factors for atheroma and develop due to thinning of the media exacerbated by hypertension. The aneurysms are generally fusiform in shape and often extend for several centimetres along the aorta. Aneurysms >6 cm in diameter are likely to rupture, so it is recommended that these are replaced by prosthetic grafts because replacement after rupture carries a high mortality.

Arteries are not fixed structures but living tissues that require continual maintenance of the ECM. In atherosclerosis, the thickened intima can reduce diffusion and so lead to ischaemic damage in the inner third of the media; hypertensive arteriosclerosis of the aortic vaso vasorum produces ischaemia of the outer media. Both result in SMC loss, reduced ECM synthesis, accumulation of amorphous proteoglycans and fibrosis, which is termed 'cystic medial necrosis'. Atheromatous plaques also contain macrophages with increased MMP expression, which leads to ECM degradation in the underlying media. Rarer conditions also produce aneurysms. Ehlers–Danlos syndrome type IV has abnormal type III collagen, which is an important component of the ECM. Marfan's syndrome has an abnormal amount of the scaffolding protein fibrillin, which indirectly affects

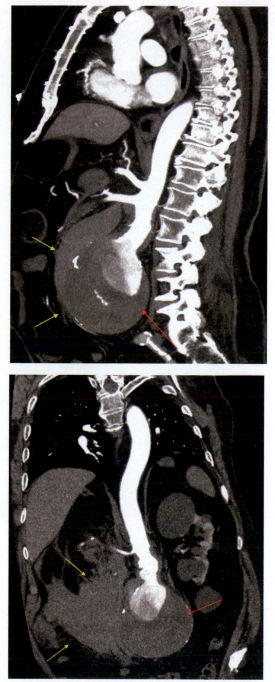

Figure 8.18 Aortic aneurysm: CT in a 79-year-old man who presented one night with severe back and abdominal pain. The CT images show a leaking abdominal aortic aneurysm with thrombus in the wall of the aneurysm (red arrow), and blood leaking into the periaortic soft tissues (yellow arrows).

Chapter 8: Atherosclerosis and hypertension

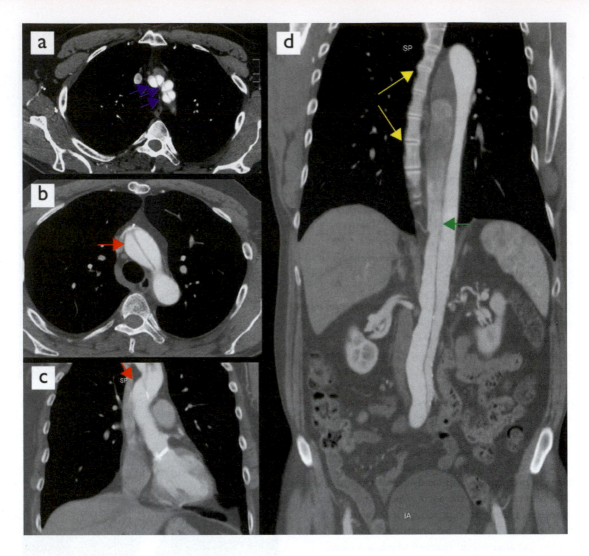

Figure 8.19 (a–d) Computed tomography (CT) scan image series in a 34-year-old man with Marfan syndrome which illustrates a life-threatening complication of the disease: aortic dissection. The dissection raises an intimal flap which can be identified on cross-sectional imaging as a low-density line running through the origin of the brachiocephalic, carotid and subclavian arteries (a; blue arrows), the aortic arch (b,c; red arrow) and the descending thoracic and abdominal aorta (d; green arrow). Note the aortic valve replacement (c; blue arrow) a consequence of aortic regurgitation and the scoliosis of the thoracic spine (d; yellow arrows), all features of Marfan syndrome.

the elastic tissue in the media through altering the activity of transforming growth factor β (TGF-β).

Aortic dissection usually occurs in the 40- to 60-year-old group and affects men more commonly than women, although it does occur in pregnant women, possibly because of generalised hormonal actions, which soften connective tissue. Patients complain of sudden severe pain in the centre of the chest, similar to that felt in myocardial infarction, but this often radiates to the back and moves as the dissection

History

Antoine Marfan (1858–1942)

Antoine Marfan was a French paediatrician whose major contribution is not the syndrome named after him but his studies on pulmonary tuberculosis, which recognised the immunity conferred by healed local lesions; this became known as Marfan's law and inspired Calmette to develop the BCG (Bacille Calmette–Guérin) vaccination. He was the pioneer of clinical paediatrics in France, specialising in infections and nutrition.

'In medicine it is always necessary to start with the observation of the sick and to always return to this as this is the paramount means of verification. Observe methodically and vigorously without neglecting any exploratory procedure using all that can be provided by physical examination, chemical studies, bacteriological findings and experiment, one must compare the facts observed during life and the lesions revealed by autopsy.'

progresses. The first event in aortic dissection is a tear in the intima, so that blood enters the media and tracks down between the middle and outer thirds of the media. The tear often occurs in the ascending aorta and is thought to be due to shearing forces on the intima because of turbulent blood flow. Any hypertension will exacerbate both the turbulence and the forces splitting the media. Once the blood begins to track along the media, it can travel in either direction and rupture back into the aorta, or it can rupture out into the peritoneal cavity, pericardial sac or pleural cavity. Rupture outwards is catastrophic and common. Rupture into the aorta is rare but has a good prognosis

and will produce a double-barrelled aorta. Extension of the dissection will occlude the mouths of any tributaries that become involved, and this commonly affects the coronary, renal, mesenteric, iliac or cerebral vessels. Dissection of the vertebrobasilar vessels can also occur (Fig. 8.19).

Aneurysms secondary to inflammation will include those due to syphilis, arteritis and infection. Syphilitic aneurysms tend to occur in the ascending arch of the aorta where they are ideally situated to cause mischief. Those that arise close to the aortic valve ring lead to dilatation of the ring and hence to aortic incompetence, the result of which is overload of the left ventricle and cardiac failure.

Aneurysms may rupture into the trachea or oesophagus to produce haemoptysis (coughing up blood), haematemesis (vomiting blood) or death. Any cause of aortic expansion within the chest can produce difficulty in breathing or swallowing due to compression, persistent cough due to irritation of the recurrent laryngeal nerves or problems of bone erosion. Fortunately, untreated syphilis is now less common in the western world and these complications are rare.

Aneurysms secondary to vasculitis, such as polyarteritis nodosa, tend to occur in the renal and mesenteric vessels where they lead to local ischaemia. Patients, therefore, may present with renal failure or with intestinal infarction and peritonitis, all of which have a significant mortality. Aneurysms secondary to infection are called mycotic aneurysms. Such aneurysms tend to be 'saccular', i.e. the wall is weakened in a particular focus that 'blows out' to form a sac. Most of the other conditions we have mentioned cause more diffuse weakening of the arterial wall, which dilates to form a 'fusiform' or spindle-shaped aneurysm.

Chapter 8: Atherosclerosis and hypertension

CHAPTER 9
CIRCULATORY FAILURE

If you think back to Chapter 1, you will recall that one of the most common causes of cell damage in any part of the body is hypoxia, which is a lack of oxygen because of either poor perfusion (reduced blood flow) or reduced oxygen in the blood as a result of lung or blood disorders. In Chapter 7, we looked at the chronic vascular disorders that narrow the arteries and, in Chapter 8, we considered the causes of acute blockage of blood vessels. In this final chapter on circulatory disorders, we pull this all together as circulatory failure and its effects.

ACUTE CIRCULATORY FAILURE – SHOCK

Acute circulatory failure is called shock. The 'shocked' patient is desperately ill and requires intensive treatment both to correct the condition that has produced the circulatory collapse and to cope with the widespread ischaemic damage resulting from shock. By definition, the patient will have hypoperfusion of many tissues. Blood pressure may be low but need not be, because the patient may either have compensated by increasing peripheral vasoconstriction to keep the pressure normal or may have had a high blood pressure that has now dropped. There may be pallor, cold extremities, sweating and a tachycardia: the first two signs due to poor perfusion and the other two resulting from the attempt to compensate, which includes the release of adrenaline.

What has happened to precipitate this disastrous state? Well, logically, there will be a sudden generalised poor perfusion if the pump fails or if there is insufficient blood, so-called cardiogenic and hypovolaemic shock. Abrupt heart failure may result from myocardial infarction, arrhythmias and cardiac tamponade, whereas hypovolaemic shock follows fluid loss due to haemorrhage, severe burns, diarrhoea or vomiting. Shock after a pulmonary embolism mimics cardiogenic shock, but the heart is normal and the reduced output is because the left atrial filling has dropped. A rather special but clinically very important form of shock is 'septic shock' due to overwhelming infection, especially that caused by Gram-negative bacteria that have endotoxic lipopolysaccharides (see page 72). Here the pathogenesis is complicated because of the varied effects of the bacterial products on endothelial cells, platelets and leukocytes, which leads to a veritable web of interactions, resulting in disseminated intravascular coagulation (DIC) and reduced blood volume because of vasodilatation and increased vascular permeability (Fig. 9.1).

Whatever the cause of shock, it has initial stages that are reversible and final stages that are irreversible, not dissimilar to cell damage (see page 51). In the initial stage, homeostatic mechanisms kick in, work adequately and tissue perfusion is maintained

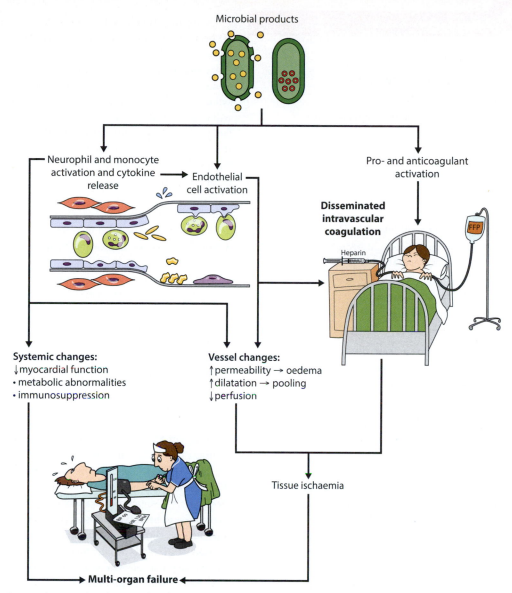

Figure 9.1 Pathogenesis of septic shock. A similar series of events can occur in systemic inflammatory response syndrome (SIRS) due to overwhelming activation of immune mechanisms from non-infective causes.

(*non-progressive stage*). Eventually these compensatory mechanisms may be overwhelmed and tissues become hypoxic and their functioning deteriorates (*progressive stage*). Finally there is an *irreversible stage* in which tissue damage cannot be recovered even if the circulation is restored. The effects on individual organs are covered on page 280. The prognosis of shock depends on the cause and the general health of the patient. More than 90% of previously healthy patients with acute hypovolaemic shock will survive if treated promptly. Those with septic or cardiogenic shock do less well. Septic shock kills around 20% of affected patients. Many with cardiogenic shock may also have some irreversible underlying heart problem that has suddenly become worse and will kill them.

Chapter 9: Circulatory failure

Key facts

Types of shock

Cardiogenic: pump failure due to:

    Myocardial infarction

    Cardiac tamponade (external pressure)

    Arrhythmias

    Pulmonary embolism

Hypovolaemic: loss of blood volume due to:

    Trauma

    Burns

    Diarrhoea and vomiting

Septic: complex interaction of microbial products and inflammatory mediators in severe infections most commonly:

    Gram-positive septicaemia

    Gram-negative endotoxin shock

    Fungal sepsis

    Superantigens, e.g. toxic shock syndrome

Anaphylactic: allergic reaction involving IgE-mediated hypersensitivity

Neurogenic: loss of vascular tone with spinal cord injury or anaesthetic

CHRONIC CIRCULATORY FAILURE

Chronic circulatory failure is most commonly due to chronic heart disease. In the developed countries, this is usually the result of ischaemic heart disease (as discussed in Chapter 8), whereas, in other parts of the world, damage from rheumatic fever is still important. Around 6 million Europeans have heart failure and as many as one in six people aged >85 years are affected. Let's start with the clinical scenario of a myocardial infarction.

CLINICAL SCENARIO: MYOCARDIAL INFARCTION

A 63-year-old man presented to accident and emergency complaining of chest pain. He had had central chest pain for 1–2 hours, and the pain radiated down the left arm and into his neck. He was feeling nauseated and also complained of shortness of breath. He had a history of hypertension for the last 5 years and had been receiving treatment. He also smoked 25 cigarettes a day and was obese. His father died at the age of 55 of a 'heart attack'.

On examination he was found to have a pulse rate of 40 beats/min, blood pressure of 110/80 mmHg, and he was in cardiac failure. An electrocardiogram (ECG) suggests a myocardial infarction and he was found to be in complete heart block. This man has acute coronary syndrome and, as house physicians and residents, you will encounter this sort of situation with alarming regularity. The first priority is to make the correct diagnosis, recognise the type of acute coronary syndrome and instigate urgent treatment. Acute coronary syndrome takes one of three forms:

1 Unstable angina
2 ST-elevation myocardial infarction (STEMI)
3 Non-ST-elevation myocardial infarction (NSTEMI).

Myocardial infarction is diagnosed when two of the following criteria are present:

- Prolonged cardiac pain at rest and unresponsive to glyceryl trinitrate (GTN)
- Characteristic ECG changes
- Detectable T or I troponin or creatine kinase CK-MB isoenzyme in the blood 12 hours after the onset of symptoms.

As urgent therapy is crucial, it is not sensible to wait 12 hours for the rise in serum markers of myocyte damage and so, on admission, acute coronary syndromes are divided into those with and those without ST segment elevation. ST segment elevation suggests that there is transmural ischaemia, and these patients require urgent reperfusion therapy (percutaneous coronary intervention [PCI] and/or thrombolysis). For those with non-ST segment elevation, anti-*thrombotic* treatment is appropriate with a combination of anti-platelet agents (aspirin and a platelet membrane ADP-receptor antagonist) and anticoagulants. In high-risk patients, early coronary angiography to assess the state of the vessels and revascularisation, if indicated, are used.

Figure 9.2 shows the risk of death from admission to 6 months after discharge for patients with different acute coronary syndromes. It should be remembered that approximately 20% die before they reach the hospital.

In our patient, the cardiac markers were raised after 12 hours and so he is classified as having a myocardial infarction. With the increase in sensitivity of tests for troponin markers, it is appreciated that more patients have myocyte damage.

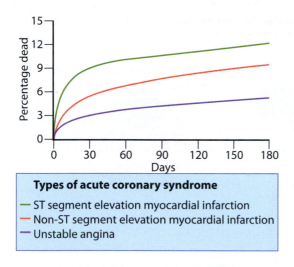

Figure 9.2 Risk of death from admission to hospital to 6 months after discharge, (Redrawn from Fox *et al.* (2006) *BMJ* **333**, 1091. With permission from BMJ Publishing Ltd.)

WHAT MECHANISMS LEAD TO MYOCARDIAL INFARCTION?

So what is the pathological sequence of events that leads to a myocardial infarction and what are the complications that may arise?

In myocardial infarction, the cardiac muscle cells die because of a lack of nutrients, most importantly oxygen. Generally, this results from poor blood flow to the myocardium because of narrowing or total occlusion of one or more coronary arteries. The extent of the infarction will depend on the amount of collateral flow, the metabolic requirements of the cells and the duration of the insult. Atheroma of the coronary vessels accounts for most cases but rarer causes include vascular spasm, emboli, arteritis and anaemia.

We have stated in Chapter 8 that plaques can be stable or unstable. The stable plaques narrow the coronary arteries so that blood flow is insufficient for even a moderate increase in cardiac work, such as walking upstairs, and the patient will complain of chest pain on exercise that is relieved on resting. This is called angina and occurs because the myocardial cells become ischaemic, but the damage is reversible. Unstable plaques may not be producing any clinical problems until an 'acute' event occurs, when the fibrous cap of the plaque splits so that blood can reach the soft necrotic centre. This can distort and enlarge the plaque

but, most significantly, the plaque contents activate the thrombotic cascade. Platelets and fibrin will aggregate to block the lumen and the platelet constituents (thromboxane A_2, histamine and serotonin) may worsen the situation by promoting spasm in the vessel wall. It is not known why the plaque fissures but it may be influenced by macrophage activity in the soft atheromatous centre, by vasospasm in the wall, by bending and twisting of the vessel as the heart contracts, or by altered distribution of stresses on the wall. It is often stated that coronary artery stenosis is not likely to produce clinical symptoms unless the cross-sectional area is reduced by 75%. This is true for long-standing fibrosed areas of atheroma, but most plaques that fissure to produce occlusion are fairly small and have an abundance of soft lipid. You will recall that soft plaques are more likely to fissure than hard, fibrous plaques.

Vasospasm is an elusive mechanism for a pathologist to identify because there will be nothing to see *post mortem*. However, it may be seen on angiography in some patients with angina or infarction, and occurs principally in areas damaged by atheroma. It is potentially of great therapeutic importance because it may be influenced by drugs. Nitric oxide (NO; originally called endothelin-derived relaxing factor) may be important in vasospasm either through reduced local production, reduced responsiveness of the smooth muscle cells or early neutralisation. Vasospasm-related angina is often called Prinzmetal's angina or variant angina. It generally responds to GTN and calcium channel blockers.

PATTERNS OF INFARCTION

Occlusion of a single vessel, as described above, will produce a regional infarct, occupying the segment of myocardium that is normally supplied by a particular coronary artery (Fig. 9.3). The infarct may involve a variable thickness of the myocardial wall, but, when it involves the full thickness of the wall, it is referred to as a transmural infarction and generally produces ST elevation myocardial infarction (STEMI) and Q waves on an ECG. Of transmural infarctions, 90% result from thrombosis complicating atheroma. Myocardial infarction is much more common in the left ventricle and interventricular septum, but approximately 25% of posterior infarctions will extend into the adjacent right ventricle or even into the atria. The patterns can vary because of collateral circulation or natural anatomical variation (Fig. 9.4). Approximately 40% are due to left

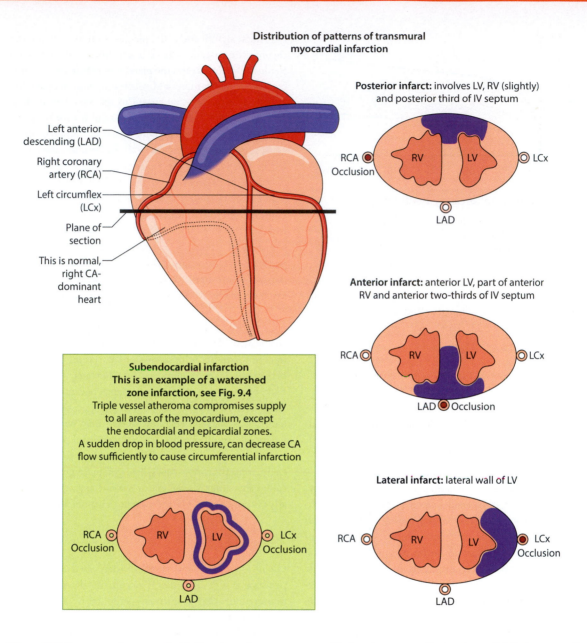

Figure 9.3 Patterns of myocardial infarction. Acute transmural infarction generally produces ECG changes of ST elevation and Q waves (STEMI). Acute subendocardial infarction generally has ST depression and no Q waves (NSTEMI).

anterior descending artery occlusion, 15% left circumflex artery, and 30% proximal right coronary artery. The posterior descending artery supplies the posterior aspect and comes off the RCA in 90% of people. Obstruction to the left main coronary artery is generally immediately fatal.

The other important pattern of myocardial damage is the subendocardial infarction. The pathogenesis of this type of infarction is different from the regional infarction because there is generally widespread atherosclerosis in all coronary vessels but no specific occlusion. The subendocardial region is the most vulnerable part of the myocardium for two reasons: first, any collateral supply that is developed tends to supply the subepicardial part of the myocardium; second, the subendocardium is under the greatest tension from the compressive forces of the

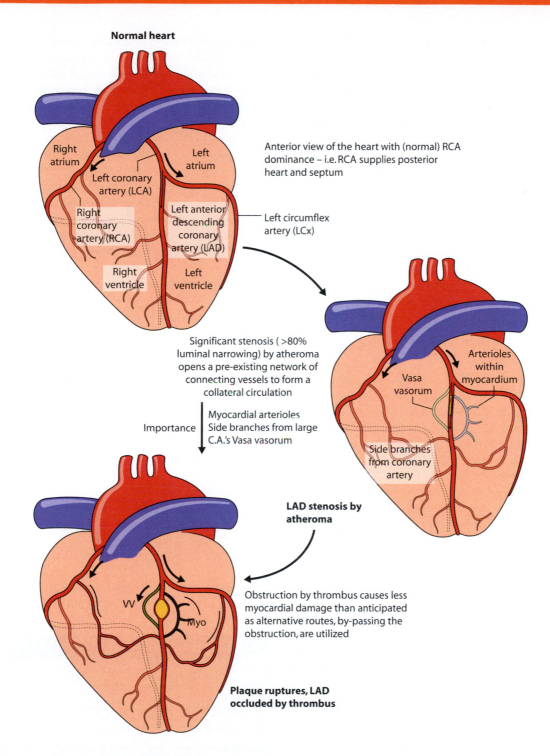

Normal heart

Right atrium

Left atrium

Left coronary artery (LCA)

Right coronary artery (RCA)

Left anterior descending coronary artery (LAD)

Left circumflex artery (LCx)

Right ventricle

Left ventricle

Anterior view of the heart with (normal) RCA dominance – i.e. RCA supplies posterior heart and septum

Significant stenosis (>80% luminal narrowing) by atheroma opens a pre-existing network of connecting vessels to form a collateral circulation

Importance

Myocardial arterioles
Side branches from large C.A.'s Vasa vasorum

Vasa vasorum

Arterioles within myocardium

Side branches from coronary artery

LAD stenosis by atheroma

VV

Myo

Obstruction by thrombus causes less myocardial damage than anticipated as alternative routes, by-passing the obstruction, are utilized

Plaque ruptures, LAD occluded by thrombus

Figure 9.4 The development of a collateral circulation in the heart. In many patients with coronary artery atherosclerosis, luminal narrowing occurs sufficiently gradually for the heart to adapt, by opening alternative circulatory paths. This leads to less than the expected amount of damage should a bypassed segment of coronary artery undergo sudden complete obstruction by thrombosis.

Chapter 9: Circulatory failure

myocardium and, hence, most likely to be ischaemic. Normally, blood will flow into the myocardium when the aortic root pressure exceeds the left ventricular cavity pressure, as occurs during diastole. Generalised reduction in myocardial perfusion results from any combination of coronary stenosis, reduction in aortic root pressure, increase in left ventricular cavity pressure, myocardial thickening and shortening of diastole.

Subendocardial infarction is much less common than transmural infarction. It is confined to the inner half of the myocardium and may be regional or circumferential. On an ECG, it generally does not have ST elevation or Q waves and is clinically designated NSTEMI. A very thin layer of subendocardial muscle remains viable because it receives nutrients and oxygen from the ventricular luminal blood. It should be noted, however, that even a transmural, regional infarct probably begins in the subendocardial region and then spreads to the rest of the wall.

WHAT ARE THE APPEARANCES OF INFARCTION?

Let us consider the clinical example described earlier in which the patient's ECG showed him to have infarction and complete heart block. If he had died within a few hours, a postmortem examination would have revealed a thrombus within the right coronary artery. This artery supplies the posterior wall of the left ventricle and the posterior third of the interventricular septum. Ischaemia of the septum would explain his complete heart block because this would damage the conduction pathway. No macroscopic abnormality would be seen in the myocardium, because the infarction would be only 6 hours old. If the patient had died at 24 hours, the infarcted area would either appear pale or be red–blue due to the trapped blood. Later the dead myocardium becomes pale yellow, softened and better defined with a rim of hyperaemic tissue at the periphery. Over the next few weeks, the necrotic muscle is replaced by fibrous scar tissue and this is usually complete by 6 weeks. The exact time course depends on the size of the infarct and any complications that may occur (Fig. 9.5).

WHAT COMPLICATIONS MAY OCCUR?

Our 63-year-old man was in cardiac failure and complete heart block, two of the most common complications of myocardial infarction (Fig. 9.6). First, we consider arrhythmias: this may be a type of heart block,

ventricular tachycardia or bradycardia, ventricular fibrillation or asystole. Arrhythmias are responsible for many cases of sudden death after a myocardial infarction and their prompt diagnosis is of crucial importance in the management of these patients. The arrhythmias occur because of either ischaemia or death of the specialised conducting tissue of the heart, or are due to the interruption of the conduction of impulses within the damaged myocardium. A damaged atrioventricular node, for example, may lead to complete heart block whereas damage to the conducting fibres within the ventricles will produce left or right bundle-branch block. Damaged myocardial fibres may also be electrically unstable (irritable) and so initiate abnormal impulses, which may terminate in ventricular fibrillation.

The second complication mentioned was cardiac failure. His cardiac failure could be due to complete heart block, so restoring normal sinus rhythm will be important in his treatment. Cardiac failure may also occur because of extensive death of muscle cells in the left ventricular wall, or because they have been 'stunned' by a short period of ischaemia and are temporarily unable to contract, but may recover over a few days.

The mainstay of treatment is to reperfuse the heart muscle but there are also hazards associated with this because various changes occurred called *reperfusion injury*. There are five main features of reperfusion injury:

- Mitochondrial dysfunction
- Myocyte hypercontraction
- Free radical-related damage
- Leukocyte aggregation
- Platelet and complement activation.

The last two lead to microvascular damage and limit restoration of blood flow. Ischaemic mitochondria can

Dictionary

Glossary of cardiological terms

Tachycardia: increase in pulse rate

Bradycardia: abnormally slow heart rate

Arrhythmia: abnormal cardiac electrical rhythm

Asystole: absence of cardiac electrical activity

Fibrillation: uncoordinated and ineffective muscle contraction

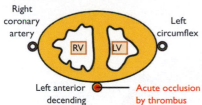

0–12 hours: *Potentially reversible*

Gross: Nil, but NBTZ positive 2–8 hours
Light microscopy (LM): Nuclear pyknosis, vague
      loss of striations, scanty polymorph (PMN)
      infiltrate 8–12 hours

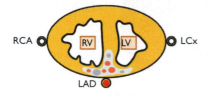

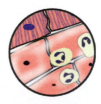

12–24 hours: *Ischaemic damage*

Gross: Blotchy, pale, slightly soft
LM: Increase in PMNs, obvious loss of
       striations, coagulative necrosis of
       myocytes

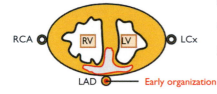

1–3 days: *Necrosis and inflammation*

Gross: Mottled pale infarct, red hyperaemic
       border
LM: As above, more marked mainly PMN
       infiltrate and early capillary ingrowth,
       particularly at periphery

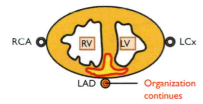

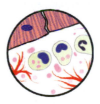

4–7 days: *Removal of debris, early*
               organization

Gross: Depressed, soft, yellow infarct, prominent
       hyperaemic edge
LM: As before, with increased macrophages
       phagocytosing debris from dead myocytes,
       peripheral granulation tissue formation

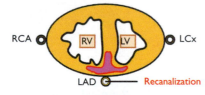

7–14 days: *Organization*

Gross: 'Bruised' look: red/purple colour,
       increasingly firm as granulation tissue
       forms
LM: Decreased inflammation as dead tissue is
       cleared, granulation tissue replaces
       damaged area

2–6 weeks: *Scar formation*

Gross: Infarct becomes firm and white and
       eventually contracts, the LV wall is
       thinned
LM: Capillaries and fibroblasts are replaced by
       acellular fibrous scar tissue

Figure 9.5 Myocardial infarction: changes with time.

Chapter 9: Circulatory failure

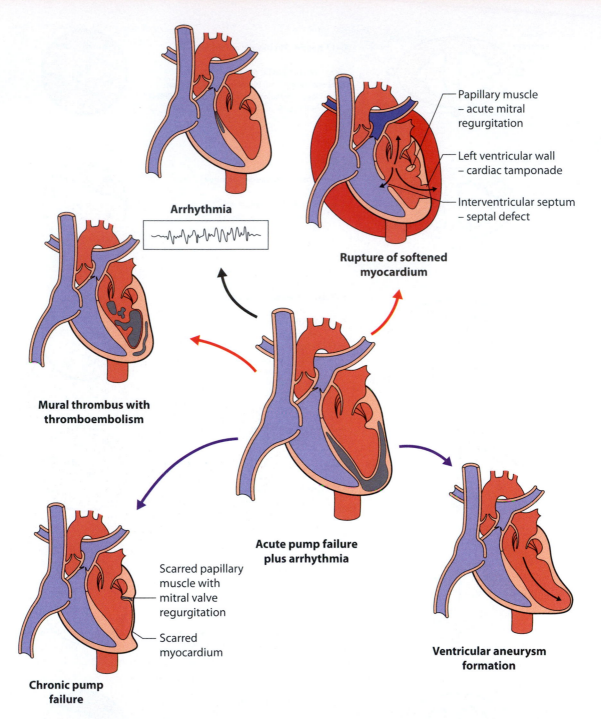

Figure 9.6 Patients who survive an acute myocardial infarction may develop subsequent cardiac problems. Those problems that tend to occur acutely are arrowed red, and more chronic problems are arrowed purple. Arrhythmias may occur at any time.

Radiology

The investigation of heart disease with imaging

There are 'many ways to skin a cat' and this saying applies aptly to imaging diseases of the heart based on what the clinician is looking for, the imaging modalities available to the investigating doctor and the costs of the investigation:

- The chest radiograph: provides some information about the size of the heart, the size of specific cardiac chambers and the presence or absence of heart failure. Calcification may be seen as a marker of myocardial, pericardial or valvular disease.

- The echocardiogram: provides a real-time ultrasound image of the beating heart and uses Doppler to measure flow, providing information on the contractility of the chambers of the heart valve function and the presence of pericardial disease.

- Coronary calcium scoring (Fig. 9.7): a CT study of the heart used as a screening test in asymptomatic patients (possibly with risk factors for cardiovascular disease). The coronary artery calcium score is calculated using a combination of density and volume of calcification and the value can be compared against data in a matched patient population to give an assessment of risk. It has a false-negative rate and should not be used alone in patients with symptoms of cardiac disease.

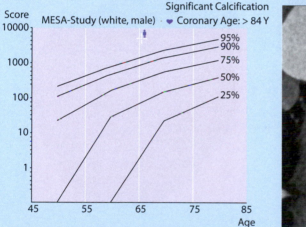
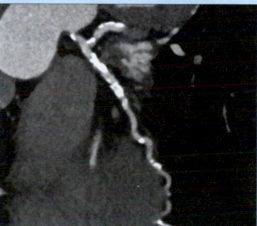

Figure 9.7 Coronary Artery Calcification Scoring: a CT screening test which gives a score for the degree of calcification in the coronary arteries. The calcification is a marker for atherosclerotic plaque. This is used in effect as a screening test in asymptomatic people (who have risk factors such as hypercholesterolaemia, hyperlipidaemia, family history, obesity, diabetes or hypertension) or people with non-specific chest pain to establish their risk for having clinically significant coronary artery disease and therefore requiring further investigation. The 66-year-old patient illustrated above has a Calcium score >1000 placing him above the 95th percentile for a patient of his age (and in effect having a coronary age older than an 84-year-old) placing him at high risk of coronary artery disease and at least one significant coronary artery stenosis. The image on the right shows significant calcified plaque in the D1 branch of the Left Anterior Descending artery.

- The technetium-99m-labelled Sestamibi scan to assess myocardial perfusion: a nuclear medicine technique using an intravenous radioisotope. This is used in the patient with possible angina to detect ischaemic changes in the myocardium. The patient is imaged at rest and then stressed, with either a treadmill or

(Continued)

(Continued)

intravenous coronary vasodilators (dipyridamole, adenosine or dobutamine). Ischaemic myocardium is identified by the presence of perfusion defects, which infer diminished blood flow.

- The stress echocardiogram assesses changes in cardiac wall motion following induced stress (with a treadmill or coronary vasodilators). Similar to the technetium scan, it can be used in the investigation of symptomatic patients to determine whether symptoms are due to myocardial ischaemia, and in patients with known coronary stenosis to determine whether the lesions are functionally significant.

- Coronary artery CT (Fig. 9.8): used predominantly in symptomatic patients who are of low-to-intermediate risk of coronary stenosis. CT of the heart is performed with intravenous contrast, and images are reconstructed of each coronary artery with detailed information about the presence of plaque, degree of stenosis in the artery and the anatomy of the coronary arteries. This is a relatively non-invasive method of imaging coronary artery disease and may help determine which patients should proceed to catheter angiography, cardiac surgery or conservative management. It is highly specific in ruling out coronary artery disease as a cause of symptoms. It is also very useful in evaluating patients with previous coronary grafts as a guide to anatomy and patency before angiography.

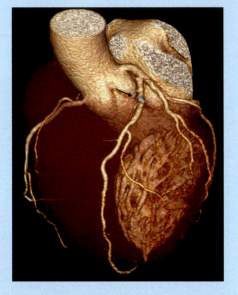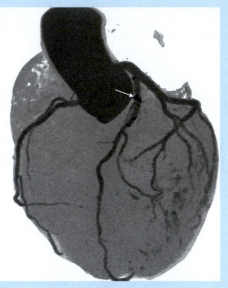

Figure 9.8 Coronary Artery CT involves the use of intravenous iodinated contrast and is less invasive compared to Catheter Angiography. This is used predominantly in symptomatic patients and often done at the same sitting as a Coronary Calcium Score when the score is high. The images show a significant stenosis caused by calcified plaque in the circumflex artery.

- Catheter angiography (Fig. 9.9): the most invasive and therefore most risk-laden imaging study with a catheter introduced into the arterial tree via a femoral artery puncture. This is negotiated to each of the coronary arteries and iodinated contrast is injected to obtain a coronary angiogram. The technique is also used to perform therapeutic angioplasties (dilatation of a narrowed artery) and stent insertions. There are now also catheter techniques to treat septal defects ('hole in the heart') and replace the aortic valve. The angiogram is also required in planning for bypass surgery.

- MRI: used to assess cardiac tumours, and congenital cardiac and aortic disease. This modality is the gold standard for assessment of left and right ventricular function and left ventricular hypertrophy. MRI perfusion techniques are also used to assess the presence of viable ischaemic myocardium after a myocardial infarct. MRI is expensive and technically challenging, and is reserved for situations where the combination of echocardiography, CT and angiography does not provide all the information needed. It is unsuitable for most patients with pacemakers.

Part 3: Cardiovascular disorders

- Positron emission tomography (PET): used to distinguish between viable and non-viable myocardium after a myocardial infarction. Also used in the investigation of cardiac sarcoid.
- Gated blood pool scans: nuclear medicine technique to measure left ventricular function. Used in oncology to follow patients treated with potentially cardiotoxic chemotherapeutic agents. Also used to identify intracardiac shunts.

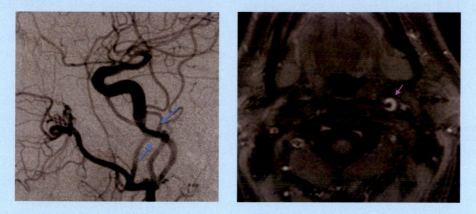

Figure 9.9 A catheter angiogram in another patient on the left shows concentric narrowing of the lumen of the internal carotid artery (blue arrows). The same patient has an MRI scan on the right showing the bright clot in the subintimal layer of the internal carotid artery (purple arrow). During a dissection, the intima is torn and blood tracks into the intima. It can break back into the lumen thus forming a vessel with two channels with an intimal flap between the two channels. The patient would be at risk of the dissection propagating intracranially, distal thrombosis or embolism. The blood in the subintimal layer can clot, as in the above patient, and produce a narrowing of the lumen and restrict flow to the brain.

swell and rupture, releasing contents that cause myocyte apoptosis. Hypercontraction of myofibrils causes cytoskeletal damage and potential cell death related to the rise in intracellular calcium levels during the ischaemic period. Free radical damage occurs as in other tissues (see page 52).

If a papillary muscle is damaged then mitral valve incompetence will produce cardiac failure. Initially, the papillary muscle is likely to be intact but incapable of contraction. After 4–5 days, the infarction has softened and the muscle may rupture, allowing the valve leaflet to prolapse, i.e. to float upwards into the left atrium. Similar softening occurs in infarcted tissue in the left ventricular wall so that it may rupture. This occurs in transmural infarction (i.e. full thickness) but not in subendocardial infarction. Within 24–48 hours of transmural infarction, the damaged ventricular muscle stretches, i.e. it becomes thinner, and is liable to aneurysm formation or rupture. This is often referred to as 'infarct expansion' but it should be appreciated that the amount of tissue damage is not increasing. It is merely a stretching of the damaged area. Rupture may take place in the interventricular septum, which creates a ventricular septal defect (VSD), or through the ventricular wall so that the blood leaks into the pericardial cavity, producing a haemopericardium, which inhibits the normal action of the heart, so-called cardiac tamponade. Generally, either of these complications is fatal and is most likely to occur 5–7 days after a myocardial infarction.

The body's immune system responds to the infarction so that the pericardial surface overlying the infarcted area usually becomes inflamed (pericarditis) by the second or third day. In most cases, this is self-limiting, but the friction between the pericardial surfaces produces a pericardial rub that may be heard through a stethoscope. Similar changes occur on the endocardial surface of the infarct which, in combination with stasis, predispose it to mural thrombosis. Whenever there is thrombosis, there is a risk of embolism. In this case, these would be systemic emboli affecting organs such as the brain or kidneys.

Chapter 9: Circulatory failure

Finally, the healed and fibrotic wall may balloon out to produce a cardiac aneurysm, which itself can be a site of thrombus because of stasis.

Extreme circulatory failure, as in shock, is a consequence of decreased cardiac output due to pump failure or blood volume loss. Lesser degrees of poor tissue perfusion can occur through any combination of pump failure, blood abnormalities and peripheral vascular disease. We have discussed heart and vessels quite extensively and should now turn our attention to the blood.

ANAEMIA

It is beyond the scope of this book to describe the whole of haematology but we can provide a framework for thinking about blood disorders and help you to appreciate that abnormalities in the blood can manifest as symptoms and signs in almost any organ, i.e. it is a common mechanism of disease.

The key components of blood (excluding platelets and clotting factors) are the fluid, the red cells with their haemoglobin for oxygen transport, and the white cells with their defence functions. What can go wrong with each of these? The simple answer is too much, too little or the wrong sort. Abnormalities of white cells are discussed in Chapters 4 and 5 (inflammation) and Chapters 11 and 12 (malignancy). Derangements of fluid balance are best looked up in a physiology or renal book. This leaves us to consider the red cells, but first a brief overview may help.

WHAT IS ANAEMIA?

The modern haematologist, instead of describing in English what he can see, prefers to describe in Greek what he can't.

Richard Asher, 1959 (in *The Lancet*)

A few key words or parts of words will help us through the maze of blood disorders. Any word ending in '-aemia' relates to the blood (e.g. polycythaemia, hypoalbuminaemia) just as words ending in '-uria' relate to the urine (e.g. haematuria, blood in the urine; anuria, not passing urine). A prefix of 'hypo-' indicates too little, 'micro-' means too small, 'hyper-' is too much and 'macro-' is too big. *Leuk*aemia literally means white blood but has become synonymous with a malignancy of blood cells. Anaemia strictly means a lack of blood, which is not really correct because there is a reduction in red blood cells rather than an absence. The term is now used to indicate a reduction in haemoglobin concentration in the blood. So what are the possible causes?

The red blood cells contain haemoglobin, which is important in the transport of oxygen. The haemoglobin level may drop because the number of red cells is low (reduced red cell mass) or the content of haemoglobin is reduced. Red cell numbers may be reduced because of impaired production, acute or chronic blood loss through bleeding, or a reduced lifespan for a variety of reasons.

Anaemia can be classified according to its cause (Tables 9.1 and 9.2) or according to the appearance of the blood (Figs 9.10 and 9.11). It is important to know both because the first investigation of a pale patient is to perform a full blood count which will detail the

Table 9.1 Morphological classification of anaemia

Type	MCV	MCH	Common causes
Normocytic/normochromic	Normal	Normal	Anaemia of chronic disease; chronic renal failure
Microcytic/hypochromic	↓	↓	Iron deficiency; β-thalassaemia trait; anaemia of chronic (severe) disease
Macrocytic (megaloblastic)	↑	Normal	Folic acid deficiency; vitamin B_{12} deficiency
Macrocytic (non-megaloblastic)	↑	Normal	Liver disease; alcohol ingestion; hypothyroidism
Leukoerythroblastic anaemia	Normal	Normal	Replacement or infiltration of marrow

Megaloblastic refers to abnormal maturation of erythroid cells detectable on examination of the marrow.
MCH, mean corpuscular haemoglobin; MCV, mean corpuscular volume.

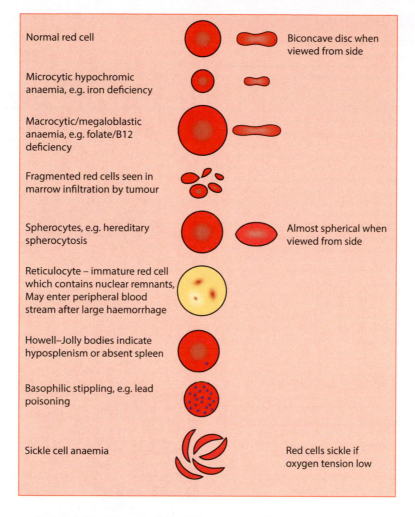

Normal red cell — Biconcave disc when viewed from side

Microcytic hypochromic anaemia, e.g. iron deficiency

Macrocytic/megaloblastic anaemia, e.g. folate/B12 deficiency

Fragmented red cells seen in marrow infiltration by tumour

Spherocytes, e.g. hereditary spherocytosis — Almost spherical when viewed from side

Reticulocyte – immature red cell which contains nuclear remnants, May enter peripheral blood stream after large haemorrhage

Howell–Jolly bodies indicate hyposplenism or absent spleen

Basophilic stippling, e.g. lead poisoning

Sickle cell anaemia — Red cells sickle if oxygen tension low

Figure 9.10 The morphology of red blood cells in different types of anaemia.

haemoglobin level in the blood, the number of red blood cells, the size of the red cells (mean corpuscular volume) and the haemoglobin in a red cell (mean corpuscular haemoglobin). If the haemoglobin level indicates that the patient is anaemic, then you can use the MCV and MCH to classify the anaemia on cell size and haemoglobin concentration and also look up the possible causes in Table 9.2. Let us try it for an imaginary patient. Remember '-cytic' refers to the cell and '-chromic' refers to the haemoglobin concentration.

A 21-year-old woman goes to see her general practitioner for a prenatal health check because she wishes to become pregnant for the first time. The doctor advises her about the risks of smoking and alcohol on the fetus, performs a physical examination and finds no abnormalities. Her blood pressure is normal and

he takes a blood sample for further analysis. The blood result and normal values are shown in Table 9.3.

When the patient returns to the surgery the following week, the doctor explains that she has a minor problem because she is anaemic and probably has iron deficiency. How do you classify her anaemia?

Hopefully, you have concluded that our woman has small red cells with a reduced haemoglobin concentration, i.e. a microcytic/hypochromic anaemia. To search further for a cause, you need more information. The most common causes of anaemia can be identified through a combination of the following:

- Reticulocyte count
- Morphological appearance of the cells in the peripheral blood

Chapter 9: Circulatory failure

Table 9.2 Causes of anaemia

Decreased red cell production

Defective haemoglobin production
   Iron deficiency
   Anaemia of chronic disease
   Myelodysplasia
Defective DNA synthesis (megaloblastic anaemia)
   Vitamin B12 deficiency
   Folic acid deficiency
Anaemia of elderly people (low iron, vitamin B12, testosterone)
Stem cell failure, e.g. aplastic anaemia
Bone marrow replacement, e.g. infiltration by malignant disease
Inadequate erythropoietin stimulation, e.g. chronic renal failure
Other nutritional and toxic factors
   Scurvy
   Protein malnutrition
   Chronic liver disease
   Hypothyroidism

Increased red cell destruction, i.e. haemolytic anaemias

Intrinsic defect of erythrocytes
Congenital
   Haemoglobinopathies, e.g. sickle cell anaemia and thalassaemias
   Membrane defects, e.g. hereditary spherocytosis
   Enzyme deficiency, e.g. glucose 6-phosphate dehydrogenase deficiency (G6PD)
Acquired, e.g. paroxysmal nocturnal haemoglobinuria
Extrinsic cause for haemolysis
Immune-mediated
   Autoimmune haemolytic anaemia
   Haemolytic disease of the newborn
   Blood transfusion-related haemolysis
   Drug-induced immune haemolytic anaemia
Direct acting
   Infections, e.g. malaria
   Snake venom
   Physical trauma, e.g. microangiopathy
   Hypersplenism

Blood loss

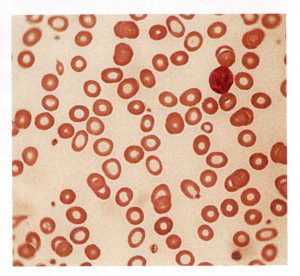

(a)

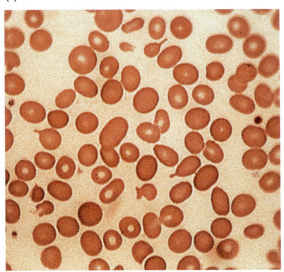

(b)

Figure 9.11 (a) Microcytosis and hypochromia in a case of iron-deficiency anaemia; (b) oval macrocytes in a case of pernicious anaemia.

- Haemoglobin electrophoresis
- Serum iron
- Bone marrow examination
- Serum ferritin (serum iron and iron-binding capacity were formerly used)
- Serum vitamin B_{12} and folate levels
- Antibody screens (e.g. parietal cell and intrinsic factor antibodies).

Dictionary

Glossary of haematological terms

Anaemia: reduction in haemoglobin in blood

Pancytopenia: reduction in all blood cell types

Neutropenia: reduction in neutrophils

Thrombocytopenia: reduction in platelets

Polycythaemia: increase in red cells

Thrombocythaemia: increase in platelets

Leukocytosis: increase in white cells

Leukaemia: malignant haematopoietic cells in blood

Aplastic marrow: no haematopoiesis in marrow

Hypoplastic marrow: reduced haematopoiesis in marrow

Hyperplastic marrow: increased haematopoiesis in marrow

Leukoerythroblastic: red and white cell precursors in peripheral blood may indicate marrow replacement by fibrosis, tumour, abscesses,

Red cell changes

Normocytic: normal cell size

Macrocytic: increased mean corpuscular volume (MCV)

Microcytic: decreased mean corpuscular volume (MCV)

Normochromic: normal haemoglobin (Hb) concentration

Hypochromic: Decreased mean corpuscular haemoglobin (MCH)

Poikilocytosis: variation in shape

Anisocytosis: variation in size

Howell–Jolly bodies: nuclear remnants in delayed maturation

Table 9.3 Normal peripheral blood values and those from a patient

	Normal	Female patient
Haemoglobin (g/dL)	Male, 13.5–17.5	8.2
	Female, 11.5–15.5	
Erythrocytes ($\times 10^{12}$/L)	Male, 4.5–6.5	4.7
	Female, 3.9–5.6	
Haematocrit (PCV) (%)	Male, 40–52	31
	Female, 36–48	
MCV (fL)	80–95	69
MCH (pg)	27–34	24.5
Reticulocytes ($\times 10^9$/L)	25–125	96

Reticulocytes are newly released red cells that are slightly larger than mature red cells and have a more basophilic (blue) cytoplasm on routine Giemsa staining. Approximately 1% is the normal level (giving a count of 25–125 $\times 10^9$/L) and an increased number indicates increased red cell turnover. Normally, the reticulocytes become mature red cells in the marrow, and it is only when demand for red blood cells exceeds supply that these immature forms are released in significant numbers. Our woman has a normal reticulocyte count so the most likely diagnosis is iron-deficiency anaemia, which is a common finding in premenopausal and pregnant women and easily treated with iron tablets. The doctor is not likely to investigate any further unless the pregnant woman has other problems, but we digress to discuss the causes of iron deficiency and iron overload. Iron deficiency is important to understand because it is common. Iron overload is worth remembering as an example where homeostasis cannot be achieved through increased excretion. Think about it! You drink too much water and the kidney responds. Take too much salt or vitamin C and you excrete what you do not need. However, there is no controllable excretion method for iron. Is this a unique situation? No, just consider calorie intake!

IRON-DEFICIENCY ANAEMIA

Iron is an important element with well-known roles in haemoglobin, myoglobin, cytochromes and various enzyme systems in the cells. Approximately 80% of the body's iron is in one of these functional forms whereas the remaining 20% is stored as ferritin or haemosiderin. Iron is absorbed in the upper small intestine and transported in the blood bound to the glycoprotein

Chapter 9: Circulatory failure

Key facts

Causes of iron-deficiency anaemia

Inadequate intake
Deficient diet
Malabsorption
    Generalised malabsorption, e.g. coeliac disease
    Post-gastrectomy (rapid gastrojejunal transit)
Hypochloraemia due to proton pump inhibitors

Increased iron loss
Reproductive tract
    Heavy menstruation
    Pregnancies and miscarriages
Gastrointestinal tract
    Oesophageal varices
    Peptic ulcer disease
    Chronic aspirin ingestion
    Hookworm infestation
    Haemorrhoids
    Tumours
Miscellaneous
    Epistaxis
    Haematuria
    Haemoptysis
Repeated venesection for blood transfusion

Increased demand for iron
Early childhood
Pregnancy and lactation
Erythropoietin therapy

transferrin. There is no control over the excretion of iron, and iron loss from the body is through loss in secretions, exfoliated cells and menstrual blood. There is some control over the absorption of iron with between 10 and 20% of dietary iron being absorbed. This does not leave much of a safety margin so iron deficiency is the most common cause of anaemia and occurs because of an imbalance between absorption and loss. In developed countries, an average diet contains about 15 mg of iron and the daily requirement for *absorbed* iron is 0.5–1.0 mg for men and 0.7–2.0 mg for women. This means that any reduction in dietary iron, problem with absorption or increased requirement for iron will lead to deficiency (see Key facts box above). Iron balance is particularly precarious in premenopausal women because of menstrual blood loss. In 50 mL of whole blood there is about 25 mg of iron, which would require an extra 250 mg of iron in the diet to be back in balance.

There are no immediate problems for the person developing iron deficiency because red cell production continues, but the iron stores in the marrow become depleted. Once the iron stores are inadequate, red cells are still produced but they are small (microcytic) with too little haemoglobin (hypochromic), and the patient will become tired and lethargic, breathless on exertion and appear pale. She may also have problems due to the effects of iron deficiency on epithelial cells if she does not receive treatment but remains chronically iron deficient. This complication, however, is rare in most countries. The mucous membranes of the mouth, tongue, pharynx, oesophagus and stomach become thin (atrophic), which may cause difficulty in swallowing (dysphagia) and produce mucosal webs in the upper oesophagus. Fingernails become spoon shaped (koilonychia) and split easily, and the thinned stomach wall does not produce a normal amount of acid. This combination of problems in severe iron deficiency is called the Plummer–Vinson syndrome and is cured by giving iron.

MEGALOBLASTIC ANAEMIA

The next woman in our prenatal clinic has Crohn's disease, which is an inflammatory disease that can affect any area of the gastrointestinal tract. The terminal ileum is commonly involved and this can result in anaemia due to vitamin B_{12} deficiency. Her blood results are shown below:

Haemoglobin	5.0 g/dL
Erythrocytes	1.7×10^{12}/L
Haematocrit (packed cell volume or PCV)	21%
MCV	128 fL
MCH	33 pg
Reticulocytes	40×10^9/L

The mean corpuscular volume (MCV) is increased and the mean corpuscular haemoglobin (MCH) is normal, consistent with a macrocytic anaemia. We are going to concentrate on the subset of macrocytic anaemias that have abnormal erythroid maturation in the bone marrow resulting in large precursors and called 'megaloblastic'. Megaloblastic anaemia is most commonly due to vitamin B_{12} or folate deficiency (Table 9.4). Both are co-factors for the conversion of

Table 9.4 Causes of megaloblastic anaemia

Vitamin B$_{12}$ deficiency

Inadequate diet
  Strict vegans excluding milk, eggs and cheese
Absorption problem
  Intrinsic factor deficiency
    Pernicious anaemia
    Total and subtotal gastrectomy
  Terminal ileal disease
    Crohn's disease
    Surgical removal
Competition by microorganisms
  Bacterial overgrowth in blind loops
  Fish tapeworm infection

Folic acid deficiency

Inadequate diet
  Malnutrition
  Chronic alcoholism
Absorption problem
  Generalised malabsorption
  Tropical sprue
  Gluten-induced enteropathy (coeliac)
Increased demand
  Early childhood
  Pregnancy
  Erythroid hyperplasia in severe haemolytic anaemias
Use of folic acid antagonists
  Anticonvulsants, e.g. phenytoin
  Anticancer drugs, e.g. methotrexate

Defects of DNA synthesis

Congenital enzyme deficiency, e.g. orotic aciduria
Acquired enzyme deficiency
  Alcohol
  Therapy with hydroxyurea

Key facts

Clinical features of megaloblastic anaemia

- Slow onset
- Gradually progressive
- Mild jaundice due to excess breakdown of RBC in ineffective erythropoiesis
- Glossitis
- Angular stomatitis
- Mild malabsorption due to epithelial abnormalities
- Purpura due to thrombocytopenia
- Excess apoptosis
- Sterility
- Increased melanin

deoxyuridine to deoxythymidine, an essential step in the synthesis of DNA.

Working down our list of investigations, we first ask for a reticulocyte count and a peripheral blood film. The laboratory tells us that there are few or no reticulocytes but the red cells contain Howell–Jolly bodies, which are nuclear remnants due to delayed maturation. The red cells are large (macrocytic) and of variable shape (poikilocytosis) and the neutrophils are hypersegmented (most normal neutrophils have three or four lobes, whereas hypersegmented cells have more). These are all features of megaloblastic anaemia. The next step is to measure the serum vitamin B$_{12}$ level and serum and red cell folate levels. If the folate level is low, we have a diagnosis of folate deficiency, the cause of which should be identified by taking a good history. If the vitamin B$_{12}$ is low, we can investigate further and need to understand a little more about vitamin B$_{12}$ absorption. Vitamin B$_{12}$ is absorbed in the terminal ileum as a complex bound to intrinsic factor. Intrinsic factor (IF) is produced by the parietal cells in the stomach. This means that you need an adequate diet and a normal stomach and terminal ileum to avoid vitamin B$_{12}$ deficiency. We assume that our woman with Crohn's disease will fail to absorb the vitamin B$_{12}$/IF complex because she has a diseased terminal ileum, but how could we prove this? We could demonstrate that her bone marrow is vitamin B$_{12}$ deficient by *injecting* some vitamin B$_{12}$ and observing an almost immediate increase in reticulocytes.

The most common cause of vitamin B$_{12}$ deficiency in Britain is pernicious anaemia (Fig. 9.12). This is aptly named because patients with megaloblastic anaemia can have a very severe anaemia and may even die without treatment. Pernicious anaemia or Addison's anaemia is when the vitamin B$_{12}$ deficiency is due to autoimmune damage to the stomach, resulting in chronic atrophic gastritis in which an inflamed and thinned stomach mucosa fails to produce adequate IF or acid. This disease occurs predominantly after the age of 50 years and is more common in men. As well as severe anaemia, the people with this condition may have neurological problems such as subacute combined degeneration of the

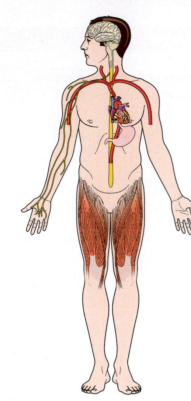

**Clinical features of vitamin B12 deficiency
due to autoimmune damage to the stomach**
Severe megaloblastic, macrocytic anaemia
Peripheral neuropathy
Subacute combined degeneration of the cord
• demyelination of posterior and lateral columns
• paraesthesia
• loss of position and vibration sense
• ataxia
• weakness
• spasticity
Increased risk of gastric carcinoma

Figure 9.12 Clinical features of pernicious anaemia (vitamin B$_{12}$ deficiency due to autoimmune damage to the stomach).

cord and segmental demyelination of peripheral nerves. These are due to the lack of vitamin B$_{12}$ and do not occur in folate deficiency. They are also at an increased risk of stomach carcinoma. *Helicobacter pylori* infection may initiate an autoimmune gastritis, which presents in younger people as iron deficiency and in elderly people as pernicious anaemia.

 Read more about abnormal full blood count in Pathology in Clinical Practice Case 22

HAEMOGLOBINOPATHIES

Our next patients in the antenatal clinic know all about anaemia because they have a family history involving anaemia as the result of an inherited abnormality with their haemoglobin. These are called haemoglobinopathies and it is possible to do genetic testing for them (Fig. 9.13).

SICKLE CELL DISEASE

Patients with this recessive haemoglobin disorder may present clinically with abdominal pain, joint pains, cerebral symptoms, renal failure and cardiac failure, which result from thrombotic and ischaemic damage. This occurs because the red cells 'sickle', so altering their shape and occluding capillaries. The red cells have an abnormal haemoglobin which, under hypoxic conditions, polymerises and alters the cell's shape.

In 1949, Pauling analysed the haemoglobin from patients with sickle cell anaemia and discovered that its mobility on electrophoresis differed from normal haemoglobin. He called it haemoglobin S (HbS). Later, family studies suggested that the gene for sickle cell haemoglobin was an allele of the normal gene on chromosome 11 for the β chain of the haemoglobin molecule, i.e. an alternative gene at the same locus on the chromosome. The difference between the normal haemoglobin gene and the sickle cell gene is a change in one base-pair: GAG becomes GTG. This causes valine

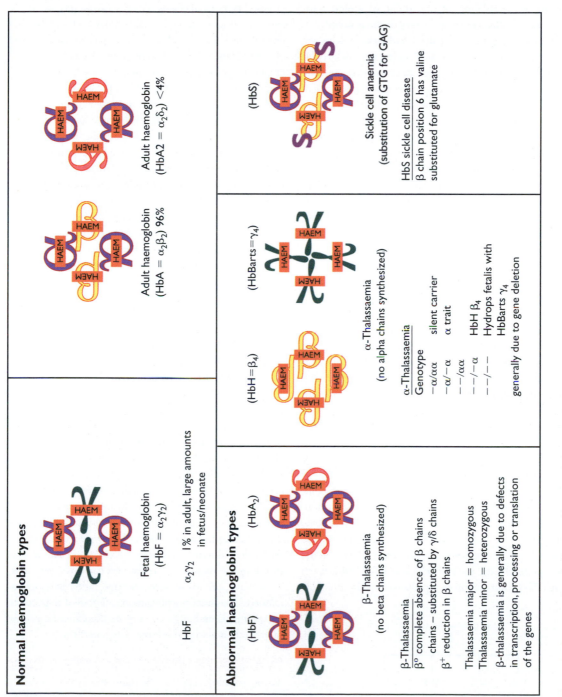

Normal haemoglobin types

(HbF)

Fetal haemoglobin
(HbF = $\alpha_2\gamma_2$)

$\alpha_2\gamma_2$ 1% in adult, large amounts
in fetus/neonate

Adult haemoglobin
(HbA = $\alpha_2\beta_2$) 96%

Adult haemoglobin
(HbA2 = $\alpha_2\delta_2$) <4%

Abnormal haemoglobin types

(HbF)   (HbA$_2$)

β-Thalassaemia
(no beta chains synthesized)

β-Thalassaemia
β⁰ complete absence of β chains
chains – substituted by γ/δ chains
β⁺ reduction in β chains

Thalassaemia major = homozygous
Thalassaemia minor = heterozygous

β-thalassaemia is generally due to defects
in transcription, processing or translation
of the genes

(HbH = β_4)   (HbBarts = γ_4)

α-Thalassaemia
(no alpha chains synthesized)

α-Thalassaemia
Genotype
−α/αα   silent carrier
−α/−α   α trait
−−/αα
−−/−α   HbH β$_4$
−−/−−   Hydrops fetalis with
               HbBarts γ$_4$
generally due to gene deletion

(HbS)

Sickle cell anaemia
(substitution of GTG for GAG)

HbS sickle cell disease
β chain position 6 has valine
substituted for glutamate

Figure 9.13 Normal and abnormal haemoglobin types.

Chapter 9: Circulatory failure

to replace glutamic acid in position 6 of the β chain. That's it – a point mutation changing just one nucleotide leads to the translation of one different amino acid, which entirely changes the property of the protein!

Fortunately, genes are paired and people who are heterozygous (i.e. one normal allele, one sickle cell allele) do not usually have any problems unless they become unusually hypoxic (e.g. during a surgical operation). They have a mixture of the normal and abnormal haemoglobins.

For practical purposes, we can regard sickle cell disease as an autosomal recessive disorder. How should we counsel a healthy pregnant woman who has a family history of sickle cell disease? The problem lies in deciding which members of the family are carriers of the gene because two people with sickle cell trait (heterozygotes) are likely to produce one healthy child, one sick child and two carriers (Fig. 9.14).

Techniques have been developed to analyse the DNA, which are particularly useful in prenatal diagnosis for testing the fetus before it has switched on to full production of the β chains. It is not possible to detect the abnormal β chains in fetal red blood cells because the fetus is relying on haemoglobin produced from α and γ chains, i.e. HbF (see below). However, it is possible to remove a small piece of placenta (chorionic villous sampling) for DNA analysis, relying on the point mutation to alter the binding of specific oligonucleotide probes or interfere with restriction enzyme digestion.

OTHER HAEMOGLOBINOPATHIES

Sickle cell disease is not the only haemoglobinopathy, although it is one of the most common. Haemoglobinopathies can be due to an abnormal chain being present or one type of chain not being produced. Haemoglobin is composed of two pairs of (i.e. four) polypeptide chains, each of which is linked to a haem group. The haem group is a protoporphyrin molecule

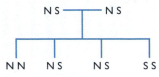

Figure 9.14 Possible outcomes for the offspring of two people with sickle cell trait. N, normal gene; S, sickle cell gene.

chelated with iron and able to carry oxygen. For the moment, we shall concentrate on the polypeptide chains. The normal chains are called alpha (α), beta (β), gamma (γ) and delta (δ). All normal haemoglobin has one pair of a chain combined with a pair of another type of chain. Hence there is haemoglobin A ($\alpha_2\beta_2$), haemoglobin A_2 ($\alpha_2\delta_2$) and haemoglobin F ($\alpha_2\gamma_2$). HbF predominates in the fetus and HbA in the adult.

Besides sickle cell disease, the other major group of haemoglobinopathies is the thalassaemias. In thalassaemias, the chains are normal in structure but not enough are produced. This condition is common in people originating from the Mediterranean, Africa and Asia, and is due to a wide variety of underlying genetic changes. The severity of the person's symptoms depends on the chain involved and whether they are homozygous or heterozygous for the abnormality. All normal haemoglobin has α chains, so complete absence of α chains is incompatible with life and an affected fetus will be oedematous (hydropic). In the absence of any α chains, the other chains do their best to produce a haemoglobin molecule by combining together as Hb Barts (γ_4) and HbH (β_4). This is the position with deletion of all four genes and, predictably, the severity decreases with the addition of each α chain (see Fig. 9.10). This group of conditions is called α-thalassaemia.

Around 95% of adult haemoglobin is HbA, composed of α and β chains, so the other important disease is β-thalassaemia with absent or reduced β chains, which attempts to compensate by producing HbA_2 and HbF using γ and δ chains, respectively.

The main clinical problems in thalassaemia are due to haemolysis of the red cells with the abnormal haemoglobin. These cells have a reduced life span and are removed in the reticuloendothelial system, particularly the spleen and marrow. Here the red cell components are broken down for reuse or excretion. The iron is stored in the tissues as ferritin and haemosiderin, and, if excessive, can cause tissue damage called haemochromatosis. The protoporphyrins are degraded to produce bile pigments and any excess gives the patient a yellow tinge to their skin and sclerae, i.e. jaundice.

SIGNS AND SYMPTOMS OF ANAEMIA

What would have happened to our women if their anaemia had not been discovered at prenatal testing? Let us discuss the effects of anaemia first. The anaemic patient has too little haemoglobin and hence a potential problem with the transport of oxygen to the tissues. The demand from the tissues will depend on the person's level of activity, so that an anaemic person may be asymptomatic when sitting at a desk but would have problems running a marathon; pregnancy can be regarded as a 9-month marathon (with a sprint finish!). An anaemic person can often compensate for the reduced amount of haemoglobin in the blood by pushing the blood round faster, i.e. increasing the cardiac output. The pregnant woman has the problem that in a normal pregnancy cardiac output needs to increase by around 30%, so compensation for anaemia may not be possible. So what will suffer? The mother will suffer with breathlessness, tiredness, weakness, and possibly dizziness or fainting. The baby will suffer because nutrition through the placenta may be inadequate, leading to a 'small-for-dates' baby lacking the normal stores of nutrients transferred between mother and baby in the last trimester.

If we think about patients in general with anaemia, their problems will depend on how low the haemoglobin is, how quickly the anaemia has developed and how well they can compensate. Older people are usually less able to adapt and, in addition to problems of shortness of breath, weakness, lethargy, palpitations and headaches, may have symptoms of cardiac failure, angina, intermittent claudication or confusion (Fig. 9.15).

ORGAN DAMAGE DUE TO POOR PERFUSION

Whatever the cause of poor perfusion, the effect at tissue level is similar, with a reduced delivery of oxygen and nutrients so that cells' normal functions are

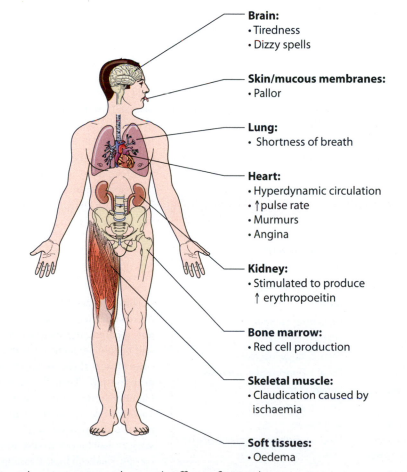

Brain:
• Tiredness
• Dizzy spells

Skin/mucous membranes:
• Pallor

Lung:
• Shortness of breath

Heart:
• Hyperdynamic circulation
• ↑pulse rate
• Murmurs
• Angina

Kidney:
• Stimulated to produce ↑ erythropoeitin

Bone marrow:
• Red cell production

Skeletal muscle:
• Claudication caused by ischaemia

Soft tissues:
• Oedema

Figure 9.15 Presenting symptoms and systemic effects of anaemia.

Chapter 9: Circulatory failure

disturbed. The details of reversible and irreversible cell injury were discussed in Chapter 1, so here we will concentrate on the clinical effects. The most important organs for immediate survival are the heart and brain, so the body has mechanisms for shunting blood from other tissues to protect these two organs. Frequently the kidneys and lungs will be underperfused and sufficiently damaged to be the immediate cause of death.

 Read more about supporting haematopoiesis in Pathology in Clinical Practice Case 36

ACUTE RESPIRATORY DISTRESS SYNDROME

The lungs are fairly resistant to short periods of ischaemia but, if prolonged, the patient may develop acute respiratory distress syndrome (ARDS). For oxygen to reach the alveolar blood, air must move in and out of the lungs and be able to diffuse across the alveolar septa. In 'shock lung' there is severe oedema, which both reduces the lung compliance and impairs alveolar gas diffusion – this means double trouble and a mortality rate of around 40%.

The probable sequence of events (Fig. 9.16) is that the 'shock' causes the production and release of IL-8 by pulmonary macrophages, which promote neutrophil aggregation and activation in the lung. The neutrophils

Key facts

Organs damaged in a patient with shock

Organ	Damage
Kidney	Acute tubular necrosis
Lung	ARDS
Heart	Ischaemic damage
Brain	Watershed infarcts
Liver	Fatty change/necrosis
Adrenal	Focal haemorrhagic necrosis
Pancreas	Pancreatitis
Stomach	Erosive gastritis
Duodenum	Ulceration
Small and large bowel	Haemorrhagic gastroenteropathy/ infarction

produce arachidonic acid metabolites, such as thromboxane, which cause pulmonary vasoconstriction, oxygen-derived free radicals, which injure the endothelial and epithelial cells, and lysosomal enzymes, which digest local structural proteins.

Key facts

Key causes of ARDS

- Shock
- Diffuse pulmonary infection, especially viral
- Aspiration pneumonitis
- Cardiac surgery involving extracorporeal pumps
- Inhalation of toxins or organic solvents

The damaged alveolar capillary endothelial cells are leaky, which leads to interstitial alveolar oedema and fibrin exudation. The damaged alveolar epithelial cells, particularly the type I pneumocytes, desquamate to form the characteristic hyaline membranes in combination with surfactant and protein-rich oedema fluid. These are the same as the hyaline membranes in neonatal hyaline membrane disease; in both situations they indicate severe epithelial injury with lack of surfactant, which leads to collapse of alveolar air spaces (atelectasis) and so further reduces compliance and gas transfer. Most patients who survive regain normal lung function after a year, but some develop chronic fibrosis. ARDS is one of the complications of the systemic inflammatory response syndrome described on pages 139 and 142.

RENAL DAMAGE

Impaired renal blood flow results in acute tubular necrosis, a major cause of acute renal failure. This is not apparent immediately in a 'shocked' patient but will become evident once the 'shocked' state is under control and there is no circulatory reason for poor urine output. Then the patient will be noted to have oliguria (urine output of 40–400 mL/day; normal, 1500 mL/day), salt and water overload, a high plasma potassium and urea, and a metabolic acidosis. At this stage a renal biopsy would show numerous foci of tubular epithelial cell loss, affecting any area of the nephron, and epithelial 'casts', i.e. dead epithelial cells, present

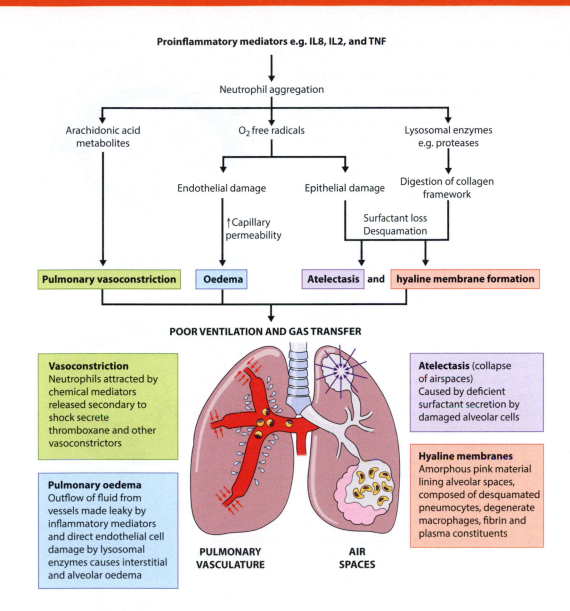

Figure 9.16 Pathogenesis of adult respiratory distress syndrome (ARDS): an initiating event causes shock. Acute inflammatory mediators damage vascular endothelium, the alveolar walls and lining epithelium, and cause pulmonary vasoconstriction, oedema, collapse and the formation of membranes over the alveolar surface.

in the tubular lumina. The important clinical point is that the patient can make a complete recovery if appropriately managed, e.g. by dialysis, and the cause of the shock rectified. After a few days, the tubular epithelium will regenerate and the urine volume will increase, often to above normal values because the tubules are unable to concentrate the urine, and there may be

excessive loss of water, sodium and potassium – the so-called diuretic phase. Slowly the tubular epithelium returns to normal and reasonable renal function is restored. The picture can be complicated by the addition of an inflammatory reaction in the damaged tissue and also intrarenal vasoconstriction as a consequence of sublethal endothelial injury

BRAIN AND CARDIAC DAMAGE

Despite the body's best efforts to protect the heart and brain, these organs may become underperfused.

Damage to the brain may be mild or devastating. The neurons are most vulnerable to ischaemia, particularly the large Purkinje cells of the cerebellum and the pyramidal cells in the hippocampus. A short episode of hypoperfusion may not cause any irreversible neuronal damage, or the number of neurons damaged may be too few to produce any clinical effect beyond temporary confusion. However, prolonged ischaemia will result in infarction, which most commonly affects the 'watershed' areas at the junctional zones between the main arterial territories (Fig. 9.17). This may result in severe permanent cerebral damage or coma and death. It is important to remember that the 'watershed' effect operates in many organs if there is poor perfusion. In the heart, this is the subendothelial zone because the endocardium is nourished by direct diffusion from the blood in the cardiac chambers, and the outer myocardium is supplied by arterioles penetrating from the outside. In the gut, areas such as the splenic flexure of the colon are at the boundary between arterial supplies and vulnerable to poor perfusion.

The changes described above relate to shock, i.e. acute circulatory failure. In chronic heart failure there are various compensatory mechanisms that attempt to maintain organ perfusion. There is a combination of autonomic, neurohormonal, immunological and haemodynamic alterations. A key pathway involves the renin–angiotensin–aldosterone system (RAAS), which is activated as renal perfusion falls and helps to maintain cardiac output. In the long term, however, it causes myocardial damage through apoptosis and fibrosis, increased risk of arrhythmias and an enhanced response to sympathetic nervous stimulation.

The sympathetic nervous system is stimulated by arterial baroreceptors reacting to the reduced cardiac output and results in raised levels of circulating noradrenaline. In the short term, this increases heart rate and myocyte contractility to improve cardiac output but, in the long term, it can increase the strain on the heart and accelerate left ventricular remodelling and dilatation.

Why is this important? It is because there are drug treatments that can block these pathways and improve survival. So we use angiotensin-converting enzyme (ACE) inhibitors and angiotensin-receptor blockers, aldosterone antagonists and β blockers, frequently in combination. Unfortunately this benefits the heart but

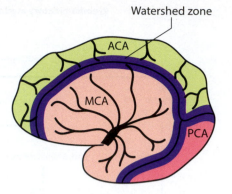

Lateral aspect of left cerebral hemisphere

Key: MCA Middle cerebral artery
ACA Anterior cerebral artery
PCA Posterior cerebral artery

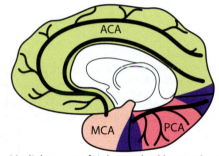

Medial aspect of right cerebral hemisphere

Figure 9.17 Watershed zones: the brain, heart and colon are particularly at risk of ischaemic damage because they receive oxygenated blood from the most peripheral branches of supplying arteries with no territorial overlap. This supply can be compromised by hypotension, often due to shock. The regions affected in the brain are shown here. In the heart, subendocardial infarction may occur at the boundary between the peripheral myocardial blood supply from the coronary arteries and direct diffusion from the endocardial side (see Fig. 8.2). In the colon, the splenic flexure lies at the boundary between the inferior mesenteric and superior rectal.

can worsen the kidney damage, so that deteriorating renal perfusion can also lead to anaemia of chronic disease, due to reduction in erythropoietin production. Chronically raised circulating noradrenaline is also thought to contribute to cachexia, loss of fat and muscle wasting.

STROKES

As we reach the end of the section on cardiovascular disorders, it is a good moment to discuss 'strokes'. To understand 'strokes', you need to pull together the topics we have covered in these three sections — namely, thrombosis, embolism, atherosclerosis, hypertension, aneurysms and circulatory failure. So what is the clinical picture? Very variable is the answer because different areas of the brain perform very distinct functions and the patient's symptoms and signs will depend on the location of the damage.

First, a definition: *a stroke is a sudden loss of some cerebral function due to a vascular lesion*. It is usual to exclude haemorrhage caused by trauma (subdural and extradural haemorrhage) and global loss of function, as might occur with generalised hypoxia or hypoperfusion (e.g. brain damage secondary to a cardiac arrest). If the loss of function lasts for less than 24 hours, it is called a transient ischaemic attack (TIA).

Second: are strokes a significant cause of death and disability? Approximately 10% of deaths are due to stroke, which puts it in the top five causes of death. In the USA, there are approximately half a million cases of stroke each year. Roughly half will die and, of the remainder, half will have permanent significant disability. Around a third of strokes occur in people aged <65 years. Thus, any intervention that can reduce the incidence or severity of strokes can have a major impact.

Third: what is the most common type of stroke? Strokes occur when cerebral tissue is deprived of its blood supply. This is most commonly ischaemic in nature (≥80%) due to blockage of an artery by thrombus (≥50%) or embolus (≥30%). You already know about the causes of thrombosis and the importance of unstable atheromatous plaques (see page 239) and the types and sources of emboli (see page 225). Most other strokes are due to haemorrhage, which can be intracerebral (about 10%) or subarachnoid (about 5%), and are associated with aneurysms and arteriovenous (AV) malformations (see page 252). Around 80% of intracerebral haemorrhages occur in the presence of hypertension. The most common causes are listed in the box below.

Improved imaging of the brain and a greater understanding of the importance of protein-folding abnormalities (proteinopathies) in neuropathology have identified another distinct cause of haemorrhagic strokes, which is cerebral amyloid angiopathy. This affects the medium and small meningeal and cortical vessels, in which

> ## 🔑 Key fact
>
> **The most common causes of spontaneous intracerebral haemorrhage**
>
> - Small vessel disease: associated with hypertension
> - Amyloid angiopathy: older patients
> - AV malformation
> - Arterial aneurysm
> - Cavernous malformation
> - Venous thrombosis
> - Clotting factor deficiency
> - Neoplasm
> - Vasculitis

amyloidogenic peptides deposit in the wall, weaken the vessel and can cause focal haemorrhage affecting specific lobes of the cerebral cortex. The protein type is usually amyloid b peptide, which is the same as that involved in Alzheimer's disease.

CEREBRAL BLOOD SUPPLY AND STROKES

The clinical effect of a disease process will depend on how well the body can respond to the insult. With vascular problems affecting the brain, this is influenced by the vascular anatomy and so it is time to revise some basic facts. You will recall that some vessels are end-arteries and blockage will lead to infarction of a clearly defined tissue segment. In the brain, the small cerebral vessels are end-arteries but the others are not and, to a greater or lesser extent, may compensate for blockage through being part of a collateral circulation.

The brain receives blood through two internal carotid arteries and two vertebral arteries. The two vertebral arteries join to become the basilar artery, which, with the two internal carotid arteries, feeds the circle of Willis at the base of the brain. If the circle of Willis is anatomically normal and free from significant atheroma, it can be an effective collateral route, such that one internal carotid artery can be totally blocked without causing damage. Blockage of the vertebrobasilar system is, however, less easy to compensate for.

The blood supply to the brain has two other unusual features. One is the effect of brain swelling (cerebral oedema), and the other is the ability to respond to and protect the brain from the effects of high or low blood pressure (autoregulation). The brain is encased in a hard, non-elastic structure (the bony skull), which has some fibrous internal dividing walls (tentorium cerebelli and falx cerebri) and one significant opening (foramen magnum). If a part of the brain increases in size, a so-called space-occupying lesion, then it presses on adjacent structures and may interrupt the blood supply directly or by herniation, e.g. tentorial herniation can stop flow in the posterior cerebral or superior cerebellar arteries. The swelling that accompanies infarction can cause this (Fig. 9.18).

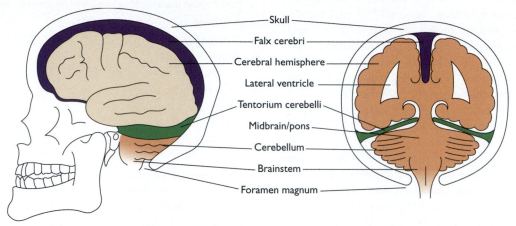

Skull
Falx cerebri
Cerebral hemisphere
Lateral ventricle
Tentorium cerebelli
Midbrain/pons
Cerebellum
Brainstem
Foramen magnum

(a) Sagittal section of a skull with a slightly shrunken brain, revealing part of the falx cerebri and tentorium cerebelli

(b) Coronal section through the skull

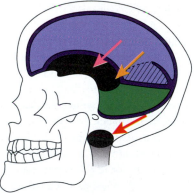

(c) Sagittal section of a skull with no brain, to show dural folds and possible sites of herniation

There are three main sites at which herniation of the brain can occur in response to a space occupying lesion in the brain

Beneath the FALX CEREBRI (pink arrow)

Through the TENTORIUM CEREBELLI (yellow arrow)

Through the FORAMEN MAGNUM (red arrow)

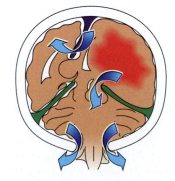

(d) Coronal section through a head containing a space occupying lesion in one cerebral hemisphere

The **falx cerebri** separates the cerebral hemispheres. Superior and inferior sagittal veins run within the fixed and free borders respectively. The corpus callosum passes beneath the falx to unite the hemispheres. The cingulate gyrus may be displaced laterally. The free edge is usually also displaced and the ventricles compressed.

The **tentorium cerebelli** separates the cerebellum from the cerebral hemispheres. The midbrain passes through the space anterior to the free border and may be displaced inferiorly.

The **foramen magnum** provides the exit hole for the spinal cord, which is sheathed in tough dura mater in the spinal canal. Tonsillar herniation occurs when the cerebellar tonsils are pushed through the skull and death by 'coning' occurs when the vital centres in the brainstem are compressed.

Figure 9.18 Sites of herniation in the brain.

Autoregulation refers to changes in the resistance of cerebral arterioles in response to changes in blood pressure. This is believed to be able to compensate for systolic blood pressures as low as 50 mmHg (as in 'shock') or as high as 160 mmHg in normal individuals, and is an example of effective homeostasis. However, this mechanism is disturbed after strokes, head injury or anaesthesia.

WHO IS MOST LIKELY TO HAVE A STROKE AND HOW CAN STROKES BE PREVENTED?

You have already worked through the risk factors and preventive measures for coronary artery disease (see page 242) and you should not be surprised to learn that they are similar for cerebrovascular disease. That is the advantage of understanding the key mechanisms of disease – once the principles are appreciated they can be applied to other situations. Combining the factors for thromboembolism, hypertension and atherosclerosis will give you the main pointers for stroke, e.g. there is correlation between non-haemorrhagic stroke and raised serum total cholesterol, low-density lipoprotein (LDL)-cholesterol and triacylglycerols, and lowered high-density lipoprotein (HDL)-cholesterol, i.e. the factors important in atherosclerosis. The treatment is, predictably, a combination of lifestyle, diet, exercise, smoking cessation, statins, anti-thrombotic agents and antihypertensive therapy. For the acute event, prompt revascularisation is important but may be difficult to achieve.

Ischaemic strokes are predominantly a consequence of acute events in unstable atheromatous plaques, with the important factors being the same as those for acute coronary syndromes, i.e. fibrous cap stability, size of lipid core and degree of inflammation. Ruptured carotid plaques are more likely than smooth plaques to be associated with future *coronary* events, which suggests that plaque instability is a systemic phenomenon.

That brings us nearly to the end of this section on circulatory disorders and we hope we have convinced you that an understanding of the physiology and pathology is fundamental to the practice of clinical medicine. In the words of Sir William Osler (Fig. 9.21), one of the greatest physicians of all time:

A man cannot become a competent surgeon without the full knowledge of human anatomy and physiology, and the physician without physiology and chemistry flounders along in an aimless fashion, never able to gain any accurate conception of disease, practising a sort of popgun pharmacy, hitting now the malady and again the patient, he himself not knowing which.

Radiology

The investigation of stroke with imaging

There are three components to the investigation of stroke: making the diagnosis of stroke on imaging, using imaging to direct acute treatment of stroke and investigating the causes of stroke.

Making the diagnosis of stroke: in most hospitals, stroke patients will have a CT scan as the first investigation. In accident and emergency, CT of the acute stroke may be normal or have very subtle findings that may be missed. It does, however, exclude a haemorrhage in the brain, and helps exclude other important causes of neurological dysfunction, such as a post-traumatic subdural collection or an intracranial neoplasm. The patient may also have a CT perfusion study. This is a rapid dynamic CT scan through the brain while the patient receives a bolus of iodinated intravenous contrast. This in effect produces a map of cerebral perfusion in the brain. This not only helps in showing the core area of infarction but also demonstrates the surrounding ischaemic penumbra. The patient may, depending on the facilities available, go on to have an MRI scan using diffusion-weighted imaging (DWI) (Fig. 9.19). This is highly sensitive to the motion of water molecules in the body and identifies areas of restricted water movement in the brain when water enters cells after an infarction (cytotoxic oedema), and becomes trapped or diffusion restricted. In fact the DWI scan picks up ischaemic changes or infarction within minutes, and is the most rapid way of making a diagnosis of cerebral infarction. The radiologist at the same time can also employ MRI perfusion to derive cerebral perfusion maps.

Using imaging to direct the management of acute stroke: many clinical trials have shown a benefit to the patient from having intravenous or intra-arterial thrombolytic treatment, but only if the diagnosis is made within 4–6 hours of the onset of stroke. If the patient is lucky enough to get to accident and emergency in

(Continued)

(Continued)

this time frame, then administration of a thrombolytic agent may reverse or improve the patient's neurological deficit. The CT scan is helpful in excluding other causes of neurological deficit and showing the presence or absence of acute haemorrhage in the infarct (which is a contraindication to thrombolysis). CT or MRI perfusion help in showing the anatomy of an ischaemic infarct with a core of dead tissue (the

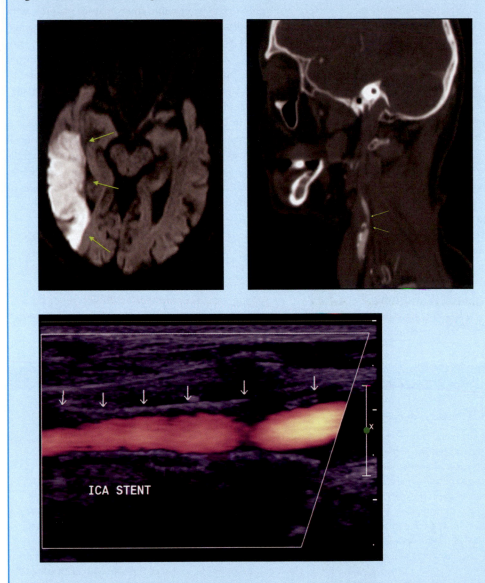

Figure 9.19 Top left: MRI showing a diffusion-weighted image (DWI), which is the most sensitive sequence used to pick up early changes of cerebral infarction (green arrows). This technique is sensitive to water motion and the high signal represents trapped water within dead or dying cells (as a consequence of cytotoxic oedema). Top right: the CT angiogram shows a stenosis of the proximal right internal carotid artery (green arrows), which predisposes the patient to a stroke in the right hemisphere. Bottom: an ultrasound scan shows the bright stent placed within the internal carotid artery to expand the stenosis and normalise flow in the artery. Although this does not treat the acute infarct, it is a preventive measure against further strokes in the territory of the right internal carotid artery.

infarct that cannot be recovered), and the surrounding ischaemic penumbra (the tissue at risk) which is potentially salvageable (Fig. 9.20). Other therapeutic manoeuvres that are used and being investigated include neurointerventional intra-arterial clot aspiration and removal.

Investigating the causes of stroke: this involves looking for the source of embolism or thrombosis and includes Doppler ultrasound of the carotid and vertebral arteries to look for ulcerated plaque or haemodynamically significant stenosis, which may benefit from surgical intervention. Echocardiography is helpful in excluding a cardiac source for a cerebral embolism. In young patients in particular, CT or MR angiography is often used to exclude a dissection of the carotid or vertebral arteries as a cause of stroke.

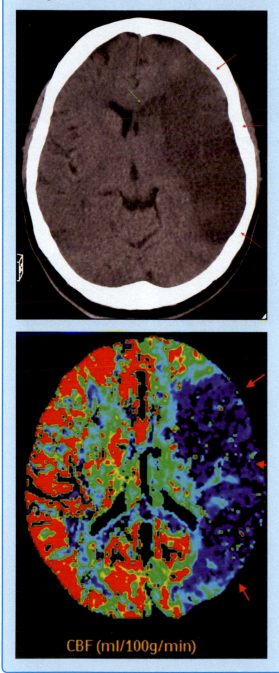

CBF (ml/100g/min)

Figure 9.20 Top: CT scan in the 24 hours after the onset of a right hemiplegia showing low density in the territory of the left middle cerebral artery – the sign of an acute infarct (red arrows). Note the mass effect or compression of the frontal horn of the ventricle (green arrow) due to the cytotoxic oedema within the infarct. Bottom: a cerebral perfusion map, generated during the scan, confirms the impaired perfusion to the same territory (red arrow). CT perfusion or MRI perfusion is increasingly used in the acute assessment of stroke in identifying potentially reversible ischaemic areas ('the ischaemic penumbra') of brain that border the irreversibly dead infarcted brain ('the infarct core'). These ischaemic areas may respond to reperfusion measures such as thrombolysis. The CT study is helpful in excluding a haemorrhagic infarct because this would contraindicate thrombolysis. Reperfusion measures such as thrombolysis are highly dependent on the patient receiving treatment within 3–6 hours of the stroke. This is the important rate-limiting step in preventing many stroke patients receiving reperfusion therapy.

Chapter 9: Circulatory failure

History William Osler (1849–1919)

Sir William Osler was especially renowned for transforming medical ewducation and clinical training. In 1888, he accepted a post at the new John Hopkins Hospital in Baltimore and became one of the famous 'Hopkins four', the others being William Welch, chief of pathology, Howard Kelley, chief of obstetrics and gynaecology, and William Halstead, chief of surgery. Between them they revolutionised the medical curriculum based on Osler's book *The Principles and Practice of Medicine*, published in 1892. Osler was particularly interested in the relationship between the teacher and student, teacher and teacher, and teacher and patient. The quotation at the end of page 289 is from his book, *Counsels and Ideals from the Writings of Sir William Osler*, 1905.

Osler was born at Bond Head, Ontario, Canada. His parents were English missionaries who had migrated to Canada. He was the youngest of nine children. Having qualified from McGill University, he spent the next 2 years travelling around Europe, the longest period being spent with Sir John Burdon-Sanderson at University College, London. He returned to McGill and later Pennsylvania before moving to take up the Regius Chair of Medicine at Oxford in 1905, where he succeeded Burdon-Sanderson.

Osler's name is associated with three medical conditions: Osler's nodes (tender, red swellings on the palms and fingers in bacterial endocarditis), Osler–Vaquez disease (polycythaemia rubra vera) and Rendu–Osler–Weber disease (recurrent haemorrhages from multiple telangiectasias in skin and mucous membranes).

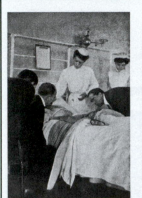
Inspection

Palpation

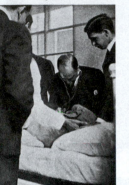
Auscultation

Contemplation

SNAPSHOTS OF OSLER AT THE BEDSIDE
From snapshots taken by T.W.Clarke

Figure 9.21 Sir William Osler. (Reproduced with permission from the Wellcome Library, London.)

Case study: circulatory failure

Clinical

A 65-year-old man complains of transient loss of vision in his right eye. He had two episodes in the previous month, each lasting for approximately 7–10 minutes. Two months ago, he also had transient slurring of his speech.

Examination

- Blood pressure 180/110 mmHg
- Displaced apex beat and ejection systolic murmur
- Carotid bruits present

- No peripheral pulses palpable below the femorals
- Chest radiograph confirmed cardiomegaly but with no evidence of pulmonary oedema
- An ECG showed atrial fibrillation and features of an old anterior infarction

- An echocardiogram demonstrated thrombus within the left atrium

- Carotid Doppler studies indicated moderate carotid artery stenosis
- Urine analysis: no glucose detected

- Blood tests: serial cardiac markers, normal
- Urea and creatinine: slightly raised.

Management and progress

It was decided that he should be treated with antihypertensive agents to reduce his blood pressure and anticoagulants to reduce the risk of further thrombosis/embolism. However, within 24 hours, he developed a right-sided weakness with hemiplegia and a right extensor plantar response. He died without regaining consciousness.

Pathology

Transient loss of vision or speech with full recovery is called a transient ischaemic attack. Generally it results from an embolus lodging in a small cerebral vessel and then being displaced or lysed. The most common sites of origin are the heart or the carotid vessels.

He is hypertensive with an enlarged heart, i.e. left ventricular hypertrophy in response to increased workload.

The bruits indicate turbulent flow, which is a result of stenosis and/or irregularities of the vessel wall due to atheroma.
Absent peripheral pulses indicate widespread arteriosclerotic disease.
The presence of pulmonary oedema would have indicated cardiac failure.
In atrial fibrillation, atria do not contract. The resultant stagnation of blood predisposes to thrombus formation.
Thrombus within the left atrium can be thrown into the systemic circulation. These can pass through the carotids to lodge in the cerebral vessels.
Doppler studies detect turbulent flow. It is the electronic equivalent of the bruit.
A simple test for diabetes: people with diabetes are at high risk of atheroma and may also have sudden temporary loss of consciousness.
Test for myocardial infarction.
Mild renal impairment probably due to hypertension and atheroma.
The signs are of upper motor neuron damage involving the motor and sensory pathways with loss of consciousness. He has sustained a large left-sided cerebrovascular accident. The neural pathways cross, hence left-sided lesions give right-sided signs.

Postmortem findings

Cardiovascular system: left ventricular hypertrophy. Atheroma in all three coronary arteries with an old anterior infarction. No evidence of recent infarct and no vegetations. A small amount of thrombus present in the left atrium. Extensive atheroma in aorta and carotids with narrowing of the mouth of the renal arteries.

(Continued)

(*Continued*)

Central nervous system: a large haematoma in the region of the left internal capsule. Extensive atheroma in the cerebral vessels.

Genitourinary tract: small scarred kidneys showing ischaemic damage.

Clinically, the stroke could have been due to an embolus but, in his case, it was a result of rupture of a microaneurysm on the lenticulostriate branch of the middle cerebral, due to hypertension. Such aneurysms are called Charcot–Buchard aneurysms.

Read about a complex case of circulatory failure in Pathology in Clinical Practice Case 45

PART 4

CELL GROWTH AND ITS DISORDERS

INTRODUCTION
A BRIEF HISTORY OF CANCER

Does early death come
As a punishment?
Or
Does it come too late,
For those who are tortured
By incurable pain?
Is death really cruel?
Or
Is it merciful?

Gitanjali (1961–77)

Gitanjali, as beautiful as the poem by Rabindranath Tagore, died at the age of 16 of cancer. To many people, cancer is a disease that appears suddenly, takes a tight grip, progresses relentlessly, and causes a slow and painful death.

In 1731, Lorenz Heister, a German surgeon, wrote: 'The name *Scirrhus* is given to a painless tumour that occurs in all parts of the body, but especially in the glands, and is due to stagnation and drying of the blood in the hardened part. ... When a scirrhus is not reabsorbed, cannot be arrested, or is not removed by time, it either spontaneously or from maltreatment becomes malignant, that is, painful and inflamed, and then we begin to call it *cancer* or *carcinoma*; at the same time the veins swell up and distend like the feet of a crab (but this does not happen in all cases), whence the disease gets its name; it is in fact, one of the worst, most horrible, and most painful of diseases.'

Is this pessimism still justified in the twenty-first century?

It is easy to forget that everything that comes to life dies: death is an integral part of life. There is a huge amount of large-scale and impersonal death portrayed on television but few children in our societies ever see or discuss death at the individual and personal level. The obsession of society to seek all kinds of 'pleasure' has led to a world that is fearful of death (a final and definitive deprivation of pleasure) and hence the word 'pain' is seldom far from the word 'death'. The mortality statistics paint a fascinating picture. The death rate from all cancers is higher than from heart disease, but is lower if cardiovascular disease is considered as a whole (heart disease including strokes), yet cardiovascular disease does not usually generate such intense dread.

Interesting? Despite the overlap in risk factors, perhaps death from cancer is seen as 'unfair' and therefore painful, but death from a heart attack is seen as self-induced by over-eating and smoking and therefore 'deserved'?

A BRIEF HISTORY OF CANCER

Johannes Müller, a German microscopist, established that tumours were made of cells (1883). This laid the foundation for his pupil, Rudolf Virchow, who divided tumours into 'homologous' and 'heterologous'. The homologous group resulted from proliferation of cells already present and were generally benign, whereas the heterologous group showed a change in the character of the cell and were generally malignant. Virchow, however, failed to recognise the mechanism of metastasis which was later described by Billroth (1856) and von Recklinghausen (1883).

Many investigators have looked for causes of cancer. One of the most famous was Percival Pott who, in 1775, identified that scrotal cancer in chimney sweeps was related to chronic contact with soot. Occupational exposure to industrial tar and paraffin was recognised by von Volkmann, in 1875, as causing cancers, and many such associations have since been described. The word 'association' is pertinent here because most of the time, when we talk about 'causes', we mean 'risk factors'. Although it is possible to definitively demonstrate that an organism causes a disease (see Koch's postulates – page 166), it is much more difficult to show a causal relationship in cancer, and most of the implicated factors increase risk to varying degrees rather than being an absolute necessity for the development of the tumour.

Recently, advances in molecular and cell biology techniques have facilitated the investigation of these associations at the level of DNA, RNA and protein. The information has highlighted the complexity and heterogeneity of cancer cells and the three-dimensional networks that regulate the development and progression of the disease. Hence we are now in a position to attempt to answer some of the fundamental questions relating to the control of normal growth and differentiation, and how these mechanisms go wrong during the process of neoplasia (Fig. 1).

In this section, we consider the benign disorders of cell growth and the premalignant changes that are clues of early cancer and important in screening programmes. We go on to look at the symptoms that occur in cancer, how it is diagnosed, and which features are important for prognosis and treatment. Then we turn to the aetiology (causes of cancer) and the pathogenesis (natural history) of tumours, and, finally, how they behave and what treatments are available. Perhaps at the end, you will be able to ask yourself again whether our social and cultural views about cancer, death and pain are justified.

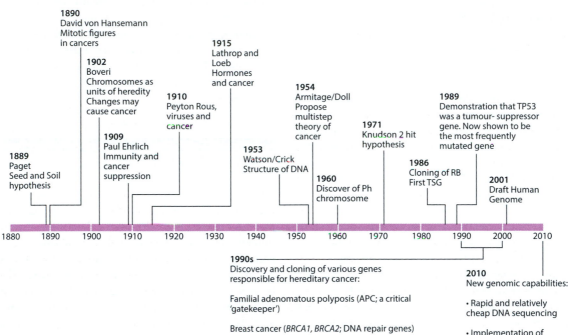

Figure 1 Cancer timeline.

CHAPTER 10
BENIGN GROWTH DISORDERS

Cells have to adapt to any changes in nutrient supply or workload in order to survive and continue performing their cellular function, i.e. to maintain homeostasis.

These adaptations take place at both the cellular and the subcellular levels. We discuss these adaptations with an emphasis on the changes that are important in pathology, and we consider the clinical situations in which they are encountered. The changes that we discuss are: hyperplasia, hypertrophy, atrophy, metaplasia, dysplasia and benign neoplasms.

CLINICAL SCENARIO: PROSTATIC DISEASE

A 73-year-old man visited the urology clinic complaining of difficulty with micturition. He passed urine 15–20 times a day and several times during the night (nocturia). The stream of urine was poor and he found that, on some occasions, it dribbled, causing embarrassment. The urologist detected an enlarged prostate on rectal examination and ultrasound confirmed an enlarged prostate with a volume of 70 mL. The patient had a transurethral resection of the prostate (TURP) of part of his prostate to improve the urine flow.

Some of you may be wondering why an enlarged prostate obstructing urine flow through the urethra should result in increased urinary frequency. The reason (as indicated in Fig. 10.1) is that the enlarged median lobe protrudes into the bladder to produce a dam behind which some urine stagnates. This means that after micturition there is still urine in the bladder, and the patient feels the urge to pass urine again.

The stagnant urine is also prone to infection or stone formation. The poor urine flow is due to narrowing of the prostatic urethra and the 'ball-valve' effect of the median lobe pressing forward on the urethral orifice.

Read more about prostatic disease in Pathology in Clinical Practice Case 3

HYPERPLASIA AND HYPERTROPHY

Hyperplasia is defined as an increase in the *number* of cells in an organ or tissue, whereas hypertrophy is an increase in *cell size*. Often the two coexist in a tissue because some cell types are incapable of division and so must increase their size (hypertrophy) to cope with any extra work whereas other cells can proliferate to share their additional work (hyperplasia). In Chapter 5 (see page 198), we noted that cardiac and skeletal muscle and nerve cells have limited potential to replicate, whereas epithelial cells and fibroblasts do so easily. Smooth muscle cells can respond by a combination of hyperplasia and hypertrophy. This means that, in the prostate, the glandular epithelium and the fibroblastic stroma will show hyperplasia and the smooth muscle is hypertrophic and hyperplastic. Why should the prostate enlarge with age, because its workload does not increase? It is not clear why this happens, although it is presumed to be due to an over-reaction to many years of androgen stimulation.

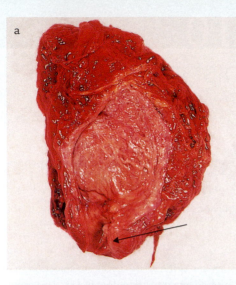

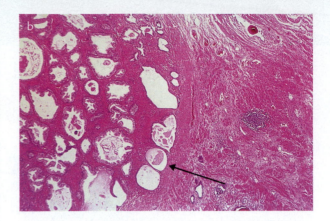

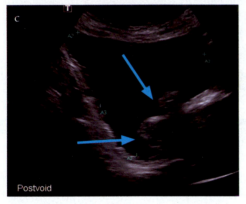

Figure 10.1 (a) Macroscopic specimen with prostatic hyperplasia and bladder showing numerous trabeculae due to urinary obstruction by a nodular enlarged prostate. (b) Photomicrograph showing nodular hyperplasia of the prostate. (c) Transpelvic ultrasound in a 70-year-old man with nocturia and frequency showing hyperplasia of the prostate with encroachment on the bladder lumen by the median lobe of the prostate (blue arrows) and a significant volume of residual urine in the post-micturition bladder.

When the hypertrophy or hyperplasia is useful, i.e. it allows the organ to cope with extra work, it is called physiological. If the enlargement does not appear to serve a purpose, then it is termed 'pathological'. Thus the prostatic changes would be pathological. Physiological hyperplasia and hypertrophy may be mediated through hormonal changes or growth factors. Pregnancy is an example of hormone-induced hyperplasia and hypertrophy that allow an organ the size of a pear (the uterus) to enlarge to accommodate a full-term baby, and also prepare the breasts for lactation. The smooth muscle cells of the uterus enlarge (hypertrophy and hyperplasia) ready for the work of pushing the baby into the world (aptly named 'labour'), and the number of glandular milk-producing cells in the breast increases (hyperplasia).

Another example of physiological hyperplasia that occurs in the body is the regeneration of the liver after a partial hepatectomy. There is a fascinating Greek myth about it. Prometheus, who was Atlas's brother, had incurred the wrath of the mighty Zeus. In anger, Zeus had Prometheus chained naked to a pillar in the Caucasian mountains where an eagle tore at his liver all day, year in, year out, and there was no end to his pain (Fig. 10.2). The reason was that the liver grew back each night! It is not something you should try yourself. The evidence suggests that the remaining liver produces various growth factors and cytokines, including transforming growth factor (TGF) α, hepatocyte growth factor (HGF) and interleukin (IL) 6, which cause an increase in mitotic activity and hence an increase in the cell number. What is remarkable is that it knows when to stop! It is believed that growth inhibitors such as TGF-β and IL-1 are involved in this process. The ancient Greeks had remarkable insights into the body's capacity to regenerate.

Figure 10.2 The Ancient Greeks appreciated the body's capacity for liver regeneration. Prometheus was punished by Zeus for bringing fire to humanity. Every night for years his liver was torn at by an eagle.

Key facts

Examples of hyperplasia and hypertrophy and their causative factors

- Hypertrophy of myocardium due to hypertension
- Skeletal muscular hypertrophy due to exercise
- Red-cell hyperplasia in bone marrow secondary to low atmospheric oxygen (living at high altitude)
- Uterine hyperplasia/hypertrophy secondary to hormonal changes of pregnancy
- Hyperplasia of epidermis and connective tissue due to release of growth factors to aid wound healing

Key facts

Causes of atrophy

Decreased functional demand
  Loss of hormone stimulation
  Decreased physical exercise
  Immobilisation of limb following fracture

Loss of blood supply
  Injury to blood vessel
  Decreased flow – atheromatous occlusion

Loss of innervation
  Transection of nerve fibres
  Infective/inflammatory disorders, e.g. polio

Developmental
  Decrease in thymic size with age
  Atrophy of ductus arteriosus

ATROPHY

Atrophy is defined as a decrease in *cell size* and/or *cell number*. Extra cells are lost through the process of apoptosis described in Chapter 1. Strictly speaking, the reduction in *cell numbers* is called involution. As this is part of normal development, it is termed 'physiological atrophy' and the classic example is the involution of the thymus gland during development. It is

distinguished from pathological atrophy, which results from an abnormal state. An example of pathological atrophy is the severe muscle wasting that may follow an episode of poliomyelitis or the muscle wasting that is commonly observed in limbs immobilised in plaster following a fracture (Figs 10.3 and 10.4).

Chapter 10: Benign growth disorders

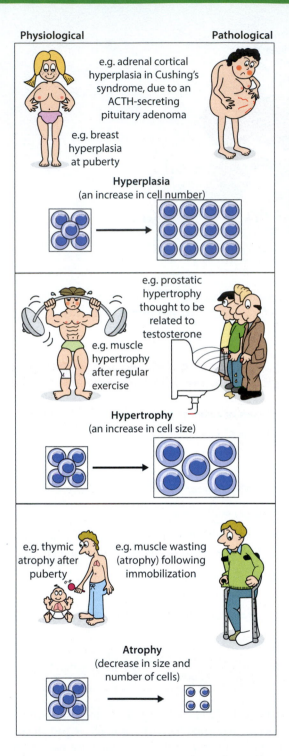

Figure 10.3 Hyperplasia, hypertrophy and atrophy may represent physiological responses to normal development or a pathological change in response to a disease process.

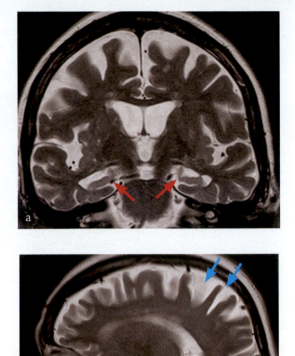

Figure 10.4 An example of atrophy but this is pathological. Note the prominence of the ventricles and the extracerebral fluid spaces, which mean that the brain has lost volume, i.e. atrophied. This happens with age. However there is quite marked loss of volume of the parietal cortex (blue arrows) and the hippocampal formations (red arrows) in the medial temporal lobes – a feature of Alzheimer's type dementia.

METAPLASIA AND DYSPLASIA

The uterine cervix serves as a useful model for this discussion of metaplasia and dysplasia. The changes in the cervix are now well documented because of national programmes designed to screen women of reproductive age to detect early changes associated with cancer.

Screening involves scraping some cells from the junctional zone of the cervix using a spatula. These are generally spread on to a glass slide, then fixed and stained by the Papanicolaou technique. Liquid-phase cytology is also increasingly being used.

The cervix has a transitional zone between the squamous epithelium of the ectocervix and the columnar epithelium of the endocervix. If there is chronic inflammation of the cervix, the columnar epithelium may be replaced by squamous epithelium, so-called 'squamous metaplasia'. Other examples include 'intestinal metaplasia' in the stomach and 'Barrett's metaplasia' of the oesophagus, where the squamous epithelium is replaced by gastric-type tissue. Metaplasia is the *conversion of one type of differentiated tissue into another type of differentiated tissue*. This is most common in epithelial tissue, although it can occur in other types of tissues such as mesenchymal tissues. Generally, it is a response to chronic irritation and is a form of adaptation that involves, for example, replacing a specialised glandular or respiratory epithelium with a more hardy squamous epithelium (Figs 10.5 and 10.6). It is worth noting that the metaplasia probably occurs at the level of the tissue-specific stem cell because the metaplastic process leads to the production of a number of cell types characterising that particular tissue. Hence the process occurs at the *tissue* level. The conversion of one *cell* into another is referred to as *transdifferentiation* rather than metaplasia. Currently, little is known about the molecular mechanisms involved in this process; however, Cdx1 and -2 transcription-factor genes are believed to be important in intestinal metaplasia. Recent evidence suggests that the conversion is a stepwise process requiring the involvement of Cdx1 and -2, as well as a number of other pathways, including Wnt, Notch and Sonic hedgehog signalling.

Metaplasia itself is generally an adaptive, benign and reversible process, but its importance lies in the fact that the stimulants and irritants causing the metaplasia may persist and play a role in carcinogenesis (Fig. 10.7).

The exfoliated cervical cells in a smear may show dysplasia, which is more worrying because it is a step on the road to an invasive tumour. The term 'dysplasia' was originally used to mean an abnormality of development. Unfortunately, it is a term that is used too loosely and this causes confusion. In pathology reports concerning the microscopy of

Figure 10.5 In smokers, the normal ciliated columnar epithelium lining the bronchus undergoes metaplasia to stratified squamous epithelium. This confers greater protection against heat damage. Metaplasia is reversible if the stimulus is removed. Metaplasia is not a premalignant condition, but areas in which metaplasia occurs are at increased risk of developing dysplasia, which is premalignant. In this example, carcinogens in the cigarette smoke may induce mutations in the exposed, rapidly dividing epithelial cells (see also Fig. 10.6).

Normal respiratory epithelium (pseudostratified ciliated columnar): ash and other particles are wafted away by cilia

Heat and toxins in cigarette smoke damage the epithelium and impair cilial action

Robust stratified squamous epithelium replaces the respiratory epithelium

tissues, dysplasia refers to a combination of abnormal cytological appearances and abnormal tissue architecture. Its importance lies in its precancerous association. However, the term is still used to describe some gross abnormalities of development encountered in neonatal pathology, such as renal dysplasia

Chapter 10: Benign growth disorders

and bronchopulmonary dysplasia, which have no pre-cancerous association.

Dysplasia in the cervical squamous epithelium involves an increased cell size, nuclear pleomorphism (variation in size and shape), hyperchromatism (increased blueness due to abnormal chromatin), loss of orientation of the cells, so that they are arranged

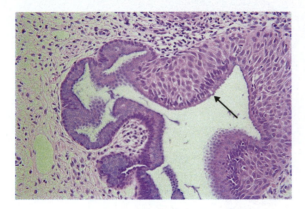

Figure 10.6 Photomicrograph showing squamous metaplasia of endocervical glands.

rather haphazardly, and abnormally sited mitotic activity (Figs 10.8 and 10.9).

Of course, these appearances are the same as those described in malignant change but they differ in *extent*. When the full thickness of the epithelium is involved, it can be called 'carcinoma *in situ*' although involvement of only the lower third is 'mild dysplasia'. Many pathologists and clinicians felt that it was inappropriate to have different names for the various stages and so the term 'cervical intraepithelial neoplasia' (CIN) was introduced. CIN I is the equivalent of mild dysplasia and describes abnormalities affecting the lower third of the epithelium, CIN II (replacing moderate dysplasia) is used for changes reaching the middle third and CIN III (replacing severe dysplasia or carcinoma *in situ*) refers to full-thickness involvement. Similar terminology can be used for changes in the squamous epithelium of the vulva (VIN) and larynx (LIN), although *glandular* epithelial changes (e.g. stomach or large bowel) are usually subdivided into mild, moderate and severe dysplasia. There is a move to simplify it further by dividing into low- and high-grade types. A potential problem with the terminology is that it implies a biological continuum from low-grade (CIN I) to high-grade lesions

Figure 10.7 Effects of stopping smoking at various ages on the cumulative risk (percentage) of death from lung cancer by age 75. (© Cancer Research Campaign 2001.)

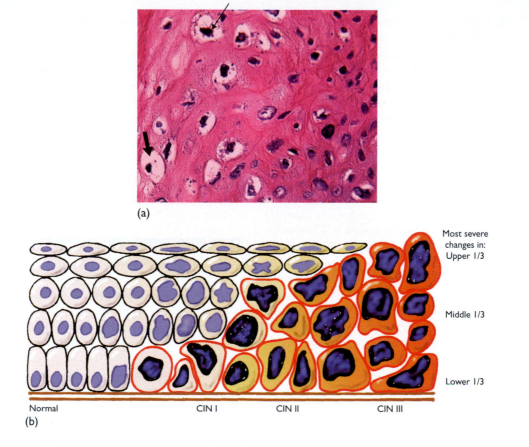

(a)

(b)

Most severe changes in:
Upper 1/3

Middle 1/3

Lower 1/3

Normal   CIN I   CIN II   CIN III

Figure 10.8 Dysplasia is a precursor lesion to squamous cell carcinoma of the cervix and is invariably associated with wart virus (human papillomavirus) infection. (a) Medium-power photomicrograph of a cervix with warty change; this is 'koilocytosis', recognised microscopically as a halo effect around the nucleus (thick arrow), which appears spiky (thin arrow). (b) Dysplasia is classified as cervical intraepithelial neoplasia (CIN) I, II or III, depending on the degree of nuclear atypia seen. Nuclear atypical changes predominate in the lower third of the epithelium in CIN I and throughout all layers in CIN III. CIN III carries a very much higher risk of progression to cancer than CIN I or II.

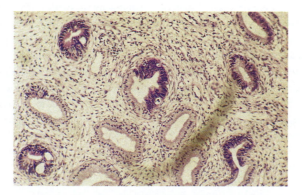

Figure 10.9 Dysplasia involving endocervical glands.

Chapter 10: Benign growth disorders

(CIN III), which may not be true. In some organ systems such as the breast, it is clear that, in most cases, high-grade *in situ* carcinoma does not develop from low-grade *in situ* carcinoma.

It should not be assumed that dysplasia is irreversible. It is believed that early stages of dysplasia may revert to normal if the stimulus is removed. However, severe dysplasia will often progress to cancer if left untreated and, for this reason, it is sometimes referred to as carcinoma *in situ*.

If severe dysplasia is cancer confined to the epithelium, what are moderate and mild dysplasia? This is a good question and without a 'correct' answer. In practice, severe dysplasia is treated as a favourable type of cancer, whereas milder degrees of dysplasia can be managed slightly less aggressively, but followed to ensure that they do not progress to more severe disease.

The concept of dysplasia fits with our current multistep theory of neoplasia (see page 353), in that it represents a stage between benign hyperplastic proliferation and overt cancer. The concept of dysplasia as a cancer in its early stages has also led to the institution of screening programmes for cervical and breast carcinomas. The logic behind this is that, if dysplastic changes precede carcinoma by several months or years, and patients with dysplasia can be identified and treated, we can reduce the death toll from that cancer. Obviously, deaths from that cancer must be fairly common to make this worthwhile and we must be confident that the 'at-risk' group is being screened sufficiently often to detect the early changes.

For example, if the progression from CIN II to invasive tumour took only 1 year, then it would be of limited value to screen patients every 3 years. Much of the interest in genetic and immunocytochemical markers of malignancy lies in the hope that they will be able to detect ever earlier precancerous changes to increase the potential benefits of such screening programmes. The move towards the use of human papillomavirus (HPV) molecular testing as the screening tool, rather than screening cytological specimens, is an example of the progress in this field (see also page 354).

There can be little doubt that the cervical screening programme has been effective in reducing mortality and there is some benefit from the breast programme, but whether they are cost-effective remains a moot point. However, there is little doubt that it has led to an improvement in the quality of radiological, pathological, molecular and clinical services. This is an important health issue and worth thinking about and discussing with your colleagues. What would you do if you were the Minister of Health?

Read more about metaplasia and dysplasia in Pathology in Clinical Practice Case 24

BENIGN NEOPLASMS

A neoplasm is defined as a *new and abnormal growth* and particularly one in which the cell division is uncontrolled and progressive. Neoplasms may be benign or malignant. We deal with malignant neoplasms in Chapter 11.

A benign neoplasm, such as an adenoma in the colon, is an uncontrolled focal proliferation of well-differentiated cells that does not invade or metastasise. Unfortunately, the term is sometimes used inaccurately and some tumours do not quite fulfil these criteria, but it will serve as a working definition. Although benign, these tumours can cause many clinical problems as discussed in the section on the local effects of tumours (page 365).

One of the most common benign tumours necessitating removal is the leiomyoma (fibroid) of the uterine myometrium, which may contribute to heavy and painful menstruation (Fig. 10.10). Benign melanocytic tumours of the skin are removed for cosmetic reasons or fear of malignant change (Fig. 10.11). Endocrine tumours are generally benign but can cause dramatic systemic problems through excessive production of hormones (Fig. 10.12; see also page 368), and some 'benign' intracranial tumours such as meningiomas can kill the patient because the skull cannot stretch to accommodate the 'benign' expansion. Remember that something that is 'benign' to the pathologist may appear 'malignant' to the patient!

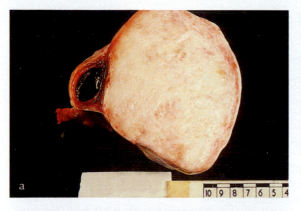

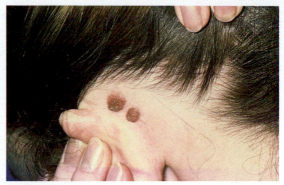

Figure 10.11 Benign naevi.

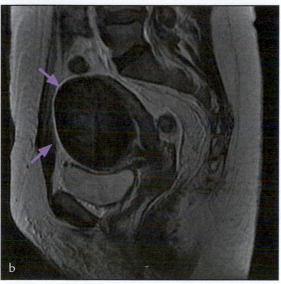

Figure 10.10 (a) Benign leiomyoma; (b) sagittal T2-weighted MRI showing a large fibroid – a benign tumour arising from the fundus and body of the uterus (purple arrows).

HYPERPLASIA AND HYPERTROPHY VERSUS BENIGN NEOPLASMS

The difference between a benign neoplasm and hypertrophy/hyperplasia is that the neoplasm, by definition, exhibits *uncontrolled* cell proliferation. This is unlike hyperplasia and hypertrophy, where the growth, due to increase in either cell numbers or cell size, is an adaptive response to a stimulus and removal of this stimulus results in regression.

Chapter 10: Benign growth disorders

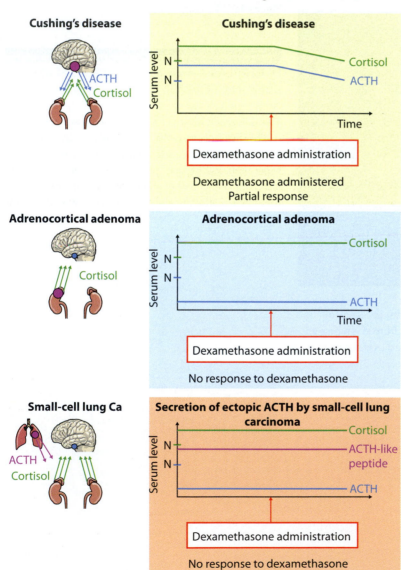

Normal

Normal

High-dose dexamethasone test
Cortisol and ACTH levels fall in response
to administration of high-dose dexamethasone

Cushing's disease

Cushing's disease

Dexamethasone administered
Partial response

Adrenocortical adenoma

Adrenocortical adenoma

No response to dexamethasone

Small-cell lung Ca

Secretion of ectopic ACTH by small-cell lung carcinoma

No response to dexamethasone

Figure 10.12 A comparison of the effects of hormone-secreting tumours on cortisol and ACTH (adrenocorticotrophic) levels in the blood, and the way in which the dexamethasone test can help to distinguish between the causative lesions. Adenoma is a benign glandular tumour; small-cell lung carcinoma is a malignant tumour of probable neuroectodermal origin.

Part 4: Cell growth and its disorders

CHAPTER 11

MALIGNANT NEOPLASMS

CLINICAL SCENARIO: BREAST LUMP

A 50-year-old woman presented to her family doctor with a lump in her left breast. She had noticed a recent enlargement in its size but the mass was not painful. She had no other medical problems but she had a positive family history, her mother having died of breast cancer 5 years previously. She had two daughters, aged 27 and 25 years, who were both well.

Her family doctor could feel a 2-cm diameter mass in the upper outer quadrant of her left breast. This was hard and poorly defined, and caused dimpling of the overlying skin. It was also fixed to the underlying tissues. The nipple and areola on that side had an eczematous appearance, but the right breast and nipple were normal. He did not find any enlarged lymph nodes in either axilla or supraclavicular fossa and no abnormalities in the rest of the body. The family practitioner suspected that this was a malignant tumour and so referred her to hospital for further investigation.

But why did the family doctor consider that this mass was malignant?

CLINICAL FEATURES OF MALIGNANT TUMOURS

It should be stressed that it was not a single criterion but a combination of factors that allowed him to draw such a conclusion. In this case, the lump was ill defined and hard, and involved adjacent tissues and skin. A characteristic feature of malignant tumours is that tongues of cancer cells infiltrate surrounding tissues, whereas benign tumours tend to grow with a smooth pushing edge.

Thus, although benign lumps are generally mobile, malignant tumours are often fixed relative to the surrounding structures. Many malignant tumours induce a proliferation of benign fibroblasts, which produce dense collagenous connective tissue. This reaction is termed 'desmoplasia' and gives the tumour its hard texture.

The lesion's size was greater than most benign lesions, although this is a variable feature. More importantly, there was a recent rapid increase in size, which often indicates malignant growth. In this example, there was one other important clue for the doctor, which is a peculiarity of some breast cancers. The 'eczema' that was noted over the nipple is referred to as Paget's disease of the nipple. This is due to carcinoma cells growing along the breast ducts towards the nipple and then into the epidermis of the skin (Fig. 11.1).

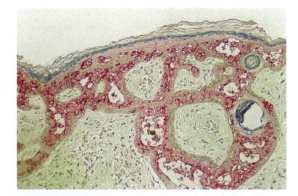

Figure 11.1 Photomicrograph of nipple stained with epithelial membrane antigen (EMA) showing tumour cells (red) within the epidermis.

History

Sir James Paget (1814–1899)

James Paget (Fig. 11.2) was born in Yarmouth, Norfolk. He was apprenticed to a surgeon at the local hospital at the age of 16 and enrolled as a medical student at St Bartholomew's Hospital, London, at the age of 20 years. In 1837, a year after obtaining his MRCS, he was appointed curator of the museum at the hospital. Paget was an excellent clinical observer and an eloquent lecturer. He is best remembered for his descriptions of Paget's disease of bone (osteitis deformans) and Paget's disease of the nipple. He was elected FRS (1851) and Surgeon Extraordinary to Queen Victoria (1858). He was created a baronet in 1871.

Figure 11.2 Sir James Paget (Reproduced with permission from the Wellcome Library, London).

Two other factors, had they been present, would have influenced the doctor: these are pain and the presence of metastasis. Many tumours, both benign and malignant, are painless, but the presence of unremitting pain is suggestive of malignancy. The presence of metastatic disease is the definitive evidence that a tumour is malignant so it is important to understand possible routes of spread, in order that the most likely sites for metastasis can be examined especially carefully. In this woman's case, there was no pain or evidence of metastatic tumour spread, so the doctor suspected that it was a localised malignant growth and the patient was referred to hospital.

EVALUATION IN HOSPITAL

Here a series of tests was performed to make a more precise diagnosis and to assess the extent of her disease; this included haematological and biochemical blood tests to look for anaemia and changes in liver function, which might suggest metastases to bone marrow and liver, mammography (radiograph of the breast) and ultrasound, a chest radiograph and bone scan to look for tumour spread. The mammography showed a 2-cm spiculated mass with linear calcification. The rest of her investigations were unremarkable. An example of a mammogram and ultrasound with a breast cancer is shown in Fig. 11.3.

Surgeons must make a definite diagnosis by obtaining some tissue from the breast lump. They have various options:

- They can insert a needle attached to a syringe to suck out some cells for examination – fine-needle aspiration (FNA) cytology (Fig. 11.4).
- They can insert a special biopsy needle which would ream out a core of tumour about 3 mm wide and 10–15 mm long (Tru-Cut biopsy).
- They can anaesthetise the patient and remove a part of it (excision biopsy) with or without an intraoperative assessment by the pathologist.
- They can anaesthetise the patient and remove the whole lump (wide-local excision or mastectomy) with or without an intraoperative assessment by the pathologist.

In this particular case the surgeon opted for FNA cytology (although in many institutions, this has been replaced by a Tru-Cut core biopsy), which showed malignant cells.

Many centres use this type of 'triple approach' of clinical evaluation, radiology (in this case mammography) and FNA cytology/core biopsy in the initial evaluation of patients. As all three investigations were positive, the surgeon went on to excise the lump and sample the sentinel lymph nodes (Fig 11.5). In due

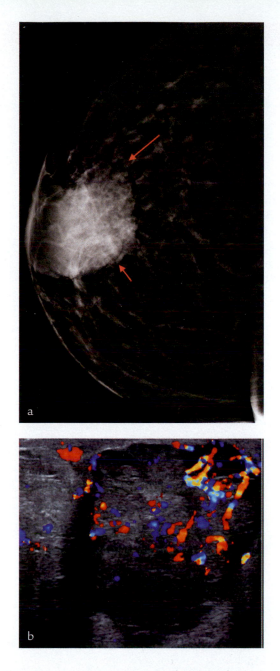

Figure 11.3 (a) A mammogram showing a large central breast mass and (b) an ultrasound scan demonstrated a hypoechoic mass with significant vascularity.

course, the surgeon received the pathologist's report on these tissues and used that information to guide his management of the patient. The report is reproduced

on page 313 (see also Figs 11.6–11.10) and we discuss its relevance for patient management and some points that it raises about the biology of tumours.

WHAT ARE THE MACROSCOPIC FEATURES THAT DISTINGUISH MALIGNANT TUMOURS?

Let us consider this report in more detail. First, there is the gross appearance which records points similar to the criteria used clinically by the family practitioner and surgeon for distinguishing malignant from benign lumps. This includes the size, the infiltrating margin and the consistency of the tumour. Other features include the presence or absence of necrosis and haemorrhage. Most importantly, there is an assessment of excision margins because incomplete excision will result in rapid recurrence and increased opportunity for spread. The report of the microscopical appearances records the pathologist's conclusion, i.e. that it is a primary malignant tumour of breast tissue. Let us consider how this conclusion has been reached.

WHAT ARE THE MICROSCOPICAL FEATURES THAT DISTINGUISH MALIGNANT TUMOURS?

Malignant tissues differ from benign tissues in that individual cells have an abnormal appearance and their arrangement is deranged. The disordered growth pattern is easy to appreciate provided that you know the normal histological appearance of that tissue.

Within the breast, the normal duct–lobular system is composed of an inner luminal epithelial and an outer myoepithelial layer surrounded by basement membrane. A crucial factor, which is often essential for diagnosing carcinoma, is that cells should have breached the basement membrane, which marks the boundary between epithelial and subepithelial tissues. Within the breast, it is possible to identify disordered growth that is still confined within the ducts, an *in situ* carcinoma.

An atypical appearance of individual cells (cytological atypia) is a rather more subtle change. The malignant cells differ not only from normal cells but also from each other; this is called pleomorphism.

Pleomorphism may involve both the nucleus and the cytoplasm. In practice, it often refers to the nucleus,

Chapter 11: Malignant neoplasms

Pathologist's report

Name: Ida Hopps
Age: 50 years
Ward: Thompson
Consultant: I.M. Surgeon
Specimen: Left mastectomy and sentinel node biopsy
Date of operation: 12/11/14

Macroscopic appearance
Specimen 1: a simple left mastectomy specimen, weighing 160 g. It consists of skin, including nipple, which measures 170 × 90 mm and covers fatty tissue with a maximum depth of 50 mm. In the upper outer quadrant, 10 mm from the nipple, there is a pale, firm, gritty mass measuring 25 × 20 × 20 mm which has an irregular, poorly defined margin. The closest excision margin (deep) is 15 mm from the mass. There is an area of erythema around the nipple.
Specimens 2 and 3: two sentinel nodes containing blue dye are submitted separately.
Specimen 4: the axillary dissection measures 80 × 50 × 50 mm and 13 lymph nodes have been identified.

Intraoperative assessment
Imprint cytology: Sp2 – no evidence of malignancy
Imprint cytology: Sp3 – groups of malignant cells identified

Microscopical appearance – synoptic report
Invasive carcinoma
Type: Ductal carcinoma – no special type
Size: 22 mm
Grade: II
(Tubule score 2; pleomorphism score 3; mitoses score 1)
Lymphovascular permeation   Present, focal

In situ carcinoma
Type: Ductal carcinoma *in situ* (DCIS)
Size: 30 mm
Site: Around invasive carcinoma and extending
                             to nipple
Grade: High
Type: Solid and comedo
Calcification: Present
Excisional margins   All clear, >10 mm for invasive and *in situ*
                             carcinoma

Lymph nodes
Site: Sentinel (Sp2 and -3) and axillary (Sp4)
Number present: 15
Number involved: 1 (sentinel node Sp3)

Nipple   Paget's disease

Immunohistochemistry
Oestrogen receptor (ER):   Positive (3+ staining in 90% of cells)
Progesterone receptor (PgR): Positive (3+ staining in 90% of cells)
HER-2:   Immunohistochemistry: negative (0)
                             In situ hybridisation: negative (2 signals,
                             diploid)

Summary
Left breast: *In situ* and invasive ductal carcinoma grade II, 22 mm (T2) with 30-mm DCIS
Complete excision
1/15 lymph node positive
ER+, PR+, Her-2−
Pathological stage: T2, N1 (stage IIB)
Reported by Dr S.P. Ecimen
13/11/14

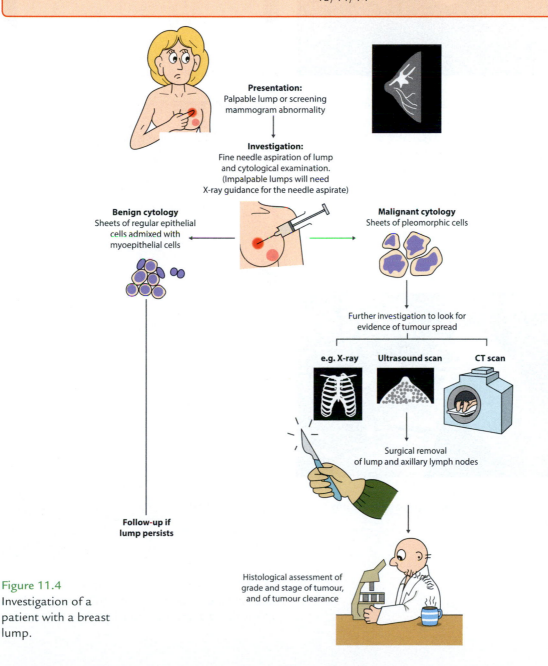

Presentation:
Palpable lump or screening mammogram abnormality

Investigation:
Fine needle aspiration of lump and cytological examination.
(Impalpable lumps will need X-ray guidance for the needle aspirate)

Benign cytology
Sheets of regular epithelial cells admixed with myoepithelial cells

Malignant cytology
Sheets of pleomorphic cells

Further investigation to look for evidence of tumour spread

e.g. X-ray   **Ultrasound scan**   **CT scan**

Surgical removal of lump and axillary lymph nodes

Follow-up if lump persists

Histological assessment of grade and stage of tumour, and of tumour clearance

Figure 11.4
Investigation of a patient with a breast lump.

Chapter 11: Malignant neoplasms

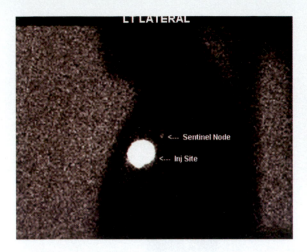

Figure 11.5 Sentinel node imaging: many forms of cancer, and in particular breast cancer, spread in a predictable fashion to regional lymph nodes. Sentinel node imaging uses a radioisotope injected around the region of the tumour to identify the very first draining lymph node. The surgeon can then remove this draining node to see if it is involved. The absence or presence of metastatic disease in the sentinel node guides further treatment, and provides staging and prognostic information.

Figure 11.6 Gross examination of mastectomy specimen.

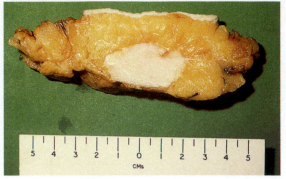

Figure 11.7 Slice of lumpectomy specimen showing spiculated tumour close to the deep margin.

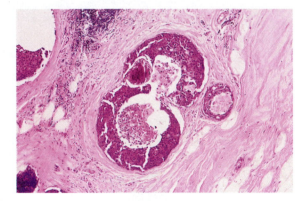

Figure 11.8 *In situ* ductal carcinoma with central necrosis and microcalcification.

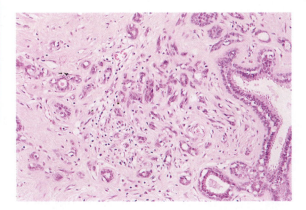

Figure 11.9 Normal breast duct (right) with adjacent tissue showing infiltration by moderately differentiated ductal carcinoma.

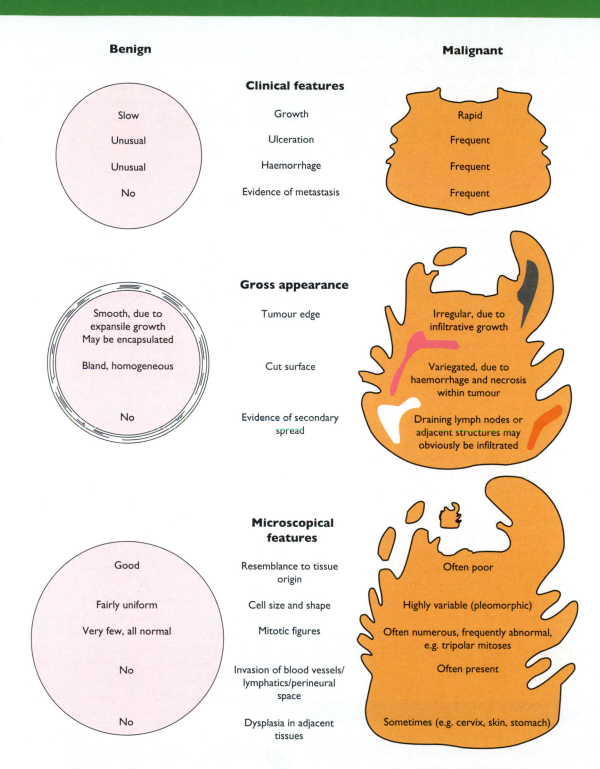

Benign   **Malignant**

Clinical features

Benign		Malignant
Slow	Growth	Rapid
Unusual	Ulceration	Frequent
Unusual	Haemorrhage	Frequent
No	Evidence of metastasis	Frequent

Gross appearance

Benign		Malignant
Smooth, due to expansile growth May be encapsulated	Tumour edge	Irregular, due to infiltrative growth
Bland, homogeneous	Cut surface	Variegated, due to haemorrhage and necrosis within tumour
No	Evidence of secondary spread	Draining lymph nodes or adjacent structures may obviously be infiltrated

Microscopical features

Benign		Malignant
Good	Resemblance to tissue origin	Often poor
Fairly uniform	Cell size and shape	Highly variable (pleomorphic)
Very few, all normal	Mitotic figures	Often numerous, frequently abnormal, e.g. tripolar mitoses
No	Invasion of blood vessels/ lymphatics/perineural space	Often present
No	Dysplasia in adjacent tissues	Sometimes (e.g. cervix, skin, stomach)

Figure 11.10 Comparison of clinical, gross and microscopical features of benign and malignant tumours.

Chapter 11: Malignant neoplasms

which may be many times the size of a normal nucleus and may show marked variability in *size* and *shape* — nuclear pleomorphism. It may also have an altered distribution of chromatin and be a darker colour in stained sections, so-called hyperchromatism. These alterations reflect the increased amount and abnormalities of nuclear chromatin, which is common in tumours because they are frequently aneuploid.

Most normal cells have a small single nucleolus. In malignant cells, there may be many nucleoli of varying sizes or, alternatively, a large single nucleolus. The position of the nucleus within the cell is often abnormal, i.e. it exhibits a loss of polarity. Thus, the nucleus of, for example, a normal colonic cell is situated at the cell's base with mucus in the cytoplasm nearer the surface, whereas the nucleus is more central in a malignant cell. There is also an increase in mitotic activity due to the increase in cell proliferation and abnormal mitotic figures may also be identified. The general appearance of the cells is altered because there is an increase in the nuclear:cytoplasmic ratio, because either the amount of cytoplasm is less or the nucleus is larger or a combination of the two.

The cytoplasmic changes vary depending on the tissue but generally involve a loss of specialised features, e.g. absence or reduction in mucin content in a colonic adenocarcinoma. Ultimately, however, all these features are only guidelines and the real test is whether the tumour behaves in a malignant fashion. Of course we cannot leave patients untreated just to see how the tumour behaves. Pathologists have learnt much about the behaviour of tumours from postmortem studies and, fortunately, tumours of similar appearance usually show similar behaviour in different patients.

Note that many of the features discussed here are the same as those used for the assessment of dysplasia (see page 302), and the distinction between grades of dysplasia and frank malignancy is based on both degree and extent of the changes.

Dictionary

Aneuploid: an abnormal number of chromosomes that is not an exact multiple of the haploid (23) number.

WHAT FACTORS INFLUENCE PROGNOSIS?

Here we are concerned with factors that influence prognosis and can be assessed routinely by histopathologists. The following are the important aspects:

- The type of tumour
- The grade of tumour
- The stage of the disease
- Tumour markers.

First, it is essential to decide whether the tumour has arisen locally or is a metastasis. Two points help make this distinction and these are whether there are precancerous changes or *in situ* carcinoma present, and whether the lesion resembles tumours known to occur at that site.

In situ carcinoma is an alteration in the cytological appearance similar to that seen in malignant tumours, but it does not show any invasion through the basement membrane. If the tumour had been entirely *in situ* then it would have an extremely good prognosis, because the lack of local invasion would mean that the tumour had no ability to extend into lymphatic or blood vessels and no possibility of metastasis. Precancerous lesions are harder to define but are changes (e.g. atypical hyperplasia, dysplasia in cervix) that have been shown in large studies to be associated with the subsequent development of cancer, and are believed to represent an early, but possibly reversible, stage of malignancy.

In most organs, there is one type of malignant tumour that is far more common than any other and this generally corresponds with the type of tissue that is proliferating in that normal organ, e.g. the breast and colon have active glandular epithelium so the most common malignant tumour at both sites is an *adenocarcinoma* composed of malignant glandular epithelium. The bladder is lined by transitional epithelium which gives rise to *transitional cell carcinoma* and the oesophagus has squamous epithelium and *squamous carcinomas*.

 Read more about factors influencing prognosis in Pathology in Clinical Practice Case 25

Remembering the normal histology can be a great help in predicting the most common tumours for a particular site.

To return to our patient, she has an adenocarcinoma that has *in situ* and invasive components. The *in situ* carcinoma tells us that it is locally arising. Therefore, this is a primary tumour of the breast. Within the breast, there are a large number of different subtypes and it is worth remembering that not only is cancer many diseases, but individual organs also have many different subtypes of cancer. Our woman has a ductal carcinoma that has a poorer prognosis than if she had a mucinous carcinoma (Figs 11.11–11.15).

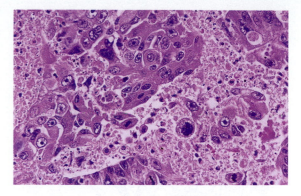

Figure 11.13 Breast carcinoma exhibiting pleomorphism, necrosis and mitotic activity.

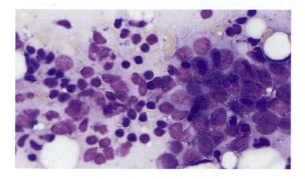

Figure 11.11 Cytological smear showing pleomorphic tumour cells admixed with smaller lymphoid cells.

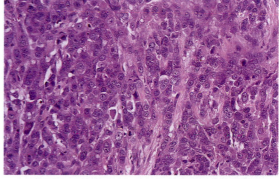

Figure 11.14 High-grade invasive ductal carcinoma.

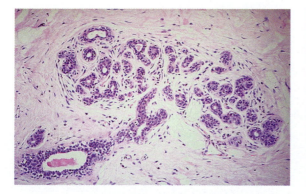

Figure 11.12 Normal terminal duct and lobule of breast.

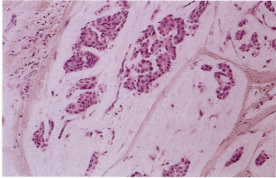

Figure 11.15 Invasive mucinous carcinoma of the breast.

Chapter 11: Malignant neoplasms

Figure 11.16

Table 11.1 Grading of breast cancer

	Score
Tubule formation	
Majority of tumour >75%	1
Moderate amount 10–75%	2
Little or none <10%	3
Nuclear pleomorphism	
Mild	1
Moderate	2
Severe	3
Mitotic count (count per 10 high-powered fields; varies with field diameter of the lens)	
0–5	1
6–10	2
>11	3

Total score	Grade
3–5	1
6–7	2
8–9	3

Grading is based on the assessment of three criteria: tubule formation, nuclear pleomorphism and mitotic count. Each is scored from 1 to 3, as shown in the text.

Although the patient had an invasive carcinoma, the tumour was not seen in lymphatic or blood vessels. The most important prognostic features are grade and stage. The grade of a tumour depends on its histological appearance whereas the stage of a tumour depends on its size and extent of spread. The histological grade is a crude measure of *how closely the tumour resembles normal tissue* combined with an estimate of its mitotic activity. There is a well-defined scoring system for breast tumours (Table 11.1), based on tubule formation, nuclear pleomorphism and the mitotic count. This divides them into three grades, with grade 1 tumours having a better prognosis than grade 3 tumours. Many organ sites have no formal grading system and so the pathologist will merely record whether the tumour is well, moderately or poorly differentiated, by assessing similar features but in a less objective way.

The stage of a tumour is a measure of the *extent of disease* and depends on pathological, radiological and clinical information. A TNM staging system (Fig. 11.16) is often used for breast carcinoma where T stands for size of primary tumour, N codes for regional node involvement and M for metastatic disease. Figure 11.17 illustrates the different pathological staging systems and links with the TNM classification.

This provides an easy shorthand for indicating the disease stage, which is helpful for deciding treatment and comparing the outcome of patients treated with

TNM system, e.g. Ca breast

T = Tumour size:
T0: impalpable
T1: 0–2cm
T2: 2–5cm
T3: >5cm±fixation to underlying muscle
T4: any size, with fixation to chest wall or skin

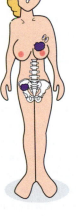

N = Lymph node status:
N1: regional nodes involved
N2,3: more distant nodal groups

M = Metastases
M0: no detectable spread
M1: metastases present (specify sites)

Dukes' staging of colorectal carcinoma

Comment: 5 year survival figures:
Dukes' A: 80–85 per cent
Dukes' B: 55–67 per cent
Dukes' C: 32–37 per cent

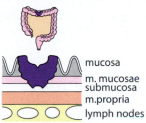

mucosa
m. mucosae
submucosa
m.propria
lymph nodes

Dukes' A: tumour confined within bowel wall; no spread through main muscle layer

Dukes' B: spread through m. propria into serosal fat, without lymph node involvement

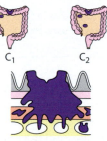

C_1 C_2

Dukes' C: tumour spread to lymph nodes.
C_1: pericolic nodes involved
C_2: involvement of higher mesenteric nodes

Cotswolds revision of Ann Arbor staging system for Hodgkin's lymphoma

Comment: the presence of 'B' symptoms, e.g. fever, drenching sweats, weight loss, adversely affects the prognosis, and is included in the stage, e.g. Stage IIA (no B symptoms), or Stage IIB.

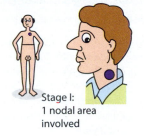

Stage I:
1 nodal area involved

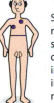

Stage II: ≥2 nodal areas on same side of diaphragm involved (no. of involved sites recorded)

Stage III: nodal areas on each side of diaphragm:
III1 upper abdo,
III2 lower abdo

Stage IV:
visceral involvement

The spleen is part of the reticuloendothelial system. Splenic involvement does not carry the same staging implications as, for instance, bone marrow or liver

Figure 11.17 Staging of tumours.

Chapter 11: Malignant neoplasms

Radiology

The use of imaging in staging cancer

There are many imaging modalities used by clinicians to stage cancer. The role of the imaging modality is to do the following:

- Delineate the anatomical extent and size of the primary tumour (T)
- Identify locoregional metastatic disease in regional lymph nodes (N)
- Identify distant metastatic disease (M)

It is often difficult to stage a cancer completely with regard to these three staging components with a single imaging modality. PET (positron emission tomography) uses FDG (fluorodeoxyglucose) – a radioisotope that is a glucose analogue and is taken up by tissues with a high use of glucose such as brain, kidney, muscles and cancers. The use of PET together with CT, and more recently MRI, has revolutionised the staging of cancer because the fused PET and whole-body CT or MR images provide an anatomical and functional map of the extent of disease in the body (Fig 14.5). This is the closest we have got to staging the patient with a 'single' imaging modality.

Unfortunately not all cancers take up FDG. It is also difficult to identify metabolically active tumours when located in metabolically active normal tissue, e.g. brain. Lastly the size of the metastatic disease is a limiting factor in its detection, i.e. the metastasis in a lymph node has to be a certain volume before it can be detected. Possibly most importantly, this is an expensive test not available everywhere.

Each type of cancer has its own unique pathway of investigation, derived from evidence-based medicine – hopefully. Historically many imaging modalities have been used in staging including the following:

- Plain films: useful in the delineation of bone tumours and tumours in the lungs.
- Ultrasound: used to identify abnormal lymph nodes and metastatic intra-abdominal disease, and to guide biopsies.
- CT: the workhorse of staging delineates both soft tissue and bone disease.
- MRI: used in malignancies of the central nervous system and as a problem-solving tool elsewhere when CT cannot provide the answer. Some cancers can now be imaged with whole-body MRI (e.g. multiple myeloma, which can be difficult to identify on a PET study by itself).
- Radioisotope imaging: this obviously includes PET but a far more commonly used test is the isotope bone scan, which targets osteoblastic activity and is used to delineate the distribution of metastatic disease in bone.

new therapeutic regimens. Obviously, assessment of a new treatment regimen must take account of the stage of a patient's disease to avoid spurious results.

The pathology report on our patient states that one of the lymph nodes contained tumour but the clinical investigation did not show distant metastases. The mapping of sentinel nodes has become routine practice for the staging of many different tumour types (see Fig 11.5). The sentinel node is defined as the first node that drains the tumour basin, hence the assessment of the node provides information about tumour dissemination and the likelihood of other nodes being involved. This has been an important development because large nodal excisions carry significant morbidity for the

patients. A complete axillary dissection can be complicated by marked lymphoedema of the corresponding arm, with resultant pain and mobility issues as well as a longer-term risk of sarcoma. As the involvement of the sentinel node also predicts for involvement of other nodes, positivity usually leads to further nodal surgery. Intraoperative assessment is therefore often carried out so that a definitive procedure can be performed at the same surgical procedure if the sentinel node is positive. In this case, an imprint cytology was done by smearing cells from the cut surface of the node, but a frozen section is also an alternative. More recently, molecular testing has also been introduced in some institutions whereby the lymph node is homogenised and assessed for the expression of cytokeratin 19 mRNA (an epithelial keratin), to evaluate the presence of epithelial cancer cells within the node.

As the tumour is this case was >2 cm and there was one positive lymph node, it is categorised as T2 (size 2.2 cm), N1, M0, which translates as stage IIB disease. This short, coded message tells the doctor that the woman is in an intermediate prognostic group.

Immunohistochemical markers for tumour-associated gene(s) and gene products are sometimes carried out to help predict behaviour and decide on treatment options. As the breast is an endocrine-responsive organ and proliferates due to stimulation by oestrogen, breast cancers may express the oestrogen (ER) and progesterone (PgR) receptors. Approximately 60–70% of breast cancers will be ER and PgR positive, and this is not only a better prognostic index but also predicts for response to anti-oestrogenic treatment (e.g. tamoxifen). *HER-2* is an oncogene (see page 338) that is over-expressed and amplified in a quarter of breast cancers, and is usually seen in high-grade cancers and predicts a poorer prognosis. A monoclonal antibody trastuzumab (Herceptin) has been developed against the *HER-2* receptor and has been shown to be effective in the treatment of advanced breast cancer. It is one of the first treatments in the category of 'targeted therapies', in which specific molecular abnormalities seen in cancers are used as therapeutic targets. This is quite different to chemotherapy, which represents general poisons to cancer (and normal) cells.

There has been a huge explosion in knowledge about the genomic alterations in cancers as a result of new technologies, including microarray expression profiling and massive parallel sequencing. It is already becoming a standard of care to test for mutations in the relevant gene(s) as part of the pathological assessment. Besides the assessment of *HER-2* amplification status in breast cancer, the assessment of mutations in *BRAF* in melanoma and colon cancer, *MSI* testing for mismatch repair genes in colon cancer, and *EGFR* and *ALK* genes in lung cancer is now more or less routine.

When the doctor talks to his patient about these results, the patient may well ask a variety of questions about her prognosis, but before we attempt to answer those questions, we should digress to discuss the classification of tumours.

CLASSIFICATION OF TUMOURS

The pathological classification of tumours is illustrated in Fig. 11.18.

You will recall that knowledge of the normal structures at a particular site can be of great help in predicting the most common tumours. In most organs, there is one particular type of malignant tumour that is more common than any other, and this generally corresponds with the type of tissue that is proliferating at that site. The stomach and colon have active glandular epithelium so the most common malignant tumour at both sites is an *adenocarcinoma*, which is composed of malignant glandular epithelium. The bladder is lined by transitional epithelium which gives rise to *transitional cell carcinoma* and the skin has squamous epithelium and *squamous carcinomas*.

As the tumour resembles part of the parent tissue, the classification is based on the assumed histogenesis, i.e. because a transitional cell carcinoma has some similarities with transitional epithelium, it is assumed to arise from it.

The broad classification divides tumours into those arising from epithelia (carcinomas), connective tissue (sarcomas), lymphoid tissue (lymphomas) and 'the rest', which includes specialised tissues such as the brain. Included in Fig. 11.18 are the benign counterparts arising from the same tissues.

At this point there needs to be a word of caution because, although this classification originated from ideas on histogenesis, it is now apparent that cells of one tissue type may 'differentiate' to resemble cells of another type (a process called metaplasia; see page 302), e.g. bronchial glandular epithelium may become squamous due to chronic irritation from smoking. A tumour arising in such a patient may hence appear squamous although the original epithelium at this site was glandular. The histogenetic approach to classification is destroyed in such circumstances. Fortunately, we only have to classify tumours according to their type of differentiation and we eliminate the problem! Thus a tumour resembling squamous cells is a squamous cell carcinoma regardless of the original tissue type. You will discover that, although rare, it is possible to get squamous carcinoma in the breast and adenocarcinoma in the bladder! Sometimes a tumour cell is very poorly differentiated so that, even to the trained

Chapter 11: Malignant neoplasms

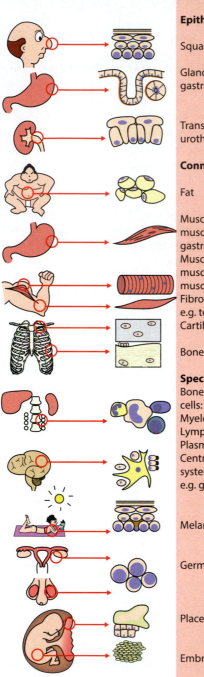

TISSUE TYPE	BENIGN	MALIGNANT
Epithelium		Carcinoma
Squamous, e.g. skin	Squamous papilloma	Squamous carcinoma
Glandular, e.g. gastrointestinal tract	Adenoma	Adenocarcinoma
Transitional, e.g. urothelium	Transitional cell papilloma	Transitional cell carcinoma
Connective tissue		Sarcoma
Fat	Lipoma	Liposarcoma
Muscle: i. Smooth muscle, e.g. wall of gastrointestinal tract	Leiomyoma	Leiomyosarcoma
Muscle: ii. Striated muscle, i.e. voluntary muscle	Rhabdomyoma	Rhabdomyosarcoma
Fibrous tissue, e.g. tendon	Fibroma	Fibrosarcoma
Cartilage	Chondroma	Chondrosarcoma
Bone	Osteoma	Osteosarcoma
Special categories Bone marrow-derived cells:	(non-systematic nomenclature retained mainly for historical reasons)	
Myeloid cells		Myeloid leukaemia
Lymphoid cells		Lymphocytic leukaemias, Lymphomas
Plasma cells	Plasmacytoma	Myeloma
Central nervous system, e.g. glial cells		Gliomas
Melanocyte	Benign melanocytic naevus	Melanoma
Germ cells	Benign teratoma	Malignant teratoma Seminoma/Dysgerminoma
Placenta	Hydatidiform mole	Choriocarcinoma
Embryonal cells		Embryonal cell tumours (may show differentiation towards tissue types, e.g. neuroblastoma) Special categories

Figure 11.18 Pathological classification of tumours.

histopathologist's eye, it does not resemble a particular type of normal cell. In this situation, special stains to demonstrate cytoplasmic or surface molecules can be helpful. Thus, the presence of intracellular mucins would suggest an adenocarcinoma and immunohistochemical stains for different intermediate filaments or lymphoid antigens would help to distinguish between a wide variety of tumours. If it is not possible to demonstrate any differentiation, the tumour is referred to as anaplastic. Some of these substances are also released into the blood, which is useful both for diagnosis and for following the patient's response to treatment, e.g. prostatic-specific antigen (PSA) levels can be measured in the blood to help screen for prostatic adenocarcinoma, although levels are also raised in some non-malignant prostatic disorders because the antigen is present on both benign and malignant prostatic cells. The β subunit of human chorionic gonadotrophin (βhCG) and α-fetoprotein (AFP) are also useful markers in patients with germinal cell tumours.

Now we must turn to the patient's questions. What causes cancer? How will it behave? What treatments are available? Will there be a lot of pain?

If we are not to be stumped by the patient, we have to understand a little more about the natural history of cancer.

Case study: oesophageal cancer

A 62-year-old man presents to his GP with a 3-month history of difficulty with swallowing. He has also lost his appetite and lost 7 kg in weight. The dysphagia is mainly for solids. Examination is unremarkable.

Question What are possible causes for his dysphagia?
Answer There are many causes including benign stricture, neurological disorders such as cerebrovascular accident and multiple sclerosis as well as benign and malignant neoplasms.

The GP sends him for an upper gastrointestinal endoscopy. A tumour is seen at 28 cm. The biopsy report reveals a moderately differentiated squamous cell carcinoma.

Question What is a carcinoma?
Answer A carcinoma is a malignant tumour derived from *epithelium*.

Question What is meant by the term 'differentiation' as applied to the tumour?
Answer Differentiation is an assessment of how closely the tumour resembles the tissue of origin. The closer the resemblance, the better the differentiation and lower the *grade* of tumour.

Question What is the significance of grade?
Answer Grade is associated with prognosis — the higher the grade, the worse the prognosis.

The patient underwent further investigations to assess if the tumour had spread.

Question What is the medical term for assessing extent of spread?
Answer Stage

🔗 Read more in Pathology in Clinical Practice Case 4

OVERVIEW OF CANCER EPIDEMIOLOGY

Cancer is one of the major non-communicable diseases, with a worldwide incidence of approximately 14 million new cases and 8 million cancer-related deaths in 2012. The incidence is expected to rise to nearly 25 million over the next 10–20 years. The increasing incidence, together with increasing age of the population, technologically driven imaging and expensive therapies, puts a significant burden on the health-care system. Not surprisingly, therefore, there is a move to improve preventive strategies in dealing with the cancer burden in our communities.

The most common cancers in men include lung, prostate, colorectal, stomach and liver cancers, whereas, in women, the most common cancers are breast, colorectal, lung, cervix and stomach. It may come as a surprise to some that about 60% of the cancers worldwide occur in Africa, Asia, and Central and South America, and these regions also account for nearly 70% of the cancer-related deaths (Stewart BW, Wild C, eds. *World Cancer Report*. Lyon: IARC, 2014).

CANCER AS A DISEASE OF GENETIC MATERIAL

The view that cancer originates within single cells due to abnormalities within their DNA is now generally accepted. The evidence comes from five main sources:

1 Some cancers have a heritable predisposition; examples include familial retinoblastoma, familial adenomatous polyposis (FAP), and familial gastric and breast cancer.

2 Many tumours exhibit chromosomal abnormalities and karyotypic studies have even identified specific changes in some tumours, e.g. 8;14 translocation in Burkitt's lymphoma.

3 A number of rare inherited disorders involve an inability to repair damaged DNA. An example is xeroderma pigmentosa, and patients with this condition have an increased susceptibility to skin cancer after damage to DNA from ultraviolet light.

4 Many chemical carcinogens are also mutagens, i.e. they have been shown to cause genetic mutations.

5 DNA recombinant technology has demonstrated that DNA from tumour cells, when transferred into a normal cell, can convert them into tumour cells of the same type.

The isolation of genes with a direct role in tumour formation (oncogenes) has firmly established cancer as a disease of genetic material. However, unlike, for example, cystic fibrosis, where mutation in one gene causes the disease, no single gene defect has been shown to 'cause' cancer; cancer genes in general should be thought of as significant contributors to the development of malignancy rather than as 'causes' of cancer. The concept of DNA changes that act as 'drivers' is discussed in Chapter 13.

We now consider the many predisposing and aetiological factors that lead to tumour formation. Although it appears that these factors must alter either DNA's structure or DNA's function in some way, the details of many of these processes remain unclear.

One of the patient's concerns will relate to what causes cancer and the risk factors that are important in his or her case.

RISK FACTORS FOR CANCER

AGE

There is a very strong relationship between age and cancer development. The advent of antibiotics, improved sanitation and good nutrition has extended people's expected life span, so that they can now achieve an age in which there is a high incidence of malignant tumours, particularly those of the breast, colon, lung, prostate and bronchus. It is postulated that carcinogens may have a cumulative effect over time, which may explain an increased incidence with age. The ability of carcinogens to induce genetic mutations is well known and the large number of cell divisions with increasing age may contribute to the neoplastic process. Cell division itself is a risk because each time the DNA is copied, there is a potential to introduce mistakes within the genome. There are elaborate DNA repair mechanisms in place to correct such errors, but mutations in genes coding for proteins involved in DNA repair will allow such errors to pass on to the next generation of cells. All these factors, together with the age-related metabolic or hormonal changes, may combine to account for the increasing incidence of tumours with age.

Within the age groups, there are some differences between the sexes. Until about the age of 14, the incidence is similar between the sexes, but, after that, it is greater in females up to approximately age 50 years. This increase is mainly due to cervical and breast cancer incidence. Males take over after about age 60 years due to the incidence of lung and prostate cancer.

Tumours are of course not just confined to elderly people, and some malignancies such as leukaemias are more common in children. The incidence worldwide in children aged 0–14 years in 2012 was 165 000 (95 000 in boys and 70 000 in girls) (World Cancer Report 2014).

GENETIC FACTORS

In the case of breast carcinoma discussed in Chapter 11, the doctor discovered that the patient's mother had died of breast carcinoma. (The patient also had two daughters who at the time were both well.) This history of malignant disease within close family members is of relevance because there are several tumour types where the risk of cancer in close family members is increased. How much the risk is increased in individual cases and with different tumours is not easy to specify but, in general, it is about two to three times the risk in the general population. Obviously, tumour development is not inevitable and many other factors such as environmental and dietary influences may modify the risk. Alternatively, the patient may also undergo prophylactic surgery which can also reduce the risk.

In some tumours, the genetic susceptibility is better understood and in two autosomal dominant conditions, familial polyposis coli and retinoblastoma, it involves loss of a tumour-suppressor gene or anti-oncogene (see Chapter 13). Familial polyposis coli (familial adenomatous polyposis) is a disorder in which individuals develop hundreds of polyps in the gastrointestinal tract. These polyps show varying degrees of dysplasia (see page 302) and, although benign, should be regarded as premalignant because practically all of these patients will develop a colonic carcinoma if the colon is not removed by the age of 25 years. Retinoblastoma is a malignant tumour of the eye that is most common in children. Of cases of retinoblastoma, 25–30% are hereditary and the rest sporadic. Both the familial and sporadic cases arise due to two mutations in the retinoblastoma gene, but the familial cases inherit one mutation through the germline cells (see later).

GEOGRAPHICAL AND RACIAL FACTORS

Geographical factors merge with environmental factors, because a geographical factor is only an environmental factor that affects the population of a particular area. This may be a sunny climate, radioactive rock formations or a carcinogen in the water supply (Fig. 12.1).

Let us discuss the increased incidence of stomach cancer in Japan compared with North America. The tumour is seven times more common in Japanese people living in Japan than in Americans living in the USA. Is this a racial difference or an effect of some climatic, geological or dietary factor that operates in Japan? To answer this we need to know the incidence in Japanese people who move to America and raise families. They will keep their racial (genetic) factors and may import their dietary factors but not their geographical factors. We find that the incidence drops in these immigrants and is halved in their first-generation offspring, but is still higher than in white Americans, so we haven't achieved a definite answer to our question.

Some reports suggest that the incidence drops further in future generations until it equals the American rate. This would appear to rule out a racial (genetic) factor and may implicate a cultural dietary change. Similar arguments apply to the interesting observation of an increase in prostate cancer among the Japanese migrants to USA.

A much easier example is the incidence of melanomas in white-skinned Australians. Here there is a *racial predisposition*, because they do not have sufficient skin pigmentation to protect them from ultraviolet light, and the *geographical factor* of a sunny climate. If the Australian emigrates at birth to a cold grey country, then his risk of melanoma drops dramatically.

ENVIRONMENTAL AGENTS

Numerous environmental agents have been implicated in the causation of cancer. Everybody knows that there is a strong association between smoking and lung cancer. This problem may affect not only the smoker but also the 'innocent bystander' who inhales exhaled tobacco smoke (passive smoking).

Asbestos exposure increases the risk of developing lung carcinoma and malignant mesothelioma of the pleura and peritoneum (Fig, 12.2). Exposure to β-naphthylamine, which may occur in the rubber and dye industries, increases the risk of transitional cell tumours of the bladder. Exposure to vinyl chloride in the plastic industry enhances the development of liver angiosarcoma (malignant tumour of blood vessels).

Hepatocellular carcinoma
Sub-Saharan black African
Far East

Malignant melanoma
Australia
Scandinavia
North America: California whites at highest risk

Gastric carcinoma
Japan, China
Brazil, Colombia, Chile
Iceland, Finland,
USSR, Poland, Hungary

Figure 12.1 Geographical variations in tumour incidence.

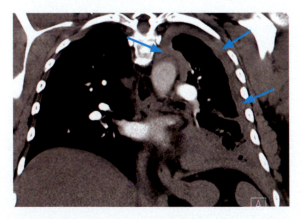

Figure 12.2 Coronal CT showing a thick rind of pleura (blue arrows) encasing the left lung and preventing it expanding fully. This is mesothelioma – a malignant primary tumour of the pleura directly related to exposure to asbestos.

One of the first examples of an environmental cancer was described in 1775 by Percival Pott, surgeon to St Bartholomew's Hospital. He had observed that chimney sweeps had a very high incidence of scrotal cancer, and correctly deduced that this was due to chronic contact with soot. In fact, Percival Pott achieved a double: he described an environmental carcinogen and an occupational cancer in one go! He is also remembered for his description of spinal tuberculosis, referred to as Pott's disease.

CARCINOGENIC AGENTS

So far we have discussed carcinogenesis under the broad headings of age, genetics, race, geography and environment. The next step is to consider what type of agent is operating (the aetiological agent) and to look at ideas on how the agent converts a normal cell to a malignant cell (pathogenesis).

It is worth remembering that, as in the case considered in Chapter 11 of the woman with breast cancer, by the time a patient presents with a tumour, a large number of cellular events and many thousands of cell divisions have already taken place. Consequently, we are looking at a growth that has been in existence for quite some time, possibly 10–15 years. Identifying the responsible aetiological factors at this stage can be extremely difficult. There are three major groups of agents involved in carcinogenesis that we need to consider:

1 Chemical carcinogens
2 Radiation
3 Infectious agents.

These groups should not be viewed in isolation. Chemicals may, for example, interact with ionising radiation or with infectious agents. Several different agents within any one group may also interact with each other. Furthermore, all these extrinsic agents may interact with endogenous or constitutional factors in the host, such as genetic susceptibility, immune status or hormonal status, emphasising that the carcinogenic process is complex and multifactorial.

Chemical carcinogens

Figure 12.3 illustrates classic experiments of chemical carcinogenesis using mouse skin, which provide

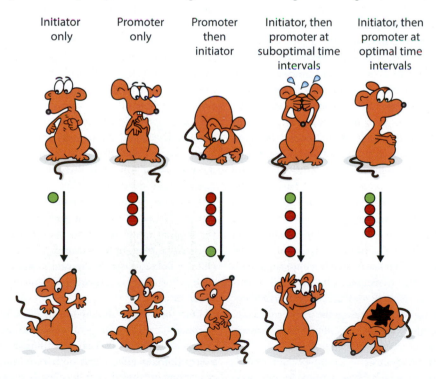

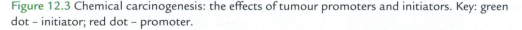

Figure 12.3 Chemical carcinogenesis: the effects of tumour promoters and initiators. Key: green dot – initiator; red dot – promoter.

the basis for the multistep theory (see page 353) and lead to the descriptions of the process of initiation and promotion. We now know that tumour development in humans is much more complex than depicted in Fig. 12.3.

Let us consider Fig. 12.3. If you apply a low dose of polycyclic aromatic hydrocarbon (initiator) to the shaved skin of the mouse and don't do any more, then no tumours will result. However, if you later apply another chemical, croton oil (promoter), to the same skin, local tumours will develop. The important points are that the initiator must be applied before the promoter, and that the promoter must be applied repeatedly and at regular intervals. There may be a long time interval between initiation and promotion, which suggests that initiation provokes an irreversible change in the DNA that is fixed by cell division. In contrast, the promoter acts in a dose-related, initially reversible fashion and appears to modify the expression of altered genes. Some chemicals (complete carcinogens) can act as both initiator and promoter, whereas others (incomplete carcinogens) fulfil only one action.

Evidence that certain chemicals are carcinogenic in humans is provided by epidemiological studies. Some chemical carcinogens occur naturally, e.g. aflatoxin B_1 is a potent hepatocarcinogen, which is a metabolite from the fungus *Aspergillus flavus*, a common contaminant of grain and other crops in the tropics. Several carcinogens occur as complex mixtures, as in tobacco smoke. Chemical carcinogens typically take 20 or more years to exert their effects, hence there is a long latency period between first encounter with the chemical and the appearance of a tumour. The dose required to induce tumours varies widely. Carcinogens act on a number of fairly specific target tissues, broadly determined by the initial routes of exposure and by subsequent patterns of absorption, distribution and metabolism. β-Naphthylamine is an interesting example. It enters the body mainly via the respiratory system, and is inactivated by conjugation with glucuronic acid. After excretion in the urine, it is activated again due to the action of urinary glucuronidase, which splits the conjugate releasing the active molecule. Its carcinogenic effects are hence confined to the urinary tract where it causes transitional cell tumours.

Chemical carcinogenesis is complex and occurs in several steps to which both genotoxic and non-genotoxic events contribute. Genotoxic carcinogens react with DNA. Various types of genetic damage will follow and, if the damage is not lethal to a cell, it will be transmitted to the daughter cells after cell division. The only protection the cell has is its array of DNA-repair enzymes, which must reconstitute the DNA before the next cell division or else the abnormality will be 'stamped' in by being transmitted to the daughter cells. Most genotoxic carcinogens undergo metabolic changes and are converted from inactive procarcinogens to activate ultimate carcinogens which bind to DNA. Some genotoxic chemicals react directly with DNA without previous metabolic activation. The conditions that determine whether a potential genotoxic chemical is activated or detoxified are very complex, but two main groups of enzymes are involved: the family of cytochrome P450-dependent monooxygenase isoenzymes, and various conjugating enzymes that catalyse the formation of water-soluble glucuronides. The process by which activated genotoxic carcinogens bind to DNA is equally complex. Once an activated carcinogen is bound to DNA, a number of consequences follow, depending on the nature and extent of the DNA damage that has been sustained. If this damage is extensive and irreversible, the cell will die. If less severe, the damage can be restored by the process of error-free DNA repair. The third possibility, mentioned earlier, is that the cell will survive with damaged DNA, which will then be passed on to the daughter cells following cell division.

Non-genotoxic carcinogens, by contrast, do not bind to DNA and do not directly damage it. They appear to act on cells in the target tissues mainly by directly stimulating cell division, or by causing cell damage and death (and thus indirectly stimulating cell division through the process of regeneration and repair). Other effects are less clearly understood, but the general mode of action of non-genotoxic chemicals can be thought of as causing disruption of normal cellular homeostasis. Some non-genotoxic chemicals, such as hormones, act through receptors on the surface of target cells.

One final point should be made. Some genotoxic chemicals exert both genotoxic and non-genotoxic effects in the target tissues. So, although genotoxic and non-genotoxic effects are both required for tumour development, they do not necessarily depend on separate genotoxic and non-genotoxic agents.

Key facts

Examples of chemical carcinogens and the associated tumour types

Carcinogen	Tumour type
Aromatic amines, e.g. β-naphthylamine	Transitional cell carcinoma of lower urinary tract (principally of the bladder)
Polycyclic aromatics, e.g. benzo(a)pyrene	Skin cancer, lung cancer
Vinyl chloride and Thorotrast	Angiosarcoma of liver
Arsenic	Skin cancers
Aflatoxin B_1	Liver cell carcinoma

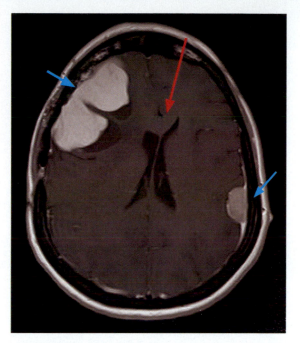

Figure 12.4 Radiation-induced meningiomas: a 39-year-old woman who had whole-brain irradiation for leukaemia 25 years previously re-presents with meningiomas (benign tumours of the meninges blue arrows). The right-sided meningioma is associated with considerable mass effect and will require surgical decompression (note the displacement of the cerebral structures across the midline – red arrow).

Radiation

Ionising radiation includes electromagnetic rays, such as ultraviolet light, X-rays and γ-rays, and particulate radiation, such as α particles, β particles, neutrons and protons. All of these are carcinogenic.

As ionising radiation passes through tissue, it interacts with atoms in its path to destabilise them. This disturbance in the electron shell of atoms may lead to chemical changes.

The precise mechanisms are still obscure. Radiation causes chromosomal breakage, translocations and mutations. Various protein molecules are also damaged and there are two principal theories to account for the observations. The direct theory states that ionising radiation directly ionises important molecules within the cell, whereas the indirect theory states that ionisation first affects water within the cell, which leads to the production of oxygen free radicals that cause the damage. Whichever mechanism operates, the end-result is that DNA is altered, analogous to the initiator effect in chemical carcinogenesis (see Fig. 12.3).

The carcinogenic effect of radiation is related to its ability to produce mutations and it is known that this depends on the type and strength of the radiation and the duration of exposure Some tissues, such as bone marrow and thyroid, are particularly sensitive to the effects of radiation and children are more susceptible than adults (Fig. 12.4).

Ultraviolet light is particularly important, because sun exposure causes vast numbers of melanomas, squamous cell carcinomas and basal cell carcinomas of the skin. Fortunately, squamous cell carcinomas and basal cell carcinomas can generally be cured by complete local excision, but melanomas metastasise early and kill. Many of the pioneers who studied radioactive materials and X-rays developed skin cancers, and miners of radioactive elements have a high incidence of lung cancers. The radiation from the atomic bombs dropped on Hiroshima and Nagasaki resulted in an increased incidence of leukaemia, especially acute and chronic myeloid leukaemia, and breast, lung and colonic cancers. In contrast, a dramatic increase in thyroid carcinomas has been described in children living in Ukraine and Belarus, who were exposed to fallout after the Chernobyl accident. Interestingly, no increase in the incidence of other types of childhood or adult solid cancers has been noted.

Chapter 12: What causes cancer?

Infectious agents

A large number of viruses, bacteria and parasites have been implicated in the aetiology of cancer (Table 12.1). We discuss briefly the following:

- Epstein–Barr virus (EBV)
- Human papillomavirus (HPV)
- Hepatitis B virus (HBV)
- Human T-cell leukaemia virus (HTLV)
- Kaposi sarcoma-associated herpesvirus (KSHV)
- Helicobacter pylori
- Schistosoma haematobium.

Table 12.1 Viruses and their associated cancers

Virus	Associated tumour
Oncovirus	
HTLV-1	Adult T-cell leukaemia/ lymphoma
Hepadnavirus	
Hepatitis B	Liver cancer
Papovavirus	
Human papillomavirus (HPV) types:	
1, 2, 4, 7	Benign skin papillomas
6, 11	Genital warts
16, 18	Cervical cancer
10, 16	Laryngeal cancer
5	Skin cancer
Herpes virus	
Epstein–Barr virus (EBV)	Burkitt's lymphoma
	Nasopharyngeal carcinoma
	Immunoblastic lymphoma
Herpes simplex 8	Kaposi's sarcoma
	Body cavity B-cell lymphoma
	Multiple myeloma

Epstein–Barr virus

There is a very strong association between EBV and the African variety of Burkitt's lymphoma (Figs 12.5 and 12.6), because over 95% of the African cases show

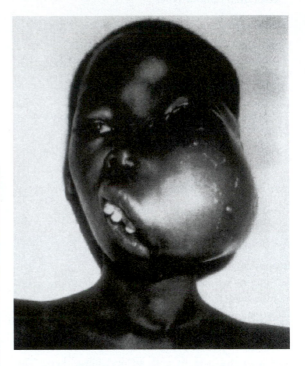

Figure 12.5 Young child with a large maxillary tumour distorting the face. This is a classic presentation of Burkitt's lymphoma.

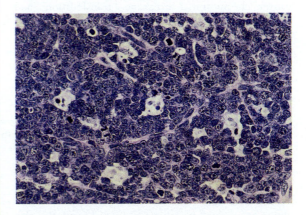

Figure 12.6 Photomicrograph showing classic 'starry sky' appearance of Burkitt's lymphoma due to apoptosis of tumour cells creating 'light holes' in a 'sky' of blue cells.

the EBV genome in the tumour cells and all the patients have a raised level of antibodies to EBV membrane antigens. Fortunately, EBV does not inevitably cause cancer, because it is a common infection in developed countries, where it causes a flu-like illness called infectious mononucleosis or glandular fever. Burkitt's lymphoma can occur without EBV and few non-African cases of Burkitt's lymphomas (20–30%) have the EBV genome. Therefore, EBV must be just one factor involved in the transformation of B lymphocytes to a B-cell malignancy.

It is interesting that the African regions where Burkitt's lymphoma is common are also regions where malaria is endemic. It would appear that malaria causes a degree of immunoincompetence that allows the EBV-infected B cells to proliferate and, hence, gives them an increased risk of mutation. Burkitt's lymphoma exhibits a specific mutation resulting in the 8;14 translocation, regardless of whether EBV is involved. This translocation moves the c-*myc* gene from its position on chromosome 8 to be adjacent to the immunoglobulin heavy chain gene on chromosome 14. The gene c-*myc* codes for proteins that control cell proliferation and the effect of this translocation is to increase its transcription, possibly because that zone of chromosome 14 is an area of frequent transcriptional activity.

Another putative association of EBV is with Hodgkin's lymphoma, a heterogeneous disorder of B cells, in which the tumour cells are often a small

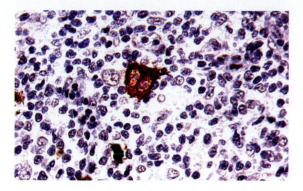

Figure 12.7 Reed–Sternberg cell in Hodgkin's lymphoma, staining positively for Epstein–Barr virus (EBV). EBV has been found in the Reed–Sternberg cells in about a third to half of all Hodgkin's lymphoma cases, although whether it truly has an aetiological role has not been conclusively established.

component of the proliferation. The classic cells are called the Reed–Sternberg cells (Fig. 12.7).

Nasopharyngeal carcinoma, although rare overall, is very common in parts of Asia (China, Singapore and Malaysia). EBV appears to be an important aetiological agent in the non-keratinising forms of nasopharyngeal carcinoma, irrespective of the geography. The EBV genome or gene products are identified in almost 10% of cases in areas with a high incidence of the disease.

History

Thomas Hodgkin 1798–1866

Thomas Hodgkin (Fig. 12.8) was a devout Quaker who was generally unsuccessful in practice because he was very reluctant to accept fees. In his early years, his major interest was pathology and he was curator of the pathology museum at Guy's Hospital and later lecturer in pathology for 10 years, but resigned after being passed over for the job of assistant physician, much to the disappointment of his students. In later years, he devoted his time to philanthropic causes, travelled widely and died from dysentery in Jaffa.

Figure 12.8 Thomas Hodgkin.

Chapter 12: What causes cancer?

Human papillomavirus

HPV is a papovavirus that has long been known to be associated with skin papillomas (warts). Its role in causing cancer was recognised during the study of a very rare disease, epidermodysplasia verruciformis, in which patients have defective cell-mediated immunity and numerous skin papillomas. These papillomas may transform into squamous cell carcinomas, which frequently contain the genome of HPV-5, -8 or -14. HPV is not a single virus but a group of around 85 genetically distinct viruses. Interestingly, some types appear to produce benign tumours whereas others predispose to malignancy. Thus HPV-16 and -18 (the most common among 13 high-risk subtypes) are implicated in squamous cell carcinoma of the uterine cervix, accounting for approximately 70% of cervical cancers worldwide, whereas HPV-6 and -11 are common in benign cervical lesions (Figs 12.9 and 12.10).

There is also good evidence that high-risk HPV, in particular HPV-16, is associated with a proportion of anal, vaginal, penile and vulvar cancers. Data have also been accumulating that HPV-16 may play a role in some oropharyngeal (head and neck) carcinomas.

HPV can be transmitted by sexual intercourse, and it is noted that there is a high incidence of carcinoma of the cervix in both those who start sexual activity at an early age and those who are promiscuous. The question is: why doesn't our immune system eradicate the virus? Many people have skin warts on their hands and feet (verrucae) as children, but appear to develop immunity so that the warts are less common in later life. HPV genome comprises a number of genes, two of which are believed to be important in malignancy. The protein products of the *E6* and *E7* genes bind to p53 and retinoblastoma protein, respectively, inactivating their function in regulating the cell cycle (see page 347).

Two types of vaccines for the prevention of cervical cancer are on the market: a bivalent vaccine targeting HPV-16 and -18, and a quadrivalent vaccine targeting HPV-16, -18, -6 and -11. The vaccines are almost 100% effective in women not previously infected and hence have to be given to adolescent girls before first sexual intercourse. Both types of vaccines appear to be safe and have been incorporated into national immunisation programmes in many developed countries. The real impact of the vaccines will not be known for several decades.

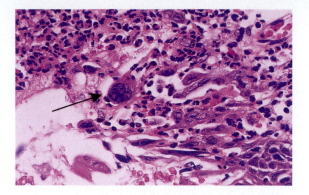

Figure 12.9 Inflamed cervical epithelium with multinucleated giant cell due to herpes simplex virus (HSV) infection. The role of HSV in cervical cancer is still debated; human papillomavirus (HPV) is of far greater importance.

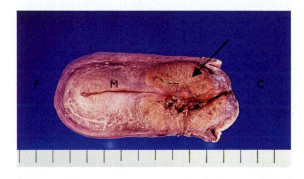

Figure 12.10 Invasive cervical carcinoma (arrow). Each division is 10 mm. C, cervix; F, fundus; M, myometrium.

Hepatitis B virus

HBV is associated with the production of a chronic hepatitis, cirrhosis and carcinoma of the liver.

In sub-Saharan Africa and south-east Asia, where infection with HBV is endemic, the infection is transmitted vertically from mother to child during pregnancy. These children, therefore, have chronic HBV infection and a high incidence of hepatocellular carcinoma (HCC) at a relatively young age (20–40 years).

The importance of HBV (Figs 12.11 and 12.12) in HCC is apparent from this sort of epidemiological work, and also from molecular biological investigation looking for integrated HBV DNA sequences. These have been identified in the hepatocytes of some patients with chronic HBV infection and some HCC tumour cells. It appears that integration of the viral genome precedes malignant transformation by several years, but to date no known oncogenic sequences have been identified. The HBV genome does contain a transactivating gene, termed 'X', which codes for a product that alters the level of transcription of other genes, including the genes in the hepatocytes. An additional mechanism that has been postulated is that HBV infection also leads to liver cell injury and regeneration due to the effect of cytotoxic T cells. It is possible that both the direct (DNA effect due to X) and the indirect (proliferation in response to immune-mediated injury) mechanisms are important in the aetiology of HCC.

Liver cell carcinomas are also associated with alcoholic liver disease, androgenic steroids and aflatoxins. Aflatoxins are toxic metabolites of a fungus, *Aspergillus*

flavus, which can contaminate food in the tropics. Aflatoxin B is thought to contribute to the high incidence of liver cancer in parts of south-east Asia and Africa. Possibly these agents act by causing damage that leads to regenerative activity and, hence, the production of proliferative nodules that are susceptible to further cellular alterations by HBV.

Human T-cell leukaemia virus-1

HTLV-1 is important because it is the only example (so far!) of a retrovirus causing a human cancer. It is implicated in adult T-cell leukaemia/lymphoma (ATLL), which is a rare tumour of the lymphoid system. HTLV-1 infection is most common in southern Japan, South America and parts of Africa, and precedes the development of malignancy by decades. It has a transactivating gene, *tat*, that increases interleukin (IL)-2 receptor expression in infected T cells, which promotes their growth. The study of retroviruses has advanced our knowledge of the role of genes in tumour biology by allowing the identification of specific transforming genes. However, to date, they have not been shown to be important in common human tumours.

Kaposi sarcoma-associated herpesvirus

Kaposi's sarcoma (Fig. 12.13) is an important vascular neoplasm that has come to prominence in HIV-infected patients. Since the early 1980s it has frequently been associated with patients who have AIDS. It is an endemic lesion in central Africa, predominantly in healthy men but also in women and children. Evidence is accumulating that this odd vascular tumour is due

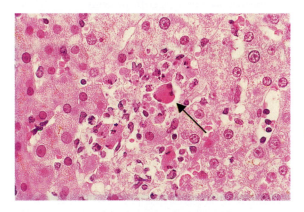

Figure 12.11 Viral hepatitis showing an apoptotic liver cell (arrow) surrounded by inflammatory cells.

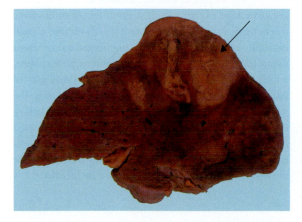

Figure 12.12 Hepatocellular carcinoma arising in a cirrhotic liver.

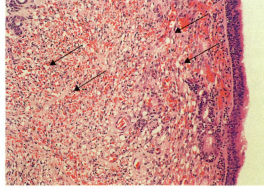

Figure 12.13 A larynx with Kaposi's sarcoma composed of slit-like channels filled with red blood cells.

Chapter 12: What causes cancer?

to a novel herpes virus. This virus has been termed 'Kaposi's sarcoma-associated herpesvirus' (KSHV) or human herpes virus 8 (HHV-8). There is strong evidence to demonstrate that this virus is linked to several other neoplasms, such as body cavity B-cell lymphoma, multiple myeloma, benign lymphoproliferative disease, angiosarcoma of the face, angiolymphoid hyperplasia with eosinophilia and multicentric Castleman's disease. Although the data are unclear, there is a suggestion that HHV-8 may play a role in some enigmatic inflammatory diseases, such as sarcoidosis. The study of HHV-8 has been very fruitful in revealing new aspects of viral carcinogenesis. The genome of this virus encodes proteins that take part in molecular mimicry of cell cycle regulatory and signalling proteins (see section on molecular genetics).

Helicobacter pylori

H. pylori, a Gram-negative bacterium, is a very common infection of the upper gastrointestinal tract. Until the seminal discovery in 1983 by Barry Marshall and Robin Warren (Nobel laureates) that peptic ulcer was a microbial disease, it was thought to be related to stress and smoking. More than 50% of the world's population is thought to harbour *H. pylori*, hence disease is not inevitable. Although gastritis and peptic ulcer are by far the most common manifestation, there is a small risk of developing gastric carcinoma. Infections can be treated using proton pump inhibitors (omeprazole) and antibiotics (clarithromycin, amoxicillin).

Schistosoma haematobium

Schistosomiasis (bilharzia) is caused by the parasite *Schistosoma* spp. Most human cases are due to *Schistosoma haematobium*, *S. mansoni*, and *S. japonicum*. It is one of the most common infectious diseases worldwide, most prevalent in developing countries such as Africa and South America. Human infection occurs due to contact with contaminated water. The manifestations are a result of an immune reaction to the schistosoma eggs (granulomatous inflammation). These eggs can be found almost anywhere in the human body, but common sites include bladder, kidneys, liver and lungs. Although chronic infections can lead to bleeding (haematuria), scarring and fibrosis, there is also a significant risk of bladder cancer (squamous cell carcinoma) and tumours of the liver.

There is an 'old wives' saying: 'Where God puts disease, he also puts a cure.' Viruses undoubtedly cause infectious disease and can be one step on the road to cancer. However, they may also provide a possible cure for disease because they may be the ideal vehicle for altering the genetic code within human cells. Ultimately, it would be best if patients with single-gene disorders could have their defective gene replaced by the correct gene. In theory, this is possible by using a retrovirus to introduce the gene, although in practice there are many problems to conquer. The most useful practical application of our rapidly expanding knowledge of the genes is in the manufacture of specific proteins.

SUMMARY

Although we started the chapter with the statement that cancer is a disease of genetic material, it should be evident that it is also a multifactorial disorder. This might appear initially as a paradox, but remember that, in any given person, the combination of a genetic make-up, the environment in which the person lives and the factors to which he or she is exposed, the longevity of the person and 'chance' will combine to produce the genetic changes (mutations) that may ultimately lead to cancer. One could argue that, at least in some circumstances, the cancer is a direct payback of an evolutionary advantage of the past. It is not difficult to envisage a genetic make-up that gives such an advantage; a good example could be decreased pigmentation in races living in the cloudy and colder northern hemisphere, which would help in obtaining adequate amounts of vitamin D from sunshine. This advantage from the genetic make-up could become a distinct disadvantage if put into the context of chance mutations, increased exposure to sunlight due to sunbathing on numerous holidays on tropical beaches, and a long enough life for the mutations to result in a cancer – a melanoma.

Case study: rapid growth and death

A 25-year-old man presents with a pigmented lesion on his back. It had an irregular margin and pigmentation and had changed recently to become much larger, now approximately 1 cm in size. It was not itchy or bleeding.

He has an excisional biopsy a week later.

The pathology was reported as: superficial spreading melanoma, vertical growth phase, Breslow thickness 2.9 mm, Clark level IV, mitoses 2/mm^2 without regression or ulceration.

Question: What are the risk factors for melanoma?

Answer: Genetic predisposition, ultraviolet radiation

Question: What are the two key prognostic features recorded in the pathology?

Answer: The growth phase – in this case 'vertical', which means that the tumour has developed the ability to penetrate deeply and be more aggressive, as opposed to horizontal growth phase, where it grows sideways, and the second feature is Breslow thickness which records how deep it has invaded.

As a result of the findings, he went on to have a wide excision and sentinel node biopsy (axilla). As the sentinel node was positive for tumour, he went on to have an axillary dissection; 9/24 lymph nodes were positive.

Over the next year, he developed numerous metastatic deposits and died.

For more details including postmortem findings, see Case 31.

Read more in Pathology in Clinical Practice Case 31

Key facts

Major aetiological factors involved in tumour formation

Age

Genetic factors

Geographical and racial factors

Environmental agents

Carcinogens: chemicals, radiation, viruses

Immunity

Chapter 12: What causes cancer?

CHAPTER 13

MOLECULAR GENETICS OF CANCER

It is now well over 100 years since Gregor Mendel first carried out his experiments with peas and almost exactly a century since Mendel's laws were 're-discovered'. It is also over a century since pathologist David von Hansemann (1890) described abnormal mitotic figures in cancer cells, and noted an association between aneuploidy (abnormal chromosomal numbers) and cancer. It is also over a century since Theodore Boveri proposed that alterations in the chromosomes were the basis of cancer formation. Remarkably, it was half a century later that the Philadelphia chromosome (discussed below) was identified.

There is little doubt that Theodore Boveri is one of the giants of twentieth-century cancer genetics. His classic experiments on fertilisation of sea urchin eggs with two sperms demonstrated the phenomenon of abnormal chromosomal segregation, and he was quick to realise that this abnormal chromosomal number may account for the unrestricted growth of tumours. In fact, he made a number of pertinent remarks relating to regulation of the cell cycle, the presence of oncogenes and tumour-suppressor genes, and genetic instability in tumours. To put it into context, Boveri published his famous monogram in German in 1914, the Philadelphia chromosome was identified in the 1960s, Knudson's two-hit hypothesis and the identification of the first dominant oncogene occurred in the 1970s, the cloning of the first tumour-suppressor gene, the retinoblastoma (*RB*) gene, was in the 1980s, the discovery of mismatch repair defects in tumours and cloning of BRCA1/2 was in the 1990s, and the publication of the first draft of the human genome was in 2000. What an incredible century (see time line on page 298)!

CYTOGENETICS

Not surprisingly, following Boveri's publications, some of the earliest indications for genetic alterations came from classic karyotypic analysis. This type of study reveals gross abnormalities at the chromosomal level. Classic examples of tumours showing such gross chromosomal abnormalities include chronic myeloid leukaemia and Burkitt's lymphoma. We have already considered Burkitt's lymphoma in the section on viruses (see page 330). We briefly consider chronic myeloid leukaemia here.

The Philadelphia (Ph1) chromosome is present in 90% of cases of chronic myeloid leukaemia and can be used as a diagnostic marker. It is produced by a reciprocal and balanced translocation between chromosomes 22 and 9. The breakpoint on chromosome 9 occurs at the locus of the *abl* proto-oncogene, and that on chromosome 22 is in the region termed 'the breakpoint cluster region' (bcr). Some recent work suggests that the *bcr* genes code for a protein kinase that could have oncogenic potential. The *abl* proto-oncogene has sequence homology with the tyrosine kinase family of oncogenes, but it is only after translocation to chromosome 22 that it produces a mutant protein with enhanced tyrosine kinase activity. This particular tyrosine kinase activity is located in the nucleus, where it is believed to influence transcription of DNA. Further examples of tumours and the cytogenetic abnormalities are shown in Table 13.1.

Table 13.1 Chromosomal alterations in human tumours

Tumour type	Chromosomal aberration	Possible action	Gene(s)
Haematopoietic tumours: translocation			
Chronic myeloid leukaemia	t(9;22)(q34;q11)	Alteration of nuclear tyrosine kinase	*ABL*
Burkitt's lymphoma	t(8;14)(q24;q32)	Cell cycle regulation	*c-myc-IgH*
	t(2;8)(p12;q24)		*Igk,c-myc*
	t(8;22)(q24;q11)		
Acute myeloid leukaemia	t(8;21)(q22;q22)		*ETO*
Mantle cell lymphoma	t(11;14)(q13;q32)		*Bcl-1-IgH*
Follicular lymphoma	t(14;18)(q32;q21)		*IgH-bcl-2*
Solid tumours: translocation			
Ewing's sarcoma/PTEN	t(11;22)(q24;q12)		*EWS-FL1*
	t(21;22)(q22;q12)		*EWS-ERG*
	t(11;22)(q13;q12)		*EWS-WT1*
Synovial sarcoma	T(X;18)		
	(p11.23,q11.2)		*SYT-SSX1*
	T(X;18)		
	(p11.21,q11.2)		*SYT-SSX2*
Malignant melanoma	t(1;19)(q12;p13)		?
	t(1;6)(q11;q11)		?
	t(1;14)(q21;q32)		?
Salivary adenoma	t(3;8)(p21;q12)		*CTNNB1*
Renal adenocarcinoma	t(X;1)(p11;q21)		*TFE3*
	t(9;15)(p11;q11)		?
Solid tumours: deletions			
Retinoblastoma	del13q14	Loss of oncosuppression	*RB*
Wilms' tumour	del11p13	Loss of oncosuppression	*WT-1*
	del11p15		?

(continued)

Chapter 13: Molecular genetics of cancer

Table 13.1 Chromosomal alterations in human tumours (*continued*)

Tumour type	Chromosomal aberration	Possible action	Gene(s)
	del17q12–21		*FWT1*
Bladder, transitional cell carcinoma	del11p13		?
Lung cancer, small-cell type	del17p13	Loss of oncosuppression	*TP53*
Colorectal adenocarcinoma	del17p13	Loss of oncosuppression	*TP53*
	del5q21	Loss of oncosuppression	*APC*
Breast cancer	del17p13	Loss of oncosuppression	*TP53*
	del17q21	Loss of oncosuppression	*BRCA1*
	del13q12–13	Loss of oncosuppression	*BRCA2*
	del16q22.1	Loss of oncosuppression	
Solid tumours: amplification			
Neuroblastoma		Cell cycle control	*N_MYC*
Breast cancer		Increased growth factor activity	CERBB2

PTEN, phosphatase and tensin homologue.

CANCER-PRODUCING GENES: ONCOGENES

The term 'oncogene' refers to any mutated gene that contributes to neoplastic transformation in the cell. Two major types of oncogenes have been identified: *dominant oncogenes* and *tumour-suppressor genes* (*anti-oncogenes*). Two other categories of genes are now known to be important: genes involved in cell death and genes involved in repair of DNA.

Some oncogenes involved in carcinogenesis are mutated versions of normal cellular genes (called proto-oncogenes, p-*onc*). The function of these normal genes is enhanced by the mutations, and hence they are referred to as 'activating' or 'gain-in-function' mutations. These genes are also known as 'dominant' oncogenes because mutation of one allele is sufficient to exert an effect, despite the presence of normal gene product from the remaining allele. It is over 40 years since such genes were discovered. This category of genes is now referred to in the literature as just 'oncogenes'.

Dictionary

Allele: alternative form of the gene found at the same locus in homologous chromosomes

In contrast, tumour-suppressor genes (TSGs) are normal genes with a function that is inactivated by mutations, hence these are known as 'inactivating' or 'loss-of-function' mutations. The genes are also known as 'recessive' oncogenes because inactivation of both alleles is required to have an effect at the cellular level. Evidence for the existence of such genes has been largely circumstantial and was based on classic genetics, cytogenetics and molecular genetics. You should be aware that the terms 'dominant' and 'recessive' refer to action at the genetic level; confusion sometimes occurs because the terminology has been borrowed from classic mendelian genetic inheritance patterns.

Programmed cell death (apoptosis) is an important component of cell regulation; and mutations in genes involved in the regulation of this pathway (e.g. the

BCL2 family) have been shown to play a role in tumour formation.

The stability or DNA-repair genes include mismatch repair (*MMR*) genes. as well as the breast cancer predisposition genes *BRCA1* and *BRCA2*. *MMR* genes repair subtle mistakes in the genome, which occur during normal cell division or due to exposure to carcinogens. *BRCA1*, on the other hand, is also involved in chromosomal stability during mitotic recombinations and segregation. These genes are fundamental to the cell because mutations in these genes allow alterations in the DNA to go unrepaired, and hence mutations in other genes occur at a higher rate. You can imagine that, if these increased mutations were to occur in cell-cycle-related genes or genes involved in apoptosis, genetic changes would be consolidated because the cell would be allowed to proceed and complete a round of division despite defective DNA. These categories of genes can act as dominant oncogenes or tumour-suppressor genes, depending on their actions within the pathways and whether loss of one or both alleles is needed to confer the effect.

Where do oncogenes come from? There are both exogenous and endogenous sources. The exogenous sources include viral oncogenes (v-*onc*), which may be introduced into cells by tumour viruses. Endogenous genes are called cellular oncogenes (c-*onc*), and these are genes that are normally present in the cell but have been altered to produce the oncogene. As mentioned above, the normal gene from which the oncogene is derived is called the proto-oncogene.

The viral or exogenous oncogenes can be divided into two types: those that show similarity to normal cellular genes and those that are completely different. This is important because viral oncogenes that resemble cellular genes are actually derived from the cell's genes. This is quite amazing when you think about it! A virus infects a cell and incorporates some of the cellular genes into its own genome. These genes, finding themselves in a new piece of DNA or RNA, become altered in their properties and are then viral oncogenes. When the virus infects another cell, it can introduce the viral oncogene (a process called transduction), which leads to altered growth of the infected cell. Retroviruses, which consist of RNA that becomes incorporated into the host DNA through the action of the enzyme, reverse transcriptase, can readily 'pick up' some host DNA and so can carry viral oncogenes derived from cellular oncogenes. Oncogenic DNA viruses generally possess gene sequences that are uniquely viral and have no homology with cellular oncogenes.

HOW CAN ONCOGENES PROMOTE CELL GROWTH?

There are a number of ways in which the function of oncogenes can be altered and this includes (1) point mutations, (2) amplification, (3) gene rearrangement/translocation, (4) deletion of part or whole of chromosome and (5) altered expression. As most

Key facts

Main differences between dominant oncogenes and tumour-suppressor genes

	Dominant oncogenes	Tumour-suppressor genes
Number of alleles in normal cells	Two	Two
Number of alleles mutated to exert effect	One	Two
Effect of mutation on the function of the protein product	Enhanced	Reduced
Germline (inherited) mutations identified, i.e. important in genetic predisposition	Rare: *RET, KIT*	Many: *TP53, RB*, etc.
Adjectives to describe mutations	Activating, gain in function, dominant	Inactivating, loss of function, recessive

of the mechanisms described apply to both types of oncogenes, they are considered here together; however, there are differences in the pattern of alterations between dominant oncogenes and tumour-suppressor genes. In contrast to dominant oncogenes, which are mutated in a consistent manner by point mutation (e.g. *ras*), translocation (e.g. *abl*) or gene amplification (e.g. N-*myc*), mutations in tumour-suppressor genes tend to be diverse in both type and position within the gene (e.g. *TP53*).

POINT MUTATIONS

These result in the substitution of one base-pair by another, e.g. substitution of G:C by A:T. The effect of the point mutation depends on its position and includes alteration of the protein structure by change in the amino acid composition and insertion of a stop codon with premature termination of the protein. The clearest example of point mutations in human tumours is found in the *ras* family (Fig. 13.1), where 99% of all *ras* mutations occur at codons 12, 13 and 61 of H-, K- and N-*ras*, respectively. They contribute to oncogenesis in many of the main types of human tumours. The *ras* mutations are encountered in both benign and malignant neoplasms. The *kit* gene encodes a transmembrane tyrosine kinase receptor that is consistently expressed in haematopoietic stem cells, mast cells, melanocytes, germ cells and interstitial cells of Cajal (ICCs). Activating germline mutations of the *kit* gene are associated with ICC hyperplasia, familial gastrointestinal stromal tumours (GISTs), cutaneous mastocytosis and cutaneous hyperpigmentation. Depending on the site where the mutation has occurred, patients will present with a combination of all or some of the above lesions.

GENE REARRANGEMENTS/TRANSLOCATIONS

This refers to the production of a hybrid chromosome due to joining of part of one chromosome to another (interchromosomal rearrangement; Fig. 13.2) or the production of a modified chromosome (intrachromosomal rearrangement), where a chromosome breaks into segments that are reformed in the wrong order. The rearrangement of DNA sequences can lead to creation of an altered gene (and product) as a result of either structural change or change in the control of transcription. Examples include 8;14 (or 8;22, 2;8) translocation in Burkitt's lymphoma and the 9;22 translocation in chronic myeloid leukaemia (CML).

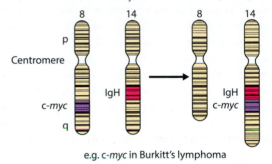

e.g. *c-myc* in Burkitt's lymphoma

Figure 13.2 Translocation.

AMPLIFICATION

The normal genome contains two copies of each gene (the two alleles). In amplification, one copy is multiplied numerous times and may result in increased mRNA and hence increased protein product (Fig. 13.3). At the level of the chromosome, these areas of amplification are seen as double minutes (DMs) or homogeneously

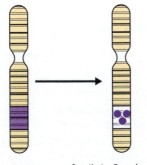

e.g. ras oncogene family in Ca colon

Figure 13.1 Point mutation.

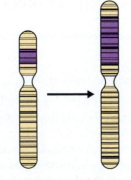

e.g. N-*myc* in neuroblastoma

Figure 13.3 Gene amplification.

staining regions (HSRs). DMs are extrachromosomal chromatin bodies without centromeres that segregate randomly during mitosis. HSRs are expanded chromosomal regions that are linked to the centromere and therefore segregate in the normal way. Most classes of oncogenes have been shown to be amplified in human malignancy. Examples include N-*myc* in neuroblastoma and *HER2* in breast cancer. Only some of these amplifications have been demonstrated to have pathological significance. N-*myc* amplification correlates with advanced stage and recurrence of neuroblastoma, and *HER2* amplification in breast cancer correlates with poor prognosis. It should be noted, however, that, in many tumours, over-expression of mRNA and protein is seen in the absence of gene amplification.

DELETIONS

These range from the loss of single base-pairs to loss of entire chromosomes. The small intragenic deletions have similar effects (abnormal protein, stop codons) to point mutations. Larger deletions will of course inactivate many genes at a time (Fig. 13.4).

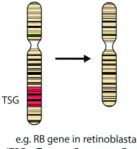

TSG

e.g. RB gene in retinoblasta
(TSG = Tumour Supressor Gene)

Figure 13.4 Deletion.

ALTERED EXPRESSION

The inactivation of a gene via deletions or intragenic mutations is now a familiar story. It has recently become apparent that some putative tumour-suppressor genes do not exhibit these common phenomena in certain tumours. Instead changes in the methylation patterns of the promoter region of the gene, or even alterations in the chromatin pattern regulated by

histone deacetylases, lead to an altered expression of some genes (i.e. alteration in the transcription of the mRNA and translation of the protein). This process is referred to as epigenetic regulation. This occurs without any mutational event in the gene; hence if the gene were sequenced, no changes would be detected. Some good examples are the inactivation of the *p16* gene on chromosome 9p21, the inactivation of the *BRCA1* gene in sporadic breast carcinomas, and the inactivation of the second allele in some rare types of gastric carcinomas (isolated or signet-ring cell gastric carcinoma). It is worth bearing in mind that, in tumorigenesis, methylation and acetylation play roles in the down-regulation of the expression of tumour-suppressor genes.

A mechanism that is emerging as a novel way in which to regulate gene expression is through the interaction with micro-RNAs (miRNAs). Unlike messenger RNA (mRNA), which is translated to proteins, miRNAs are non-protein coding and act on a target mRNA to modify the translation of its protein. It can do this either directly by suppressing the protein translation or indirectly by causing degradation of the mRNA. Micro-RNAs are short, single-stranded RNA molecules with approximately 22 nucleotides. Hundreds of miRNAs have been annotated and they are likely to act on multiple different mRNAs, creating a complex network of interactions that modify the expression patterns within an individual cell and tissue.

Recent high-throughput sequencing studies have revealed that most transcribed RNA is non-coding. Long non-coding RNAs (lncRNAs) are a diverse group of RNAs with >200 nucleotides, and are increasingly being recognised as having a role in regulatory function, adding yet another layer of complexity to the regulation of the genome.

MODE OF ACTION OF DOMINANT ONCOGENES

Normal cell growth is believed to be influenced by growth factors binding to receptors on the surface of the cell. This produces a stimulus through a 'signal transduction pathway(s)' that influences the cell's nucleus to produce instructions for proliferation. Within the nucleus itself, there are also molecules that regulate transcription and those that regulate the cell

Chapter 13: Molecular genetics of cancer

cycle during cell division. Therefore, cell growth could be stimulated due to any of the following:

- Increased growth factor production
- Increase in growth factor receptors on the cell's surface
- Abnormal growth factor receptors
- Abnormal signalling through the cascade of signal transduction pathways in the cytoplasm
- Abnormalities in the nuclear acting molecules
- Alterations to the control of the cell cycle.

Oncogenes have been identified that act through each of these mechanisms.

GROWTH FACTORS AND GROWTH FACTOR RECEPTORS

Growth factors are polypeptides that act locally to stimulate proliferation and, sometimes, differentiation.

If tumour cells produce substances that act as growth factors, then they will be continually self-stimulating (autocrine stimulation). Several classes of growth factors have been classified according to sequence homology and biological activity. The categories include epidermal growth factor family (EGF and transforming growth factor α [TGF-α]), the fibroblast growth factor (FGF) family (acidic and basic FGFs, hst and int-2), platelet-derived growth factor (PDGF), colony-stimulating factors (CSFs), interleukins and the insulin-like growth factors (IGFs). Growth factors act by binding to a receptor residing in the plasma membrane. The binding leads to activation of the receptor and signal transduction to the interior of the cell (Fig. 13.5).

When the normal cell surface receptors are activated by growth factors, the receptors usually dimerise, leading to phosphorylation of the tyrosine residues and an increase in their tyrosine kinase activity (Fig. 13.6). When the genes coding for these receptors are abnormal, there is persistent activity without receptor

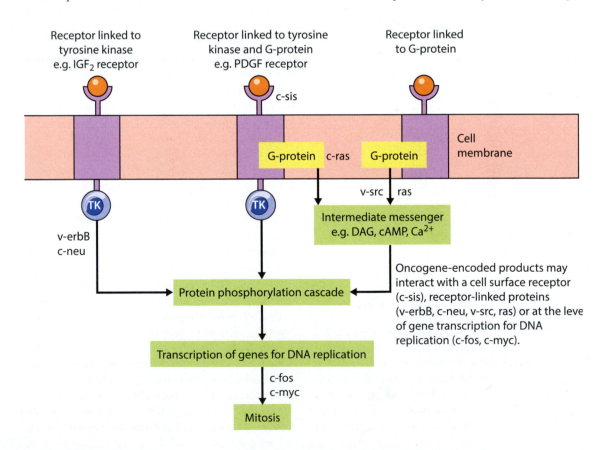

Figure 13.5 Growth factor receptors and signalling pathways: points of interaction of cellular oncogenes.

↑Growth factor
c-sis

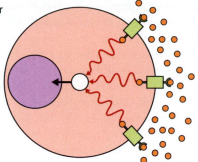

Permanent activation of receptor
v-erbB

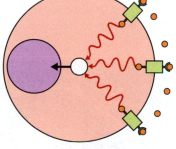

↑Growth factor receptor
c-neu

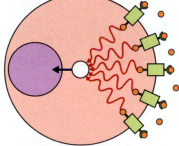

Abnormal signal transduction
c-K-ras

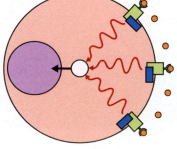

Figure 13.6 Mechanisms of oncogene activation.

binding. The growth factor receptors can be activated in this way by a number of mechanisms, including gene amplification, rearrangement and over-expression, e.g. the *HER2* gene is amplified in about 25% of breast cancers and also in a variety of other cancers such as lung. Breast cancers that over-express the *HER2* receptor are more aggressive and have a poor prognosis. Recently, a humanised monoclonal antibody against *HER2* has been developed and is used to treat patients with breast cancer in whom the tumour cells have amplification of the *HER2* gene. In other situations, the receptor may not be over-expressed but may rather have altered kinase activity, as with the mutational activation of *HER1* (EGF receptor or EGFR) in lung cancer. Some tyrosine kinases are not attached to a receptor but are anchored to the plasma membrane and participate in signalling. The *c-src* oncogene alters the activity of one of these non-receptor tyrosine kinases.

SIGNAL TRANSDUCTION PROTEINS: GTP-BINDING PROTEINS

Three mammalian *ras* genes — H-*ras*-1, K-*ras*-2 and N-*ras* — have been identified. They form one of the most important families of oncogenes identified to date. The *ras* genes usually acquire transforming activity as a result of point mutation within their coding regions. These activating point mutations are restricted to certain sites, notably codons 12, 13 and 61. The p21 product of both *ras* proto-oncogene and transforming genes is located in the inner surface of the cell membrane. It binds guanosine nucleotides (GDP and GTP) and possesses intrinsic GTPase activity. These biochemical properties resemble those of the G-proteins, which are associated with the cell membrane and implicated in the modulation of signal transduction. It is becoming clear that ras proteins function as critical relay switches that regulate signalling pathways between the cell surface receptors and the nucleus.

Overall, point mutation in the *ras* gene is the most common dominant oncogene abnormality in human tumours. It appears to play a major role in colon, pancreatic and thyroid cancers as well as in myeloid leukaemia.

BRAF is another example of an oncogene that codes for signal transduction proteins. It is a member of the Raf family. Mutations in the *BRAF* gene have been described in malignant melanoma, colorectal cancer, non-small-cell

Chapter 13: Molecular genetics of cancer

lung cancer and non-Hodgkin's lymphoma. One of the most common mutations seen is the *BRAF* $V^{600}E$ (substitution of valine by glutamate at codon 600). A number of *BRAF* inhibitors are in clinical use and have been shown to improve survival of patients with malignant melanoma. Unfortunately, in many patients, resistance to the drug also develops with recurrence of disease.

NUCLEAR ONCOPROTEINS

These oncoproteins share the feature of nuclear localisation and proven or suspected ability to bind to specific DNA sequences.

An important family of such genes involved in human malignancy is *myc*. This was originally identified as the oncogene carried by several acutely transforming retroviruses. The *myc* family consists of c-*myc*, N-*myc*, L-*myc*, R-*myc*, P-*myc* and B-*myc*. They have been isolated on the basis of homology to v-*myc* or one of the *myc* proto-oncogenes; c-*myc* is expressed in many tissues and correlates with cell proliferation. Activation due to gene amplification occurs in breast cancer and small-cell lung cancer (SCLC). In B- and T-cell lymphomas, translocation to an immunoglobulin (Ig) or T-cell receptor locus is seen.

The end-result of all the signalling pathways is the transition of the cell through the cell cycle. Here another important family of molecules is critical. These are the cyclins and cyclin-dependent kinases (CDKs). The cyclins activate the CDKs by phosphorylation and these activated kinases are critically important in allowing the various stages of the cell cycle to progress smoothly. Control of the cell cycle also involves the products of TSGs such as retinoblastoma (see later).

TUMOUR-SUPPRESSOR GENES/RECESSIVE ONCOGENES

Although 'activated' or 'dominant' oncogenes held centre stage 25 years ago, it is interesting that there was evidence for yet another type of gene long before that. The successful identification of genes, the proteins of which are physiological inhibitors of growth, stemmed from two main types of studies: (1) somatic cell hybrids and (2) genetic studies of inherited cancer syndromes. Furthermore, cytogenetic studies had

already shown that many tumours exhibit loss of DNA involving almost all chromosomal arms.

In somatic cell hybrids, the normal cell is fused with a transformed ('malignant') cell to form a hybrid cell. The main action of fusion was to produce a non-transformed state. This phenomenon of tumour suppression suggested that the normal cell must replace a defective function in the cancer cell. The first tumour suppressor to be identified was by the study of a rare familial disease, retinoblastoma (Figs 13.7 and 13.8).

Retinoblastoma, a tumour arising from the embryonal neural retina, has a worldwide incidence of 1:20 000. The tumour is of interest because it has both a sporadic and a familial form, with approximately 25–30% of the tumours being heritable. These cases

Figure 13.7 Section through the eye showing retinal detachment due to underlying tumour.

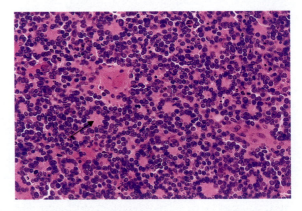

Figure 13.8 Photomicrograph showing characteristic rosettes of retinoblastoma.

tend to present earlier and develop bilateral disease. In contrast, the sporadic cases have unilateral tumours. Advancement in surgery and radiotherapy led to improved survival, and it became clear that 50% of the offspring of patients with bilateral tumours were themselves at risk of the disease. Evaluation of family pedigrees clearly shows the inherited form as mendelian dominant.

In 1971, Knudson proposed his 'two-hit' hypothesis (Fig. 13.9). He pointed out that, if cancer arises due to a series of somatic events, then it is possible that sometimes one of these changes is inherited in the germline and hence is present in every cell of the body. All the cells are therefore already one step along the pathway of carcinogenesis and this forms the basis for the dominantly inherited (mendelian) cancer susceptibility. In addition, a further mutation occurring somatically during life in the same gene would knock out the function of the gene. Hence, although the susceptibility is dominant, the action at the cell level is recessive. This therefore predicted for a class of genes that had to be inactivated or have 'loss of function' in order to provide the malignant phenotype. Knudson examined data relating to the age of first appearance of the tumour in both familial and sporadic cases, and showed that it followed the expected statistical model based on this hypothesis.

Cytogenetic studies had revealed that a few of the familial tumours showed germline deletions of chromosome 13 and careful karyotyping revealed deletions of 13q14. The retinoblastoma gene (*RB1*) was cloned in 1986. It was the first tumour-suppressor gene to be isolated. With the cloning of the gene, it could be confirmed that familial retinoblastoma is indeed due to an inactivating mutation. Further evidence for oncosuppression comes from fusion experiments. Insertion of the 4.7-kb complementary DNA sequence (cDNA) into retinoblastoma and osteosarcoma cell lines led to reversion of the tumourigenic phenotype and introduction of these cells into nude mice failed to produce tumours. Interestingly, mutations of the retinoblastoma gene and expression of the mutant protein product have also been seen in almost every tissue, despite the restricted oncogenic effects. Mutations are also found in other types of tumours such as breast carcinoma, although breast cancer does not form part of any syndrome in association with retinoblastoma.

THE MODE OF ACTION OF TUMOUR-SUPPRESSOR GENES

It is of interest that, although the mode of action of genes that are inhibitory for growth might follow the same pathways as those that promote growth, there is much less information on this topic than for dominant oncogenes.

Perhaps the best examples of molecules that act at the cell surface include the cadherins, which are cell–cell adhesion molecules. E-cadherin (an epithelial cell–cell adhesion molecule) is abnormal in a special type of breast cancer – lobular carcinoma. Mutations in the E-cadherin gene are also responsible for predisposition to inherited gastric cancer of diffuse type in a percentage of patients.

Another good example is the binding of TGF-β to its receptor, which leads to the transcription of genes that inhibit growth. The TGF-β signalling pathway is abnormal in colonic and pancreatic cancers.

A good example of a protein that regulates cell signalling is the adenomatous polyposis coli (APC) protein. The gene, which when mutated predisposes to familial adenomatous polyposis (FAP), is an important player in the development of colonic carcinoma (Fig. 13.10). The APC protein plays an important role in the signalling pathway that involves a molecule called β-catenin. This protein acts in the nucleus to increase the activity of growth-promoting genes. APC causes degradation of β-catenin, hence removing the proliferative signal. It is easy to see then that mutations in the *APC* gene will lead to increased levels of β-catenin and hence increased proliferating signal.

Not surprisingly, most information relating to the function of TSGs comes from the study of *RB* and *p53*. These molecules have an effect on nuclear transcription and regulation of the cell cycle. The regulation of the cell cycle is complex and illustrated in Fig. 13.11.

Briefly, when the cells are in the resting or quiescent stage, the retinoblastoma protein is hypophosphorylated. In this state, RB is able to bind to a transcription factor E2F and hence prevent activation of the genes that play a role in pushing the cell through division. Various growth factors including the EGF family (discussed above) activate the proteins of the cyclin family. The cyclins D/CDK-4, -6 and cyclin E/CDK2 complexes phosphorylate the RB protein, leading to release of the E2F molecule, which is then free to bind to DNA

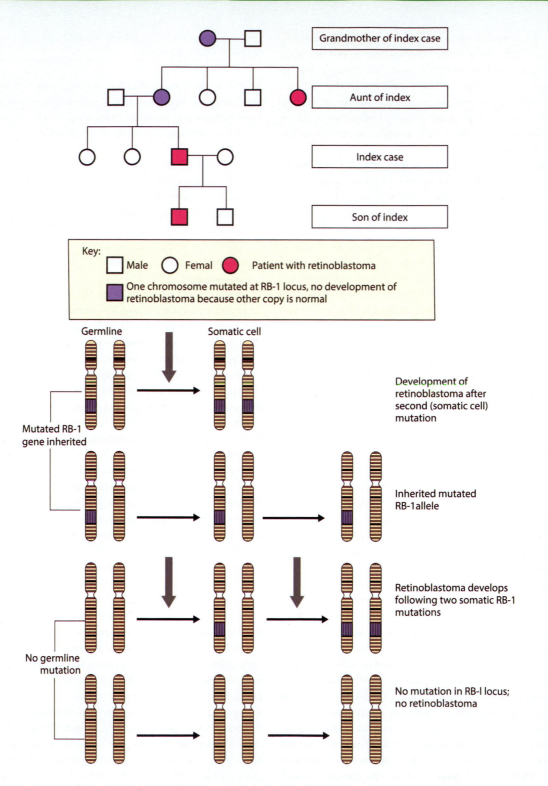

Figure 13.9 Two-hit model of retinoblastoma. Black arrow indicates inactivating mutation.

Figure 13.10 Large bowel with numerous polyps in patient with familial adenomatous polyposis.

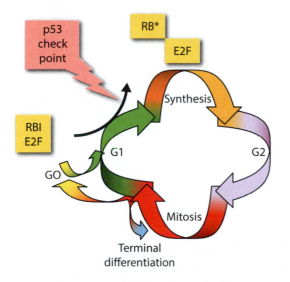

Figure 13.11 The cell cycle is largely governed by the actions of cyclins and cyclin-dependent kinases (CDK), some of which are manufactured during the cycle, others being recycled. Phosphorylation of the retinoblastoma protein (product of the *RB-1* gene) enables the transition from G1 to the synthesis phase by releasing E2F protein, which activates DNA synthesis genes. The p53 protein operates a quality control system at this checkpoint: if DNA damage or mutation is identified the defect is either repaired by DNA-repair enzymes or the cell is consigned to apoptosis. G2 is the pre-mitotic phase, during which preparation is made for mitosis. The resultant cells either re-enter the cycle, terminally differentiate and leave the cell cycle, or enter a resting phase, G0.

and activate transcription of genes involved in cell division. This process is balanced by signals that inhibit the process, with p16 playing an important role.

Unlike RB, p53 protein does not play a role in maintaining a check on the normal cell cycle. However, damage caused to the DNA by irradiation or chemicals brings *p53* into play. Levels of p53 protein rise after DNA damage, leading to two main effects: cell cycle arrest by transcription of an inhibitor of CDK, p21, or if the damage is too severe, cell death via activation of apoptosis. The cell cycle arrest is necessary and important because it allows the cell time to repair the damage

before completion of cell division. It is not too difficult to see that, if cells that undergo cell division to produce daughter cells do so without repairing the defect, then, in essence, the damage is consolidated (Fig. 13.12). This is one of the features of tumour cells. No surprise then that *p53* has been called 'the guardian of the genome' and is one of the most mutated genes in cancer.

Many TSGs have now been identified and, although the exact mechanism of action is not clear in every case, the biology is slowly being unravelled. Some examples of TSGs that have a familial predisposition are shown in Table 13.2.

Chapter 13: Molecular genetics of cancer

Normal epithelial cell – cell junctions and normal regulation of proliferation

Mutation in one cell confers freedom from normal growth control mechanisms.
It proliferates without restraint

During proliferation, new mutations occur, some of which confer an increased growth advantage and an ability to spread through basement membrane

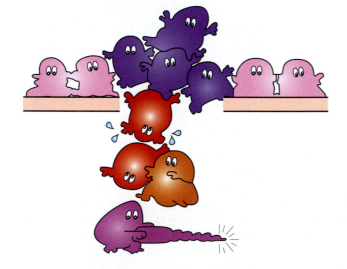

Further mutations occur. Thus although the tumour has originated from a single mutated cell (i.e. it is clonal), it is clear that the final population is mixed, each new population evolving to the advantage of the tumour. This is clonal evolution

Figure 13.12 Clonal evolution.

Table 13.2 Tumours and familial predisposition

Syndrome	Principal tumour types	Genetic locus/gene
Familial polyposis coli	Colorectal carcinoma	5q21 (*APC*)
MEN I	Pituitary, parathyroid thyroid, adrenal cortex, islet cell tumour	11q13
MEN IIa	Medullary carcinoma of thyroid, phaeochromocytoma, parathyroid tumours	10q11 (*RET*)
MEN IIb	Medullary carcinoma of thyroid, phaeochromocytoma, mucosal neuromas	10q11 (*RET*)
Von Hipple–Lindau	Haemangioblastoma of cerebellum and retina, renal cell carcinoma, phaeochromocytoma	3p25 (*VHL*)
Tuberous sclerosis 16p13.3 (TSC2)	Angiomyolipoma	9q34 (*TSC1*)
Familial retinoblastoma	Bilateral retinoblastoma, osteosarcoma	13q14 (*RB*)
Neurofibromatosis type I	Neurofibroma, neurofibrosarcoma, glioma, meningioma, phaeochromocytoma	17q11 (*NF1*)
Neurofibromatosis type II	Bilateral acoustic schwannoma, multiple meningioma	22q (*NF2*)
Li–Fraumeni	Breast cancer, sarcoma	17p13 (*TP53*)
Breast–ovarian	Breast cancer, ovarian cancer	17q21 (*BRCA1*)
Breast	Breast cancer, ovarian cancer, prostate cancer	13q12–13 (*BRCA2*)
Hereditary non-polyposis colorectal cancer (HNPCC)	Colorectal carcinoma	2p (*MSH2*)
		3p (*MLH1*)
		7p (*PMS2*)
Familial gastrointestinal hyperpigmentation stromal tumours (GISTs)	Multiple GISTs, hyperplasia of intestinal cells of Cajal, mastocytosis	4q12 (*c-Kit*)
Wilms' tumour (WAGR syndrome)	Wilms' tumour, aniridia, genitourinary abnormalities, learning disability	11p13.3 (*WT1*)
Familial cylindromatosis	Cylindromas (skin)	16q (*CYLD*)
Peutz–Jeghers	Intestinal polyps, ovarian and pancreatic tumours	19p13.3 (*STK11/LKB1*)
Cowden's	Hamartoma polyps of gastrointestinal tract, glioma, endometrial cancer, breast cancer	10q22–23 (*PTEN*)

MEN, multiple endocrine neoplasia.

DNA REPAIR, APOPTOSIS, TELOMERES, TELOMERASES AND CANCER

DNA REPAIR

If the human genome is not to fall apart as a result of exogenous (environmental chemicals, radiation) and endogenous (DNA replication) damage, it has to have efficient DNA repair and ability to execute programmed cell death (see below). There are many different types of DNA-repair mechanisms within the cell and undoubtedly the knowledge will become more complicated and refined with time. As some of the important genes involved in cancer are already being assigned to have a role in particular types of repair mechanisms, they are listed here but it is not the intention that you should be able to regurgitate the different mechanisms in detail.

The mechanisms include homologous recombinational repair (HRR), non-homologous end-joining (NHEJ), nucleotide excisional repair (NER), base excisional repair (BER) and mismatch repair (MMR). HRR, as its name suggests, relies on the homologous chromosome to provide the template for repairing the defective strand; it is therefore an accurate and error-free way of dealing with damage. In contrast, NHEJ is prone to error as no complementary strand is available to provide a template for accurate replication. The breast cancer predisposition genes, *BRCA1/2*, are involved in DNA repair through HRR. In their absence after mutations in the genes, DNA repair is still possible but it now occurs through the error-prone pathway of NHEJ, hence making the cell more susceptible to further genetic changes and further cancer formation.

The other major pathway that has been identified to play a role in tumour formation is MMR. Genomic instability plays an important role in the development of tumours in patients with hereditary non-polyposis colorectal cancer (HNPCC). Most HNPCC patients have mutations in one of the human counterparts of the bacterial mismatch repair genes *mutS* and *mutL*. Mutations in these genes leads to an inability to repair DNA mismatch (mistakes that happen during DNA replication) and hence contribute to neoplastic transformation by allowing mutations to be transmitted to daughter cells. Microsatellites are repeat units generally found within the non-coding part of the genome. They are highly polymorphic (the two alleles differ in size) and hence they provide a useful tool for the investigation of mutations within the genome. Patients with HNPCC show widespread alterations in these microsatellites and are therefore said to exhibit microsatellite instability (or a mutator phenotype). The implication of finding microsatellite instability is that the mismatch repair genes must be inactivated, otherwise the mismatch repair gene proteins would have corrected the mutations identified in the microsatellites. Hence analysis of microsatellite instability provides indirect evidence for mismatch repair gene abnormality, and patients with high levels of instability can have direct genetic testing to look for the mutations in the mismatch repair genes.

APOPTOSIS AND CANCER

After genetic damage, the cell may have an opportunity to repair the defect via the pathways described above. Occasionally, the damage is so severe that the cell is unable to do this and a set of signals is initiated that lead to programmed cell death. This is an important protective mechanism as a cell with severe genetic damage is no threat if it is dead! It is only when it manages to go through the cell division cycle with its damage and pass the alterations on to the daughter cells that problems are likely to occur.

Apart from showing increased proliferation, cancer cells also fail to undergo apoptosis and hence have an increased life span compared with normal cells. The inability of cancer cells to commit suicide is an important contributory factor to tumour growth, both at the primary site and at the sites of metastatic spread.

Many genes involved in the control of apoptosis have now been identified. The first to be identified was *bcl-2*. This gene is a member of a large family of genes, some of which are pro-apoptotic (*bax, bad*), whereas others are anti-apoptotic (*bcl-2, bcl-xL*). The mechanisms controlling apoptosis are complex. In response to DNA damage, a whole series of cascades is set up, the balance of which determines whether the cell is sent into a death programme (Fig. 13.13). In the presence of an overall death outcome, the *bcl-2* family of proteins operates via activation of a series of proteolytic enzymes called 'caspases', which chop the DNA up into fragments.

Two other genes that are also prominent in the apoptotic pathway are *p53* (discussed above) and the c-*myc* oncogene.

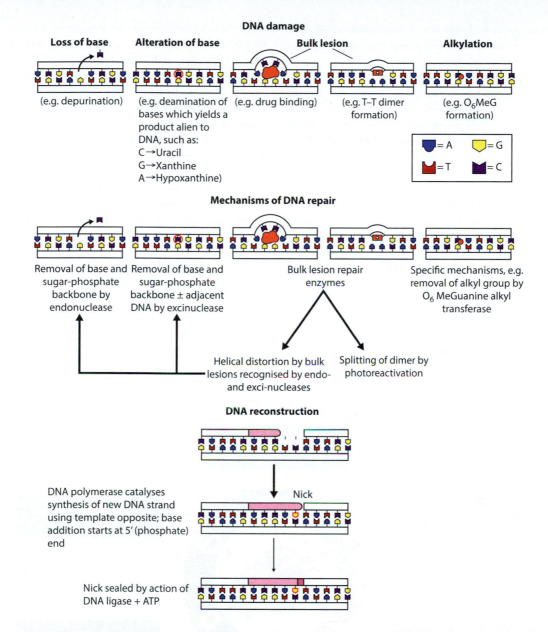

Figure 13.13 DNA damage can take several forms, some shown here. Several enzymes patrol the cell's DNA to detect DNA distortion or mismatch and effect DNA repair. Once a cell has divided, any mutation will be perpetuated if the daughter cell is capable of further division. This may lead to tumour development.

TELOMERES, TELOMERASE AND CANCER

Telomeres are a series of short tandem nucleotide repeat sequences of six base-pairs (TTAGGG), which are found at the end of chromosomes. They cap the chromosomes and protect the ends from degradation during the DNA replication that occurs as part of cell growth and differentiation. They also work as a molecular counting mechanism (an intracellular clock), because, with each round of DNA replication, the telomeres shorten until the cell reaches a 'crisis' (also known as the Hayflick limit). At this point the chromosomal ends become dysfunctional, end-to-end joining occurs and the cell is sent into apoptosis via a p53-mediated pathway.

An enzyme called telomerase plays a role in maintenance of the telomeres by preventing the erosion of telomeres that occurs during replication. This enzyme is a ribonucleoprotein reverse transcriptase composed of two components: hTERC (the RNA subunit that acts as the template for addition of new telomeric repeats) and hTERT (the catalytic protein component). Its expression is tightly regulated during development and, after this period, in humans, it is expressed in quantities sufficient to prevent telomere erosion in male germ cells, lymphocytes and stem cell populations, including basal keratinocytes. The state when the cell cannot divide anymore is called 'cellular senescence'.

It is becoming clear that, in order to create a cancer cell, two barriers must be overcome: cellular senescence and crisis, both of which limit replicative potential. This hypothesis seems to be valid for both benign and malignant tumours, at least in *in vitro* experiments. It is not difficult to postulate that cancer cells have

learnt how to circumvent the erosion of telomeres. This is indeed the case and, unlike normal, non-germ cells, most cancer cells have detectable and increased telomerase activity or develop an alternative mechanism for telomere lengthening (ALT), which is thought to involve recombination mechanisms and results in abnormal telomeres.

Several lines of evidence have supported the major role of telomerase in oncogenesis. When normal cells reach the 'crisis' (complete erosion of telomeres), lack of telomerase leads to major chromosomal abnormalities (aneuploidy and chromosomal fusions). Other connections between telomerase and other oncogenes and TSGs are currently under intense investigation.

THE HALLMARKS OF CANCER

In 2000, Hanahan and Weinberg published a seminal paper in which they described what they believe are the six fundamental hallmarks of cancer. This was subsequently updated in 2011. Figure 13.14 summarises the current understanding.

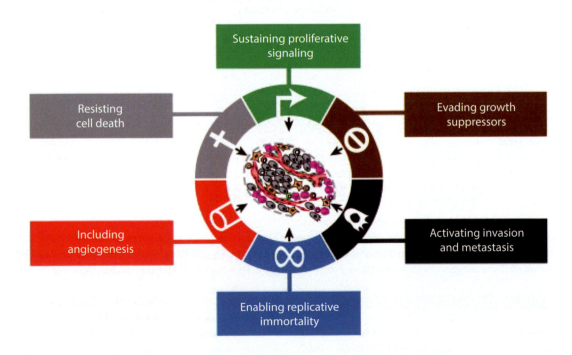

Figure 13.14 Hallmarks of cancer. (Modified from Hanahan D, Weinberg RA. Hallmarks of cancer: The next generation. *Cell* 2011;**144**:646–74.)

WHICH CELL IN THE TISSUE GIVES RISE TO CANCER?

The traditional view has been that a cell, capable of replication, undergoes a mutation in a gene that allows it to have a growth advantage. This mutated cell produces a small clonal expansion (see Fig. 13.12). Subsequent mutations over a long time period lead to further clonal expansions, which then become the reservoir for further mutations and expansions. This idea follows the traditional Darwinian model of evolution.

There has been an increasing interest in stem cell biology over the last two decades and a role for stem cells in the development of cancers has been proposed. Normal stems cells must, by definition, be able to undergo self-renewal (produce an identical copy of itself) and be able to produce daughter/progenitor cells. Progenitor cells retain some characteristics of stem cells initially, but lose these features as they differentiate into the different tissue types with subsequent cell divisions.

One of the problems with the traditional model is the knowledge that cancers take a long time to initiate and progress, and this contrasts with the relatively short life span of mature, differentiated cells. It has been proposed that stem cells, which are long lived and hence predisposed to the slow accumulation of genetic changes, may be the tumour-initiating cells. Although a huge amount of effort is currently invested in studying normal and cancer-associated stem cells, it is unclear at present whether cancers do indeed arise from normal stem cells or whether some cancer cells develop the ability to have stem cell-like characteristics as a result of transformation. There is also the idea that 'cancer stem cells' are more resistant to chemotherapy than more differentiated cancer cells, suggesting that we may also need to target this small population of cancer stem cells as part of combination therapy. This is a field in transit and worth keeping an eye on because it will change the way we think about primary and metastatic disease and how we manage it in the future.

THE MULTISTEP MODEL FOR CARCINOGENESIS

In the course of examining malignant tissues, histopathologists frequently encounter lesions that show transitions with appearances intermediate between normal morphology and frank malignancy. Occasionally, such lesions are closely associated with the invasive cancer (Fig. 13.15). This led to the suggestion that many of these lesions may be precursors of the invasive carcinoma. With the identification of dominant oncogenes and TSGs, genes involved in DNA repair and apoptosis, it became possible to investigate tumours and putative precursor lesions using molecular techniques. The study of colorectal carcinoma with its well-defined pre-invasive lesion, the adenoma, has paved the way for this type of investigation (Fig. 13.16). The results demonstrate that both activating and inactivating events are involved and it is the coordinated involvement of many of these types of alterations that are important in colon tumour formation. Furthermore, it is not just the timing of the events but also the sequential accumulation of genetic damage that is important in tumour formation. This idea that it is not one event but a sequence of genetic alterations that produces tumours is referred to as the 'multistep theory of neoplasia'.

The idea of the multistep model has been extended to almost every tumour type and there is good evidence that cancers in general develop in this way. The idea behind this type of model has also been fundamental in recommending screening programmes for cancer, the argument being that, if you pick up a 'tumour' when it is either very small or still in its early stages of development, you can cure it. Although the idea is a good one, it has also opened up a whole lot of problems. The screening programmes are identifying lesions at a very early stage in development, and the classification and natural history of many such lesions are at present unknown. Once a lesion has been removed, the patient inevitably wants

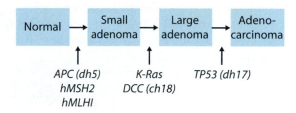

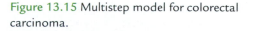

Figure 13.15 Multistep model for colorectal carcinoma.

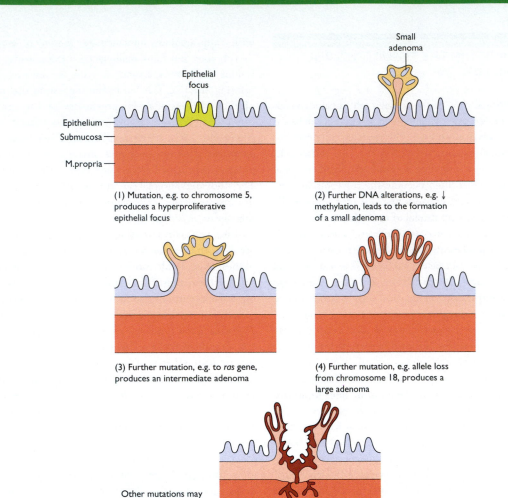

Epithelial focus

Epithelium
Submucosa
M.propria

(1) Mutation, e.g. to chromosome 5, produces a hyperproliferative epithelial focus

Small adenoma

(2) Further DNA alterations, e.g. ↓ methylation, leads to the formation of a small adenoma

(3) Further mutation, e.g. to *ras* gene, produces an intermediate adenoma

(4) Further mutation, e.g. allele loss from chromosome 18, produces a large adenoma

Other mutations may confer the ability to metastasise

(5) Another mutation, e.g. allele loss from chromosome 17, leads to the development of invasive cancer

Figure 13.16 The colonic adenoma–carcinoma sequence.

to know what it is and what its implications are. As this is not always possible to predict with accuracy, it can cause much distress, for the clinician as well as for the patient.

There has been a hope that new molecular techniques will help in understanding the biology of cancer and in particular the early lesions. So what are these techniques and how are they helping with patient management?

MOLECULAR METHODOLOGY

LOSS OF HETEROZYGOSITY

In 1971, Knudson proposed his 'two-hit' hypothesis for the presence of TSGs. The recognition that the second mutation leading to inactivation of the gene is usually in the form of a large deletion led to the development of the technique of loss of heterozygosity (LOH) by Cavenee and colleagues. The technique relies on the observation that markers (microsatellites) that are heterozygous and near the TSG would become homo- or hemizygous in the tumour compared with normal tissue. This was confirmed in the case of retinoblastoma, where markers close to the gene were seen to exhibit LOH in both sporadic and familial cancers.

Since the introduction of the technique by Cavenee, there have been numerous studies looking at LOH in many cancer types. Although patterns of LOH have been reported in some series as being of prognostic significance, this information has not yet translated into routine practice. LOH has also been used extensively to investigate precancerous lesions in the hope that patterns of LOH would help to stratify lesions into 'benign' and 'malignant'. It has been apparent that many lesions traditionally thought of as 'benign' are monoclonal and that there is considerable overlap in LOH between these lesions and those accepted as 'malignant', i.e. ductal carcinoma *in situ* (DCIS). Hence, at present, there are no robust profiles using LOH that help to definitively distinguish such lesions in clinical practice.

COMPARATIVE GENOMIC HYBRIDISATION

Comparative genomic hybridisation (CGH) is a fluorescent *in situ* hybridisation technique capable of determining changes in DNA copy number between differentially labelled test (e.g. tumour) and reference (e.g. normal) samples. Unlike LOH analysis, which gives information at a very specific locus on a chromosomal arm, CGH analysis provides information from the entire genome. This has the advantage that a single experiment can give an idea about the many changes on all the different chromosomes, data that would require hundreds of experiments using LOH analysis. Traditionally, CGH has been done using competitive hybridisation to normal metaphase chromosomes, hence resolution is poor, and is capable of detecting only high-level amplifications of around 2 Mb, and deletions of the order of 10 Mb (these represent large areas of the chromosomes with many genes). Hence, there is a compromise between global information and

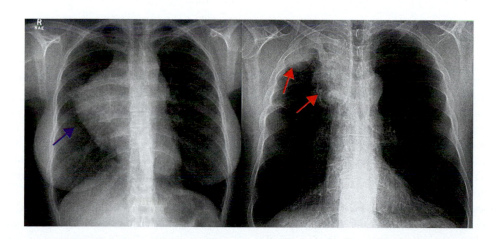

Figure 13.17 These images demonstrate carcinoma of the lung, a neuroectodermally derived tumour, common in smokers. It is renowned for its ability to metastasise widely before producing clinical symptoms and is therefore almost always fatal. New molecular techniques may assist in understanding the mechanisms by which such tumours develop, in creating early tumour detection systems and in engineering tumour-specific treatments. Left: chest radiograph, right hilar and anterior mediastinal mass (blue arrow); small-cell lung cancer. Right: chest radiograph, right apical and paramediastinal mass (red arrows) invading the mediastinum; non-small-cell lung cancer.

Chapter 13: Molecular genetics of cancer

identifying a very specific location of the change. A large number of investigators have used CGH analysis to understand the changes in DNA copy number in invasive cancer as well as precancerous lesions. Data arising from these experiments, showing precancerous lesions can have the same pattern of alterations as an invasive cancer from the same patient, strengthened the concept of the multistep model of tumour development. This method was later adapted to an array-based format (aCGH), where a large number of genetic loci (tens of thousands to several million) are spotted on to a glass slide and hybridisation is done to these loci rather than to a metaphase chromosome. This has improved the resolution of the technique enormously and this array-based technique is now used routinely in cytogenetics laboratories for the diagnosis of developmental disorders. From a cancer perspective, aCGH provides a detailed view of the DNA copy number changes across the whole genome of a tumour, including precise mapping of genes that are amplified or deleted. The integration of this type of genomic data with gene expression data (see below) is a way of identifying important cancer genes involved in driving tumorigenesis. A good example is *HER2*, a gene that is amplified and overexpressed in breast cancer and that has proved to be a very good therapeutic target.

EXPRESSION ARRAYS

As well as cataloguing the amplifications, deletions and complex rearrangements at the genomic (DNA) level, analysis of changes in the profiles of gene expression (mRNA) may give clues to the underlying molecular events in tumorigenesis and the biological 'portrait' of the tumour. The method consists of thousands of sequences of cDNA (derived from reverse transcription of the RNA) robotically arrayed on to a glass slide. The samples to be examined are labelled with different colour fluorochromes, mixed and co-hybridised to the arrays in a competitive manner, and the resulting fluorescence values reveal the relative levels of each RNA transcript in the test sample compared with the reference sample (e.g. tumour versus normal or between two different normal cell types). Mathematical algorithms are used by bioinformaticians to probe differences in expression patterns between sample sets. Expression profiles generated by cDNA arrays can reveal similarities and differences that are not necessarily evident from traditional approaches, such as morphological or immunohistochemical analysis.

The power of this technique has been elegantly demonstrated in a number of studies published since 2000. The first demonstrated that, within the morphologically homogeneous category of large B-cell lymphoma, two subtypes exist with differing patterns of gene expression. Subsequent papers have looked at other tumour types such as breast, lung and ovarian cancers, and also identified subclasses of tumour not easily identified using routine microscopy, but that exhibit different clinical behaviours, such as response to treatment or patient outcome. These analyses have been refined to show that the expression pattern of a small collection of genes (a gene signature) can be used in clinical practice to help guide patient management.

Despite the initial excitement, expression profiling has not replaced traditional histopathology yet! When the method came on the scene, the requirement was for fresh tissue but recent developments are now enabling formalin-fixed, paraffin-embedded (FFPE) tissues to be used. This will speed up its uptake because this is the normal way in which tissues are handled in pathology laboratories. There are now a number of commercially available tests that use FFPE samples, and they are used in varying frequencies throughout the world to provide additional prognostic data and stratify patients into more meaningful groups for therapy and clinical trials.

PROTEIN ARRAYS

Protein molecules, rather than DNA or RNA, carry out most cellular functions. The direct measurement of protein levels and activity within the cell may be the best determinant of overall cell function. Techniques are being developed to quantify the levels of all the proteins within a cell and compare protein levels between different cell types. Proteomic analysis, consisting of two-dimensional polyacrylamide gel electrophoresis and tandem mass spectrometry, has been used to map protein profiles in normal and tumour cells. As would be expected, the studies have highlighted differences in protein profiles between subsets of normal cells, and between normal cells and tumour cells. In a similar manner to the 'transcriptome', it is unlikely that the 'proteome' is stable over time because the cell requirements are constantly changing as it adapts to its environment. Any proteomic data are therefore likely to represent only a snapshot of the 'proteome'.

In addition, akin to work in gene expression profiling, there are now techniques capable of generating a protein expression profile, either by an array format with antibodies or by analysing protein fragments (peptides) by mass spectrometry. The use of these technologies will undoubtedly lead to the identification of new cancer-associated proteins and enable subgrouping of tumours based on their biological profile, which may aid in the stratification of patients for management purposes. These techniques also have the capability of measuring the presence of activated proteins (e.g. the phosphorylated version of the protein), which could become the gold standard in biomarker assessment – not only seeing whether a particular protein is present but also seeing whether it is functional may enhance the effectiveness of specific targeted therapies.

NEXT-GENERATION SEQUENCING

Sequencing technology is used to assess the order of the nucleotides (the bases – adenine, guanine, cytosine and thymine) within the DNA. In 1977, Sanger developed a simple method based on the principle of 'chain termination' and this method (referred to as Sanger sequencing) is still used today although usually now only for validation purposes. The drive to sequence the human genome (which was initially published in 2000) precipitated a huge revolution in sequencing technology that continues today. Collectively termed 'next-generation sequencing' (NGS) or more accurately massively parallel sequencing (MPS), this technology has recently replaced Sanger sequencing in the research setting and is beginning to impact the clinical setting too.

NGS/MPS allows a fast way to sequence a genome (or a targeted region of the genome) and, in fact, sequence multiple genomes together by introducing 'bar codes' into the process which allows the data from multiple samples to be separated (using mathematical/bioinformatics tools) after the experiment has been completed.

The technique is used primarily for sequencing DNA (to identify germline variations in familial diseases, or somatic mutations in cancer), as well as RNA (as a way of replacing gene expression arrays), but it has also been expanded to sequence miRNAs, to get DNA copy number data and to identify structural genome rearrangements (such as interchromosomal translocations) as well as methylation status. The ability to obtain such detailed profiles of a genome, and now at a relatively cheap price together with speed (days), means that NGS is ready for the clinical setting. One of the challenges lies in figuring out which mutations are 'drivers' (important in the oncogenic process and hence contributing to the tumour phenotype) and which are 'passengers' (simply there due to the instability of the genome and not directly participating in the tumour development or progression – in other words 'noise'). Over the next few years, we will increasingly see NGS data (cancer and non-cancerous diseases) being used to assess the disease process for prognostication and, more importantly, to identify mutations (drivers) and changes that may be amenable to existing and new drugs. The ability to stratify patients on the basis of an integrated genomic analysis in this way has also opened the doors for re-purposing some drugs from the past that appeared to be ineffective because they were trialled in all-comers rather than specific groups of tumours with a particular genetic change.

CONCLUSION

Pathological assessment of tissues has remained the lynchpin of diagnostic practice for over 100 years. It has become the core science of clinical medical practice, providing data for clinical management and a framework for future correlation of new markers and new therapies. With the current explosion of technology and data, it is important for pathologists and other clinical specialists to embrace and incorporate these changes into their training and practice. Molecular biologists will also benefit from a closer interaction with pathologists.

This brings us to the end of this section on the cellular events involved in producing the cancer cell. Of course, we have a long way to go before we have full understanding, but our knowledge is advancing at an exciting pace, and a whole new language of tumour terminology is emerging. For the scientist, the battle is the biology; for the clinicians and students trying to understand and apply the new knowledge, it is often the terminology!

The next question we need to address and one that will be in the forefront of the patient's mind is: How will a given tumour behave?

Chapter 13: Molecular genetics of cancer

The behaviour of a tumour can be considered under a number of headings covering how fast it will grow, whether it is likely to metastasise, which sites are affected, and what symptoms and complications the patient is likely to have.

TUMOUR GROWTH

It is often assumed that tumours grow faster than normal tissues because they expand to compress the surrounding structures. However, this does not mean that the cells are dividing more often, but that there is an imbalance between production and loss. The time taken for tumour cell division varies between 20 and 60 hours, with leukaemias having shorter cell cycles than solid tumours but, in general, tumour cells take *longer* than their normal counterparts. Cells can be in a resting phase or in growth phase, i.e. in one of the stages of mitosis. Some normal tissues have a high turnover of cells, such as the intestine, where around 16% of the cells will be in the growth fraction. In contrast, most tumours have only 2–8% of their cells actively dividing. This is important therapeutically because the cells in the growth phase are most readily damaged by chemotherapy and so tumours with a large growth fraction (e.g. leukaemias, lymphomas and lung anaplastic small-cell carcinoma) will respond better than tumours with few cells proliferating (e.g. colon and breast).

Can we predict how fast a tumour is growing? To some extent, yes. The number of mitotic figures present per unit area in a light microscopical section is a crude measure of how active the proliferation is within a tumour. Immunohistochemical markers such as Ki67 can also be used to assess the proliferation fraction within tissue sections. A tumour with a large number of cells in mitosis is likely to behave aggressively and this is why a mitotic count is one of the criteria for grading tumours (see page 317). However, the number of cells seen to be in mitosis is influenced not only by the growth fraction and the cell cycle time but also by whether they get 'stuck', i.e. the tumour cell can enter mitosis but, possibly because of an irregularity in chromosome number or the internal organisation of the mitotic spindle, may fail to complete the mitosis. Thus, on examination of a tissue section, the tumour appears to be highly proliferative but it is really 'stuck'. Tumour growth will also be influenced by factors such as the blood supply and, possibly, the host's immune response (see page 363).

It would also be wrong to assume that every cell in a tumour behaves similarly. The daughter cells of a dividing cell are identical genetically to the parent cell and are said to be a clone (clonal expansion). However, tumour cells are also prone to develop genetic instability, which results in some cells developing further abnormalities, hence resulting in the formation of multiple subclones. These may have certain survival advantages, e.g. they may have enhanced angiogenic, invasive or metastatic capabilities. This is referred to as tumour heterogeneity and it is important to consider when planning treatments because it means that some tumour cells may respond differently to particular chemotherapeutic agents. This is rather analogous to bacterial resistance to antibiotics. Just as a combination of antibiotics is most effective against an unknown organism, so a mixture of treatment modalities is often used against a tumour.

HOW DO TUMOURS SPREAD?

Well over 100 years ago, Stephen Paget (no, not the man who described Paget's disease, that was Sir James Paget), a surgeon by profession, collected postmortem records of 735 patients who had died of breast cancer and he found that most of the metastases were in the liver and brain. He concluded therefore that certain tumours were predisposed to metastasise to certain tissues. He wrote: 'When a plant goes to seed, its seeds are carried in all directions; but they can only live and grow if they fall on congenial soil.' Not surprisingly, it came to be known as the 'seed-and-soil' theory.

James Ewing, 40 years later, suggested that tumours went to particular organs not because of the seed-and-soil effect, but rather because of the routes of the blood supply to the primary organ. Using his hypothesis, organs directly in line away from the primary site would be targets for metastatic disease.

Although Ewing's idea is partially correct, it is increasingly recognised that Paget's seed-and-soil hypothesis is a better representation of what actually happens. Tumours of the colon do indeed go to the liver, which is next in line through the portal circulation, but then so do many other tumours much farther away, such as melanomas arising in the eye. We also know that organs such as the heart and skeletal muscle, despite being exposed to large volumes of blood, rarely develop metastases. In broad terms, tumour spread through lymphatics will produce metastases in the anatomically related lymph nodes whereas spread through the blood is influenced more by 'seed-and-soil' considerations, although anatomy is still of some importance.

The main routes of spread are via the following:

- Lymphatics
- Veins
- Transcoelomic cavities
- Cerebrospinal fluid (CSF)
- Arteries (Fig. 14.1).

Key facts

Possible routes of tumour spread

- Direct
- Lymphatic
- Venous
- Transcoelomic
- Cerebrospinal fluid
- Arterial

Lymphatic spread is common in carcinomas (tumours of epithelia) and the nodes that are involved first are the nodes that drain the tumour site (Fig. 14.2). Thus, knowledge about the lymphatic anatomy is useful for predicting where the tumour will spread and is the basis for many of the staging protocols (see page 319). However, lymph nodes near tumours can enlarge as part of an immune reaction, which results particularly in expansion of the macrophage compartment (sinus histiocytosis). This means that the doctor must try to distinguish between soft, mobile nodes, which are likely to be reactive, and the hard, fixed nodes that contain metastatic tumour.

Venous spread will take tumours of the gastrointestinal tract to the liver and tumours from a variety of sites to the lungs (Fig. 14.3). It is also the favoured route of spread for sarcomas (tumours of connective tissue). Some tumours may even grow along a vein, causing its obstruction, e.g. renal cell carcinoma in the renal vein. Arteries are not often penetrated by tumours but, in the later stages of metastatic spread, tumour nodules can start to develop almost anywhere, and it is likely that this happens after pulmonary metastases enter the pulmonary vein and are then distributed through the systemic arterial system. It is easy to understand how tumours that reach the pleural or peritoneal cavities can drop into the fluid and be disseminated throughout that coelomic cavity. Similarly, the CSF provides an easy route of spread for cerebral tumours, which do not generally metastasise outside the central nervous system.

One cubic centimetre of tumour can shed millions of cells into the circulation each day, so why are metastases not inevitable? Let us consider the steps required to produce a metastasis.

Chapter 14: The behaviour of tumours

Direct: e.g. mediastinal tumour surrounding and compressing the superior vena cava. This causes venous engorgement of the head, neck and arms, to produce headache and proptosis (bulging eyes due to cavernous sinus distension)

Via lymphatics: e.g. carcinoma of breast first spreads to the local axillary lymph nodes

Carcinomas generally first spread via lymphatics

Transcoelomic: e.g. gastric carcinoma cells seeding through the peritoneal cavity. Krukenberg gave his name to metastatic gastric carcinoma involving the ovaries

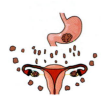

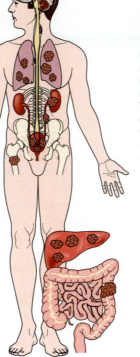

CNS spread: primary central nervous system tumours will metastasise to brain or spinal cord, but appear to be confined by the blood–brain barrier and the dural membranes

Field change e.g. transitional cell carcinoma. Urothelial tumours may synchronously arise at several sites, e.g. renal pelvis, ureter and bladder. This is not tumour spread, but a 'crop' of separate primary tumours, thought to arise due to a 'field effect', in which several sites are exposed to the same urinary carcinogens

Via the bloodstream: e.g. osteosarcomas metastasise to the lungs, or gastrointestinal carcinomas spread to the liver via the portal vein. Sarcomas generally spread via the bloodstream

Figure 14.1 Routes of tumour spread.

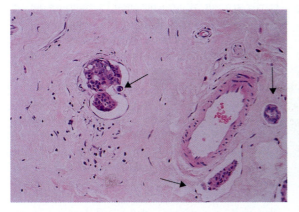

Figure 14.2 Photomicrograph of lymphatic invasion by breast cancer.

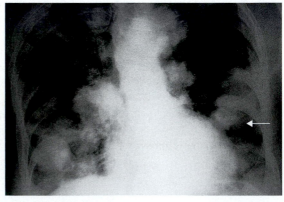

Figure 14.3 Chest radiograph showing multiple 'cannon ball' metastases.

THE BIOLOGY OF METASTATIC DISEASE

In order for metastatic disease to occur, the tumour has to first grow at the primary site and infiltrate the surrounding connective tissue, which may necessitate breaking through a basement membrane and the connective tissues nearby (Fig. 14.4). It may also have to overcome inhibitor substances to the enzymes that it produces to break down these connective tissue proteins. It is only then that it can reach the lymphatic and blood vascular channels, which are an important route for dissemination. It has to find a way of breaking the basement membrane around some vessels, attaching to

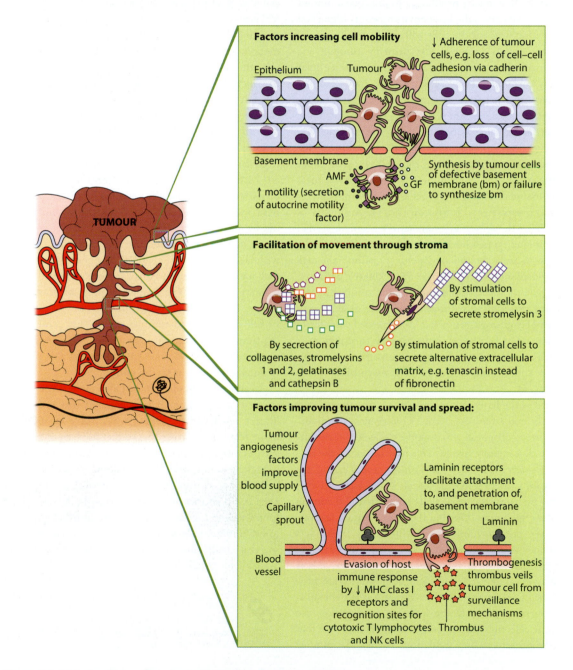

Figure 14.4 Mechanisms of tumour cell invasion and metastasis.

the endothelium and subsequently entering the channels. The vessel wall is traversed and the tumour cells must detach to float in the blood or lymph, and hope to evade any immune cells that might destroy them. Next, they must lodge in the capillaries at their destination, attach to the endothelium again, and penetrate the vessel wall to enter the perivascular connective tissue where they finally proliferate to produce a tumour deposit. Even here, the local environment is important in dictating whether the cells will grow.

What stands between the primary tumour and the vessel? First, there are a variety of extracellular matrix components to break through, for which the tumour may produce a number of enzymes. Loose connective tissue is not much of a barrier but dense fibrous areas, such as tendons and joint capsules and cartilage, can resist tumour spread. In addition, the tumour cells might acquire the ability to be more motile within this environment by developing a mesenchymal phenotype; this process is referred to as epithelial–mesenchymal transition (EMT). A number of stimuli are thought to be important in acquiring EMT: these include transforming growth factor β (TGF-β), loss of E-cadherin and oncogenic signalling through the phosphoinositide 3-kinase (PI3 kinase) pathway. The growth and dissemination of the tumour at this site are also limited without the process of angiogenesis and lymphangiogenesis. This increased vascularity is achieved by formation of new vessels (neoangiogenesis), recruitment of existing vasculature or mimicry (where tumour cells form channels resembling vascular spaces).

Next is the basement membrane so that the tumour has to be able to secrete a type IV collagenase. Tumours often have collagenases to dissolve collagen but are less able to digest elastic tissue. This may be one of the reasons why arterial walls, which contain a lot of elastic, are less readily penetrated than venous walls. Alternatively, it may be because arterial walls are thicker and contain protease inhibitors. Once in the vessel lumen, tumour cells are prey to immune surveillance by the body's lymphocytes and monocytes. Some experimental data suggest that most of these cells die; however, this is controversial and it is not clear at present what proportion of cells survive and extravasate at distant sites. Of course, apart from the immune mechanisms, the tumour cells must also resist the huge sheer forces that will be present in the circulation. Finally, the tumour

must attach to the endothelium at its destination, which may involve specific adhesion molecules (addressins) that 'home' the metastatic tumour to a particular site, analogous to the 'homing' of lymphocytes (see page 181). It is not entirely clear what dictates the site of metastatic disease. Some genes appear to predict metastases in general whereas others appear to be associated with site-specific disease.

There is considerable interest currently in understanding the factors that allow cells to colonise at the metastatic sites. Having extravasated from the vessels, the tumour cells encounter a new environment to which they have to adapt. This adaptation may be a result of beneficial mutations or simply involve a change in the expression of genes that confer advantage to growth and survival. Having survived, the growth of the cells will lead to a further clonal expansion and development of further heterogeneity, which will have an impact on the management of patients. Traditionally, we have treated patients with metastatic disease based on the phenotype of the primary cancer, but it is clear that this may not work in all patients because the metastatic disease will have evolved and changed its phenotype in ways that makes it resistant to treatments directed at the primary cancer. The understanding of metastatic disease is opening up possibilities to use new targets to kill disseminated tumour cells. This is an exciting time in metastatic cancer research; nevertheless, much remains to be done to understand this process in detail and bring it to the clinic.

An area of particular interest at present is the 'warm autopsy' programme. It is recognised that making an impact on metastatic disease will involve having fresh tissue from metastatic sites, so that modern techniques such as next-generation sequencing can be carried out to understand the underlying genomic evolution. Some patients are consenting to the autopsy immediately after death so that such samples can be harvested for future research. These samples are also being propagated in immunodeficient mice to develop xenografts, so that sufficient quantities of the tumour are available for studies.

Read more about autopsy and death certification in Pathology in Clinical Practice Cases 48 and 49

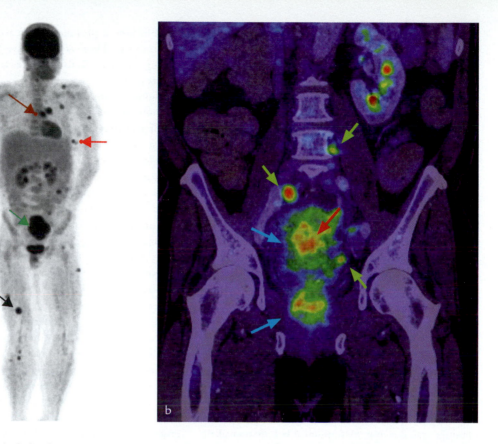

Figure 14.5 (a) Whole-body positron emission tomography (PET) scan in a 20-year-old patient with metastatic melanoma. (PET is a nuclear medicine imaging technique.) The radioactive compound used is [^{18}F]fluorodeoxyglucose (^{18}F-FDG), which is a glucose analogue taken up by metabolically active tissue. The PET scan detects areas of very high metabolic activity such as brain, muscle and brown fat, but also most cancers. PET is being increasingly used for staging malignancies and assessing treatment responses. This scan shows multiple metastatic deposits including subcutaneous deposits in the left breast (red arrow), a deposit in the femur (black arrow), a pelvic uterine metastasis (green arrow), and multiple hilar, mediastinal and mesenteric lymph node deposits (brown arrow). (b) Fused coronal PET–CT in the same patient showing FDG uptake in the primary (red arrow), diffuse peritoneal metastatic disease (blue arrows) and nodal metastatic disease (yellow arrows).

Most of our discussions about tumour cell biology have concentrated on how genetic changes enhance cell proliferation. However, we should now look at how a proliferating tumour cell differs from a proliferating normal cell. Early experiments involving *in vitro* cell cultures demonstrated that normal cells would grow to form monolayers and then stop. This was referred to as contact inhibition. If some cells from this culture were transferred to a new culture vessel (passaged), they would begin to grow again in the same way; however, normal cells would survive only about 30–50 serial passages. Cultures of proliferating tumour cells differed in that they lost contact inhibition and so could grow as disorganised multilayers and they were also immortal, i.e. although each individual cell did not last forever, the clone of cells could be passaged indefinitely. Now it is known that tumour cells may show decreased expression of E-cadherin, which normally acts as an adhesion molecule between epithelial cells, and that their increased motility may be influenced by an autocrine motility factor which some transformed culture cells release. Tumour cells can also influence the production of stroma so that tenascin may predominate, which does not bind readily to tumour cells. This sort of information suggests that it is not only changes in the tumour cells that produce local invasion and metastasis, but also the interactions of tumour cells, normal cells and stroma are important.

THE ROLE OF THE IMMUNE SYSTEM

We are all aware of the role played by the immune system in defending us against infections, so it is not surprising that questions have been raised as to whether it has any role in protection against cancer. It was Paul Ehrlich, in 1909, who postulated that, without the immune system constantly removing the 'aberrant germs', human beings would inevitably die of cancer. Many attempts were made to establish the role of the immune system in cancer and initial experiments, which transplanted tumours from one animal to another, appeared to support the concept. It was later realised that the destruction of these transplanted tumours was due not to immunity but to transplant rejection. Now, inbred mice can be used experimentally, thus avoiding the factor of transplant rejection. In certain tumours it has been shown that, if the tumour is removed from a mouse and the animal re-challenged with the tumour, the tumour is rejected. This supports the idea that immunity is involved in tumour rejection, but life is not quite so simple as we shall see.

You will recall that the cells of the immune system have to be able to distinguish between 'self' and 'non-self', by identifying specific antigens on the cell surface. Malignant tumours are derived from 'self', so if the immune system is to defend against tumours, the malignant cells must acquire antigens that differentiate them from normal cells. These are termed tumour-specific antigens (TSAs) and the whole subject has been highly controversial. In humans, such tumour-associated antigens are beginning to be defined and include differentiation antigens (e.g. CD19 and CD29), growth factor receptors such as epidermal growth factor receptor (EGFR), and intracellular proteins such as the MAGE family of proteins, which are highly expressed in melanomas. These antigens have been identified by isolating tumour-reactive T cells from patients with malignancies.

Although tumours clearly do elicit immune responses, the degree of response does not appear to be sufficient to hold the progression of the malignancy in most cases. It is conceivable that, as tumours progress, clones of cells without the antigens proliferate and escape immune destruction. A whole new area of cancer vaccine is developing rapidly and initial trials appear encouraging. Immunisation with ganglioside GM2 in melanoma appears to increase survival; vaccines against cervical cancer and liver cancer are already on the market, but time will tell whether the effects of immunisation live up to the promise suggested by initial trials.

There is also evidence from animal experiments that surface antigens are altered in some tumours induced by viruses or chemicals. Virally induced tumours in animals can display a new surface antigen (T), which is believed to be a viral peptide associated with the major histocompatibility complex (MHC). This provokes a specific cytotoxic T-cell response, and all tumours induced by a particular virus display the same antigen, regardless of the cell of origin. The obvious potential application for this lies in immunising against tumours.

Chemically induced tumours in animals (e.g. by benzopyrene) may also display new surface antigens that induce a specific immune response, but these antigens are very varied, with primary tumours in the same animal exhibiting antigenic differences, so there is no cross-resistance through immunisation.

Of course, the immune response need not be antigen-specific. Besides B and T lymphocytes, the body has at its disposal natural killer (NK) cells and macrophages. NK cells have the capacity to destroy cells without prior sensitisation as well as the ability to participate in antibody-dependent cellular cytotoxicity (ADCC). Macrophages are also involved, either due to non-specific activation or in collaboration with T lymphocytes, and can participate via ADCC or by the release of cytotoxic factors, such as tumour necrosis factor, hydrogen peroxide and a cytolytic protease.

CHAPTER 15

THE CLINICAL EFFECTS OF TUMOURS

LOCAL EFFECTS

The local effects of a tumour depend on the site and type of tumour, and its growth pattern. Some complications, such as haemorrhage, are more common in malignant tumours because of their ability to invade underlying tissues and their vessels, but it must be remembered that even a microscopically benign tumour (e.g. a meningioma on the surface of the brain) can kill the patient due to its local effects.

Local effects can complicate both benign and malignant tumours, and include the following:

- Compression
- Obstruction
- Ulceration
- Haemorrhage
- Rupture
- Perforation
- Infarction.

COMPRESSION AND OBSTRUCTION

A patient with any intracranial tumour (e.g. meningioma, astrocytoma or oligodendroglioma) may present with headaches, nausea and vomiting, because the mass growing within the closed cavity of the cranium raises the intracranial pressure. If the tumour is not removed and the pressure continues to rise, the patient will die from pressure effects on the vital respiratory centres.

A more localised example of the effect of compression is when the pituitary gland enlarges in the small cup-shaped space of the sella turcica. Local pressure will cause erosion of the bony sella and compression of the optic chiasma which sits directly above. The patient will then present with visual disturbance, classically a bitemporal hemianopia (Fig. 15.1).

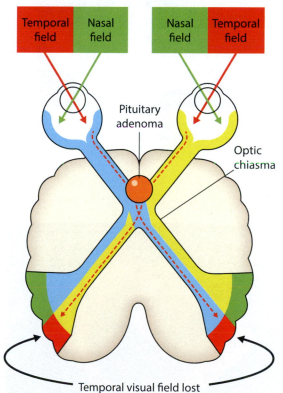

Figure 15.1 Benign tumours may have profound effects. Here, a small pituitary adenoma produced blindness in both temporal visual fields by compressing optic nerve fibres at the optic chiasma.

Compression and obstruction have been included in the same section because there is often an overlap. Compression can directly damage normal tissue, as described in the pituitary, or it may cause obstruction. This occurs, for example, when a tracheal tumour obstructs a normal oesophagus or vice versa. In the brain, compression of the brain-stem structures may obstruct the flow of cerebrospinal fluid, and a large prostate (benign or malignant) may compress the prostatic urethra. Alternatively, a tumour can grow into the lumen of the gut or into an airway, so that it produces obstruction directly (Figs 15.2–15.4).

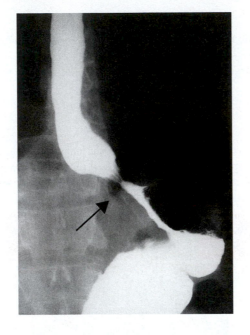

Figure 15.2 Radiograph showing constriction of lower oesophagus by oesophageal carcinoma.

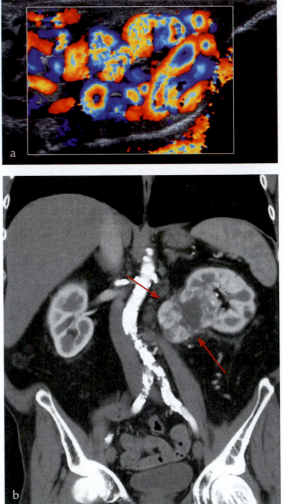

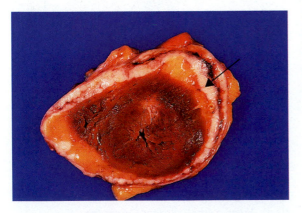

Figure 15.3 Compression of cardiac ventricles by metastatic lung carcinoma infiltrating the pericardium.

Figure 15.4 A 60-year-old male patient who presented with a tender left testicular lump. (a) The ultrasound scan demonstrates enlarged veins around the left testicle – a varicocele. This prompts assessment of the left kidney and renal vein because the gonadal vein, which may be obstructed, comes off the left renal vein. (b) The CT scan reveals a renal cell carcinoma (red arrows). The tumour has invaded the left renal vein obstructing the gonadal vein.

ULCERATION AND HAEMORRHAGE

An ulcer is defined as a macroscopically apparent loss of surface epithelium and may be benign or malignant. Ulceration of the skin will lead to a crust of dried fibrin and cells covering the area and is unlikely to produce severe haemorrhage. However, ulceration in the gastrointestinal tract, particularly the stomach and duodenum, may result in life-threatening haemorrhage or perforation. Here the absence of epithelium means the loss of an important defence mechanism, which normally protects the underlying tissue from acid and enzymes. Once the submucosa has been exposed to these agents, large vessel walls can be digested, resulting in massive bleeding (Figs 15.5 and 15.6).

RUPTURE OR PERFORATION

Rupture or perforation typically affects tumours of the gastrointestinal tract and will occur if the intraluminal pressure exceeds the strength of the wall or if the wall is eroded or weakened by tumour, ischaemia, enzymatic action, etc. (Fig. 15.7). Obviously, there is a risk of dilatation and rupture when part of the gut becomes obstructed, because the gut contents cannot follow their normal route. Rupture may also occur in closed organs, such as the ovary, because the tumour has stretched the capsule, often because of accumulation of fluid or mucin, as well as the proliferation of neoplastic cells.

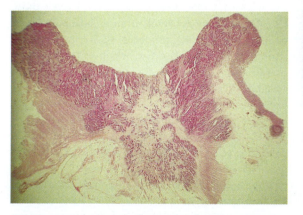

Figure 15.5 Photomicrograph showing rectal ulceration due to adenocarcinoma.

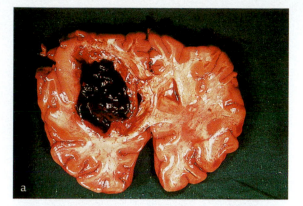

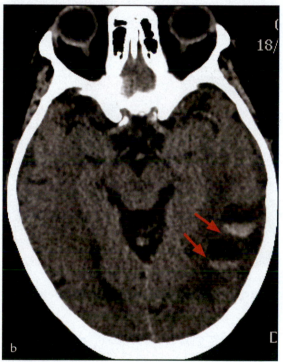

Figure 15.6 (a) Haemorrhage within intracerebral tumour; (b) intracerebral metastases in the left posterior temporal and occipital lobe The high-density fluid layering at the back of the lesions represents acute haemorrhage, which leads to a sudden increase in the size of the lesions, more mass effect, and acute headache and neurological dysfunction.

INFARCTION

Many malignant tumours will show necrosis and infarction in their central region, which is believed to result from inadequate blood supply (Fig. 15.8). In experimental models, a tumour can expand to a diameter of only 1–2 mm before it must stimulate new blood vessel formation and, in human tumours, zones of necrosis may be encountered approximately 1–2 mm from a blood vessel. Therefore, it appears that this is the maximum distance for diffusion of nutrients. Tumours attempt to solve this by secreting angiogenic factors, which stimulate capillaries to grow into the neoplasm.

Local anatomy influences the likelihood of infarction related to large vessel obstruction. The bowel, ovaries

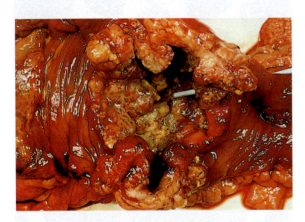

Figure 15.7 Colonic carcinoma with perforation indicated by probe.

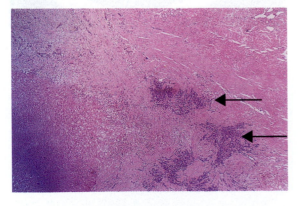

Figure 15.8 Extensive infarction in tumour, leaving small islands of viable cells (arrows).

and testes are particularly liable to torsion, i.e. twisting on their vascular pedicle which occludes the vessels.

Although we often separate the complications of tumours under the headings of local tumour and metastatic tumour, a metastasis can produce any of the local effects mentioned above. In particular, lymph nodes containing metastases can cause obstruction at crucial sites, such as the porta hepatis or the hilum of the lung.

ENDOCRINE EFFECTS

Well-differentiated tumours not only look like their tissue of origin but can also act like them. Thus tumours of endocrine organs can produce hormones that act on the same tissues as their physiological counterparts, but are not under normal feedback control.

Cushing's syndrome provides an interesting example of where different endocrine tumours can produce the same clinical problems. In Cushing's syndrome, the patient has osteoporosis, muscle wasting, thinning of the skin with purple striae and easy bruising, truncal obesity and impaired glucose tolerance. All this is the result of excess glucocorticoids. The same picture can be produced by prolonged administration of steroids to treat diseases (e.g. chronic asthma).

The adrenal produces corticosteroids when stimulated by adrenocorticotrophic hormone (ACTH) from the pituitary. The corticosteroids then provide negative feedback to the pituitary and ACTH levels drop. Cushing's disease can result from an adenoma in the pituitary gland, which produces ACTH, or a cortical tumour in the adrenal cortex, which secretes corticosteroids. Normal feedback does not operate because the adenoma cells behave autonomously. However, if very high doses of steroid (dexamethasone suppression test) are given, then the pituitary adenoma will reduce its ACTH production and endogenous steroid levels will fall, but excess steroid due to an adrenal adenoma will not be suppressed.

Some non-endocrine tumours can produce substances that have the same effects as hormones, so-called inappropriate production. One of the most common results is Cushing's syndrome when ACTH is produced by oat-cell (anaplastic small-cell) carcinoma of the bronchus, carcinoid tumours, thymomas or medullary carcinoma of the thyroid. This inappropriate and autonomous production does not suppress with high doses of dexamethasone.

History

Harvey Cushing (1869–1939)

Harvey Cushing (Fig. 15.9) was an American neurosurgeon whose greatest achievement was to reduce the mortality rate of neurosurgery from 90% to around 8%. He was a close friend of William Osler and won the Pulitzer Prize for his two-volume book *A Life of Sir William Osler*. He trained at Yale, Harvard, Massachusetts General and John Hopkins, before travelling to Europe to study and research the relationship between blood pressure and cranial pressure, and to work with Sherrington in Liverpool on the ape motor cortex. He published predominantly on classifying and removing cerebral tumours.

During the First World War, he was threatened with court martial, while serving with the British Expeditionary Forces, after a letter to his wife was intercepted by the French censors, who handed it on to the British Government. The letter contained criticism of a British surgeon.

Figure 15.9 Harvey Cushing. (Reproduced with permission from the Wellcome Library, London.)

PARANEOPLASTIC SYNDROMES

This refers to symptoms in cancer patients that are not readily explained by local or metastatic disease. Endocrine effects are generally included as a paraneoplastic syndrome if the production is inappropriate (as above), but not if the tumour arises from a tissue that normally produces that hormone.

Hypercalcaemia is a common, clinically important and complex problem with malignant tumours. In a patient with widespread metastases in bone, it may be explained as a local destructive effect of the tumour on bone, which releases calcium. However, hypercalcaemia can also occur without metastatic bony deposits and, in some cases, it appears that a parathyroid hormone-like peptide or transforming growth factor β (TGF-β) is secreted by the primary tumour, and this is most likely with bronchial squamous cell carcinoma and adult T-cell leukaemia/lymphoma.

Clubbing of the fingers and hypertrophic osteoarthropathy are also common with lung carcinoma but can occur in non-neoplastic conditions including cyanotic heart disease and liver disease (Fig. 15.10). It is not clear how it develops, or why sectioning the vagus nerve can lead to its disappearance! Equally mysterious are the skin disorders, peripheral neuropathy and cerebellar degeneration that may also occur in association with malignant tumours, particularly lung and bronchial carcinomas.

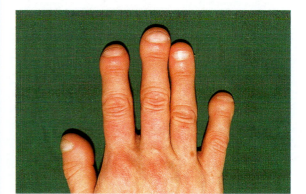

Figure 15.10 The drumstick appearance of finger clubbing is associated with hypoxia (e.g. lung cancer, cystic fibrosis, chronic pulmonary disease) and cirrhosis of the liver. Most patients with clubbing have increased blood flow through the fingers, possibly mediated by autonomic nerves (vagal transection can relieve symptoms). Such shunts may enable platelets to lodge in capillary beds and the interaction between platelets and endothelial cells, by releasing platelet-derived growth factor (PDGF), may lead to overgrowth of bone and connective tissue. There is characteristic tufting of the distal phalanges on a radiograph. Other joints may also be affected.

Chapter 15: The clinical effects of tumours

GENERAL EFFECTS

The general effects of tumours are not classed as paraneoplastic syndromes, although they are extremely common and must not be forgotten. They include general malaise, weight loss and lethargy, which are due to a combination of metabolic and hormonal influences exacerbated by any malnutrition or infection. This results in the clinical picture known as 'cachexia' (Fig. 15.11). An important chemical factor that may play a role in cachexia is cachexin, which is also known as tumour necrosis factor (TNF). This molecule is produced not by the tumour cells, but by activated macrophages. Anaemia is also common and can contribute to the general malaise. This may be a direct effect of metastatic deposits in bone marrow or an indirect effect of mediators that suppress haematopoiesis.

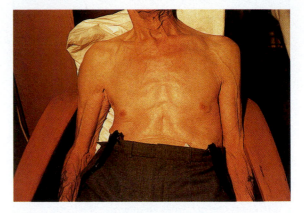

Figure 15.11 Cancer-related muscle wasting (cachexia). Similar appearances can be seen with some chronic diseases, e.g. tuberculosis and AIDS. Cachexia differs from starvation in that lean mass – skeletal muscle, especially of the shoulder girdle and head – is lost as well as adipose tissue, so there is both proteolysis and lipolysis. Cytokines such as tumour necrosis factor (TNF), interleukin (IL)-1, IL-6 and interferon γ (IFN-γ) are thought to drive the liver to produce high levels of acute-phase proteins, depleting amino acid reserves. Dietary measures alone are not effective. Eicosapentaenoic acid (EPA) in fish oils (an omega-3 fatty acid) interferes with proteolysis-inducing factor (PIF, which releases protein from skeletal muscle), found in the urine of 80% of cachectic patients. EPA does not affect lipolysis.

🔑 Key facts

Paraneoplastic manifestations of tumours

- Endocrine effects
- Cushing's syndrome
- Inappropriate antidiuretic hormone (ADH) secretion
- Hypercalcaemia
- Clubbing
- Hypertrophic osteoarthropathy
- Peripheral neuropathy
- Cerebellar degeneration
- Skin rashes

Radiology

Functional vs structural imaging

Functional imaging is a loose term applied to imaging techniques that assess a disease process by the effect it has on the function of cells, tissues or organs. Structural imaging refers to the traditional imaging techniques that depict the anatomical extent of disease (but also indirectly demonstrate the effect on function). Functional imaging techniques used widely in tumour imaging include positron emission tomography (PET)–CT, MR and CT perfusion and diffusion techniques.

Conventional PET–CT using fluorodeoxyglucose (FDG) as the isotope relies on the increased metabolism of glucose in tumour cells to allow visualisation of the extent of the disease. This modality is now widely used in the staging of many tumours as previously described (see fig 14.5). However, it has also proven to be a handy tool in assessing tumour response to oncological treatment.

Functional MRI studies have been used to map the function of the brain. The patient is challenged with a 'paradigm' such as finger tapping (to activate the motor cortex) or questions that test receptive and expressive language functions. The cortical areas responsible for these functions in the brain demonstrate changes in perfusion that are picked up by MRI, allowing localisation of function to different areas of the brain. This not only is helpful in furthering our knowledge of brain function but also has a very practical use as shown. Patients being considered for surgical resection of brain tumours have functional MRI preoperatively so that the neurosurgeon can plan a surgical approach to the tumour, avoiding vital areas of cortex and white matter tracts. The patient and neurosurgeon can also be forewarned about possible neurological deficits after surgery depending on the proximity of the tumour to particular areas of the cerebral cortex and the involvement of white matter tracts by the tumour.

MANAGEMENT OF CANCER

We discuss this under individual headings, but many patients receive a combination of management strategies.

SCREENING PROGRAMMES

The concept of the multistep model of carcinogenesis has led to the idea that mortality (Fig 15.12) as a result of the tumour may be reduced if it is identified at an earlier stage or when it is still precancerous. To identify it at this early stage, a modality that can do this is required – such as radiology (e.g. mammography), serum marker or a genetic test. Screening programmes have been instituted for cervical (cytology smear/ human papillomavirus [HPV] testing) cancer, breast cancer (mammography) and more recently for bowel cancer (faecal occult blood/endoscopy), and there is considerable debate as to whether other tumour types such as prostate cancer should also be included. The screening programmes have undoubtedly reduced mortality, although not as dramatically as had been envisaged, but it has also highlighted deficiencies in our knowledge of the classification and natural history of the early lesions, making management of patients with these 'pre-cancers' problematic. To circumvent this problem, management decisions are now made in multidisciplinary teams (comprising surgeons, oncologists, pathologists, radiologists, etc.), bringing all the expertise to bear in deciding the final strategy.

 Read more about aspects of cervical and breast screening in Pathology in Clinical Practice Cases 24 and 25

 Read more about prophylactic surgery in colorectal carcinoma predisposition in Pathology in Clinical Practice Case 37

LOCAL EXCISION

Local treatment is aimed at either achieving a cure or providing specific symptomatic relief. Cancers, such as squamous and basal cell carcinomas of the skin and cancers arising within polyps in the colon, can be cured by local excision. In tumours of the bowel, local excision may relieve an obstruction and provide good long-term remission of symptoms or even cure.

RADIOTHERAPY

Radiotherapy can be given from an external source or by implanting a small radioactive source into the tissues. Delivery schedules vary from centre to centre, but the general idea is to divide, or fractionate, the doses in order to get the maximum kill of tumour cells with the minimum damage to normal tissues. Implanted radioactive sources are very useful for providing high-dose local radiation and are particularly useful in cancers of the head and neck, where there are many vital structures close together. The combination of dosing schedule and ability to target the radiotherapy where it is needed had dramatically improved outcomes by killing the tumour and avoiding normal tissue damage.

CHEMOTHERAPY

Chemotherapy is a relatively crude but potentially effective form of treatment. In patients with haematological malignancies (e.g. leukaemia) or disseminated

Chapter 15: The clinical effects of tumours

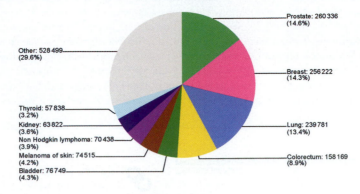

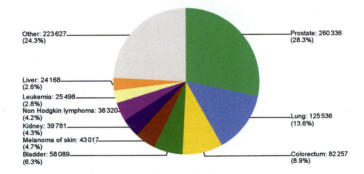

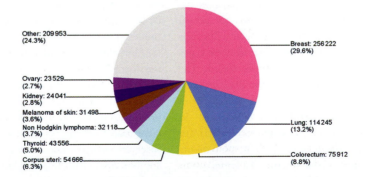

Figure 15.12 Estimated age-standardised (world) cancer incidence and mortality rates (ASR) per 100 000, by major sites, in men and women, 2012. (Reproduced from Stewart BW, Wild C, eds. *World Cancer Report*. Lyon: IARC, 2014.)

disease, surgery and radiation are not realistic options. You cannot excise a leukaemia and you cannot irradiate metastases that are widespread in the body!

In the 1950s, alkylating agents (e.g. busulfan) and anti-metabolites (e.g. methotrexate) were introduced and proved useful in the management of disseminated cancers. The main problem with such agents is that all of the body's normal tissues are also exposed to the drug, and so the challenge is to deliver enough drug to kill the tumour without killing the patient! It has also become apparent that chemotherapy is much more effective in combination with other agents than as a single agent.

ENDOCRINE-RELATED TREATMENT

This often involves giving a drug that inhibits tumour growth by removing an endocrine stimulus, e.g. many breast carcinomas have receptors for oestrogen that stimulate tumour growth. A drug such as tamoxifen (an anti-oestrogen) will block these receptors and reduce progression of the disease. An alternative approach would be to remove the ovaries, which produce oestrogen, much as the testes can be removed in men with prostatic adenocarcinomas to reduce the stimulus for tumour growth from androgens.

IMMUNOTHERAPY

DNA recombinant technology has enabled the production of cytokines in sufficient quantities for therapeutic use. The IFNs and TNFs are of particular interest. IFN-α and IFN-β have been used to treat a variety of tumours with some good effect, although it appears that they may be best used in combination with other treatments. Renal carcinomas, melanomas and myelomas have shown a 10–15% response, various lymphomas show a 40% response and hairy-cell leukaemia and mycosis fungoides have an 80–90% response rate. TNF-α has been used in the treatment of melanoma although, to date, the response has been disappointing. Lymphokine-activated killer (LAK) cells are a subset of natural killer (NK) cells which have been used in combination with IL-2 to treat renal carcinomas, and some melanomas and colorectal cancers.

As mentioned previously, attempts at specific tumour antigens are in progress and initial trials of immunisation seem encouraging. There is also considerable interest in raising monoclonal antibodies to tumour cells. The hope is that it might be possible to attach drugs to these antibodies so that they would be delivered specifically to the tumour cell – the concept of the 'magic bullet'. A lot of questions remain unanswered but the field of immunisation and targeted treatment is bound to create excitement over the next decade.

MOLECULAR MECHANISM-BASED THERAPIES (TARGETED THERAPIES)

With our increasing understanding of the mechanisms of disease, therapies directed at specific molecular pathways are rapidly emerging. These include monoclonal, antibodies and enzyme inhibitors directed at specific genetic changes.

HER2 is a tyrosine kinase-associated receptor; the gene is located on 17q. It is amplified in 10–34% of human breast carcinomas. A humanised monoclonal antibody directed against *HER2* oncogene, trastuzumab (Herceptin), has been used to treat women with primary and metastatic breast carcinomas who have HER2 over-expression and/or amplification. This has had a dramatic effect on the prognosis of patients with HER2-positive breast cancer. HER2 is also over-expressed/amplified in a proportion of other types of cancer such as of the lung and pancreas, and may have a role in those too.

Since the beginning of the 1990s, much effort has been put into the discovery of molecules that would be able to block specific kinases. The first drug that showed promising results was STI-571 (Gleevec), which was tailored to specifically inhibit the *ABL–BRC* tyrosine kinase fusion gene product. *ABL–BCR* results from a balanced translocation between chromosomes 9 and 22, and is found in more than 95% of chronic myeloid leukaemias (CMLs). Patients treated with this drug showed remarkable improvement. Not unexpectedly, STI-571 is not specific for this kinase and has been shown to inhibit other oncogenic tyrosine kinases, such as c-*Kit* and PDGFR (PDGF receptor). STI-571 seems to have a remarkable effect on neoplasms arising as a result of mutations of these genes and in particular in patients with gastrointestinal stromal tumours (GISTs) (with c-*Kit*-activating mutations). The epidermal growth factor receptor (EGFR) is also a potential target. Gefitinib (ZD1839, Iressa) has been on trial for oesophageal and colorectal cancer (CRC) and cetuximab (IMC-C225, Erbitux), a monoclonal EGFR antibody, has been approved for the therapy of CRC.

Chapter 15: The clinical effects of tumours

Vascular endothelial growth factor (VEGF) and its receptor (VEGFR) are involved in tumour angiogenesis. Bevacizumab (Avastin), a monoclonal antibody against VEGF, is being used together with conventional chemotherapy for a variety of cancers. The era of targeted therapy and the 'Holy Grail' of individualised treatments are certainly pushing ahead with speed.

GENE THERAPY

Gene therapy emerged as an alternative to conventional cancer treatments as it became possible to transfer genetic material into a cell to transiently or permanently alter its biology (i.e. to induce cell cycle arrest/apoptosis). This may be achieved by using modified, virus-encoding, specific DNA sequences (viral vector systems). The introduction of genes that encode tumour-suppressor proteins is known as therapeutic gene transfer. There are a number of drawbacks to this type of treatment which include the specificity of the virus for neoplastic cells, the side effects of the presence of the 'new gene' in normal cells and the reaction of the host immune system. The use of gene therapy for treatment had a setback when a young patient died as a result. Research continues in order to learn more about safe delivery of the vector into cancer cells and learning how to combat the different cell populations within a cancer. If the stem cell hypothesis is correct, cancers will contain slowly replicating stem cells, rapidly replicating transit-amplifying cells, as well as a variety of differentiated cells. Conventional chemotherapy may kill the dividing cells but may not have any impact on the stem cells, and use of additional gene therapy to target specific cells such as the stem cell population may be a way forward.

PALLIATIVE TREATMENT

The treatment of cancer has a wider role than merely providing a cure and cancer physicians are not interested in simply achieving a response to the administered treatment. Palliative treatment does not just refer to treatment that is given to patients in order to make them comfortable before death. It is and should be part of the oncological support given to all cancer patients and includes not only medication for the control of pain and nausea but also chemotherapy and radiotherapy for the relief of local symptoms and stabilisation of disease. This is helped by the fact that we now have a larger panel of oral drugs, allowing patients to be managed in their own homes. The term 'continuing care' is sometimes used for this multidisciplinary approach, starting with diagnosis and extending to the patient's death. The important point is that our knowledge of all modalities of treatment, including pain control, have advanced considerably in recent years and the pessimistic view that if one has cancer one must pass one's last hours either conscious, but in agony, or pain-free but unconscious is no longer justified.

CONCLUSION

Science, similar to most aspects of life, has its fashions. Much of this section has concentrated on our evolving understanding of the role of the genetic code in producing cancer. Current fashion lies very much with large-scale sequencing of all cancer genes and expression profiling to identify subsets with differing prognoses and response to treatment. Stem cells are also a hot topic, and there is a tendency to classify everything on the basis of hypothetical hierarchies of cell lineage. There is little doubt that the ability to produce differentiated tissue from stem cells to help repair, such as in cerebrovascular accidents, myocardial infarcts and traumatic neural injuries, will be a biotechnological miracle in the coming years. Whether stem cells have such a high profile in cancer biology remains to be seen.

Part 4: Cell growth and its disorders

Case study: cervical cancer

Clinical

A 38-year-old woman came to the surgery for a cervical smear.

She was single and had been living with her boyfriend for the last 6 months. She divorced her husband 2 years ago and since then had a number of casual relationships. Her first sexual contact was at the age of 16.

Four years ago her smear showed warty change, and the last one, 1 year ago, again showed extensive warty change with possible dyskaryosis. The hospital had asked for a repeat smear because the epithelial cells were obscured by inflammatory debris.

She initially ignored the recall due to social problems.

The result of the repeat smear showed warty change and severe dyskaryosis and she was referred for a colposcopic biopsy.

The cervical biopsy confirmed the above findings and she was booked in to have a cervical cone biopsy.

The results of the cone biopsy came as a shock. The report read that she had extensive squamous metaplasia with wart virus change, together with severe dysplasia between 3 o'clock and 5 o'clock, and a focus of invasive squamous cell carcinoma that was completely excised. Invasive tumour did not involve deep tissues or invade blood vessels. The dysplastic epithelium extended to the endocervical excision margin and was therefore not completely excised.

Pathology

Carcinoma of the cervix is an important cause of death and the cervical screening programme has been instituted to try to reduce this toll. The idea is that, if the disease can be picked up at an early stage, it should be possible to cure it.

The risk factors for cervical cancer include:

smoking, early onset of sexual intercourse, multiple sexual partners, a sexual partner with a history of promiscuity and infection with HPV.

Her history reveals that she had a number of risk factors, including wart virus change on her previous cervical smears.

The normal routine recall for cervical smears is 3 years, but early recall is instituted for suspicious or abnormal smears.

Severe dyskaryosis is the cytological equivalent to severe dysplasia on histological examination.

Dysplasia is a premalignant condition in which there are cytological features of malignancy, i.e. increased nuclear:cytoplasmic ratio, nuclear pleomorphism, hyperchromatism, loss of maturation and mitotic activity. Dysplasia can be graded into mild, moderate or severe. Severe dysplasia implies a full-thickness abnormality and the feature distinguishing this from carcinoma is the presence of invasion through the basement membrane.

Metaplasia, on the other hand, is entirely benign. It is a form of adaptation to injury, in which one type of epithelium is replaced by another. In the cervix, the glandular epithelium, after repeated bouts of inflammation, changes to a more resistant squamous epithelium.

Chapter 15: The clinical effects of tumours

The report was discussed with the patient and she was advised to have radiotherapy and a hysterectomy.

The examination of the hysterectomy specimen showed residual foci of severe dysplasia but no invasive carcinoma. The excision was complete and she was discharged after an uneventful recovery.

The cone biopsy is a way of performing a local excision of the cervix; the tissue removed is in the form of a cone. A suture is usually put at 12 o'clock to orient the specimen. The role of the pathologist is to map the abnormal areas, to assess the abnormality in terms of severity and to comment on completeness of excision.

This patient was 38, and still capable of having children. The decision to have a hysterectomy can be a difficult one, although she had very little choice.

The hysterectomy specimen did not reveal any more areas of carcinoma and the single focus of carcinoma was completely excised, so she should be cured. Often the ovaries are not removed, to prevent menopausal symptoms.

EPILOGUE

So what is a disease? The definition provided at the outset regarding a loss of homeostasis is, we believe, a good one. What we hope is apparent, however, is that the mechanisms involved in maintaining the balance are incredibly intricate and complex. This is the paradox of life – complexity and simplicity existing together. To maintain homeostasis – balance – seems a simple thing to do. We hardly notice the constant adaptations that the body undergoes moment to moment as we move about our daily lives: the change in body temperature and heart rate, and the flexing or relaxation of muscles as we change posture. Yet the processes that allow such a seamless transition from moment to moment are both complex and beautiful.

Imagine a cell with its 30 000–40 000 genes. These genes are transcribed into messages, which are then read and translated into proteins; these are modified to make yet more proteins, which in turn regulate the genes and messages. The pattern of gene expression is different from cell to cell depending on its function and role at that time. Imagine this whole network as a three-dimensional tube/metro system buzzing within the cell, constantly adapting to fulfil its function. Now imagine this cell as one of a hundred forming a tissue and one of thousands forming an organ and one of billions forming the body. Can you imagine a billion tube/metro networks all working together in a coordinated manner?

We have to admit that we can't! It is totally mind-boggling and, yet, this is exactly what must be happening every second, every minute of the day for the tissues, organs and body to maintain the homeostasis that allows us to function as part of a bigger 'organism' that we call life. This is not such a crazy concept; we know that our own homeostasis is to do with not just our own genes and proteins, but also the interaction of our bodies with the environment in which they find themselves. Cancer may be a 'genetic disorder' but we are all aware that the genes provide a 'predisposition', which given the right lifestyle and interaction with the environment leads to cancer. Not everyone with an inherited mutation develops cancer, but they are at increased risk, which becomes manifest given the right circumstances.

The beginning of the twentieth century marked the change in the way physicists looked at the world. The quantum theory and the theory of relativity have questioned what we mean by 'reality' and the wave–particle duality continues to show us just how simple (we all think we understand it) and complex light really is. Lest you think that this 'new' physics has nothing to do with you, the quantum physics that is used to describe light is also the same physics that applies to all atoms including those in our DNA.

The end of the twentieth century saw the emergence of a revolution in biology, the sequencing of the human genome and the ability to look at thousands of genes and their expression profiles in one experiment. This will undoubtedly change the way we look at disease and we are already into the era of targeted therapy. The technological advances are also going to change the way we diagnose disease, with new hybrid PET/CT and PET/MRI as well as functional imaging and nanotechnology providing us with tools for near-patient testing and drug delivery. Perhaps that is the key: disease, like education and life itself, should be viewed as a process rather than a defined thing with a beginning and an end.

I saw a child carrying a light.
I asked him where he had brought it from.
He put it out, and said:
'Now, you tell me where it is gone.'

Hasan of Basra

INDEX

Note: Page numbers ending in "f" refer to figures. Page numbers ending in "t" refer to tables.